Beyond Straw Men

Beyond Straw Men

Plastic Pollution and Networked
Cultures of Care

———

Phaedra C. Pezzullo

UNIVERSITY OF CALIFORNIA PRESS

University of California Press
Oakland, California

© 2023 by Phaedra C. Pezzullo

Library of Congress Cataloging-in-Publication Data

Names: Pezzullo, Phaedra C., author.
Title: Beyond straw men : plastic pollution and networked cultures of care /
    Phaedra C. Pezzullo.
Other titles: Environmental communication, power, and culture ; 4.
Description: Oakland, California : University of California Press, [2023] |
    Series: Environmental communication, power, and culture ; 4 | Includes
    bibliographical references and index.
Identifiers: LCCN 2023005939 (print) | LCCN 2023005940 (ebook) |
    ISBN 9780520393639 (cloth) | ISBN 9780520393646 (paperback) |
    ISBN 9780520393653 (ebook)
Subjects: LCSH: Plastic scrap—Environmental aspects. | Plastic scrap—
    Political aspects. | Plastics—Environmental aspects. | Plastics—Political
    aspects. | Internet and activism. | Social media.
Classification: LCC TD798 .P44 2023 (print) | LCC TD798 (ebook) |
    DDC 363.72/88—dc23/eng/20230302
LC record available at https://lccn.loc.gov/2023005939
LC ebook record available at https://lccn.loc.gov/2023005940

32   31   30   29   28   27   26   25   24   23
10   9   8   7   6   5   4   3   2   1

For all of us—the whales, the trees, and the bees, including you and me

All proceeds of *Beyond Straw Men* go to Eco-Rethink (Nakuru, Kenya) and Justice for Formosa Victims (Point Comfort, Texas, United States)

Additional podcasts and links for educators and advocates are available at: https://phaedracpezzullo.com/beyond-strawmen

The publisher and the University of California Press Foundation gratefully acknowledge the generous support of the Ralph and Shirley Shapiro Endowment Fund in Environmental Studies.

CONTENTS

ACKNOWLEDGMENTS

This book was written during a global pandemic with a world on fire. While researching and writing, I lost kin to illness and suicide, as well as community in a mass shooting at my neighborhood grocery store. My home was evacuated due to wildfire risk. I transitioned to working online with an elementary-aged child learning remotely. My child contracted (though survived) COVID before vaccines were available. Then I moved my aging and ailing parents across country into my neighborhood. Both of my parents required visits to the emergency ward in the first week. At least three times I left my mother's bedside to interview someone and return. My mother died while I slept on the floor next to her. From all these experiences and more, I do not take for granted being alive, let alone writing a book again. Surviving such tumultuous times and creating books is made possible only through the labor and the love of many, whom I acknowledge here, even as these words do not do their significance justice.

I am indebted to the people who talked with me on the podcast *Communicating Care* before it had an RSS feed: H.E. Prof. Judi Wakhungu, James Wakibia, Emy Kane, Dr. Shahriar Hossain, Michelle Gabrieloff-Parish, Nancy Bui, José Toscano Bravo, and Joe Andenmatten. At the University of Colorado Boulder (CU Boulder), Mia Ives-Rublee's podcast was recorded live, cosponsored by the Dean's Fund in the College of Media, Communication, and Information (CMCI) and the Department of Environmental Studies, with publicity support from Disability Services and Ethnic Studies. This book wouldn't be half of what it has become without those conversations. Consent to interviews doesn't mean those interviewed agree with everything I've written. My hope, however, is that their lessons are palpable. As I consider each podcast a master class from people who continue to do the work to make the world a better place despite the odds, I recommend listening to the longer conversations at https://communicatingcare .buzzsprout.com.

Part of the gift of teaching is that I constantly meet new people from whom I learn. Over the years, my undergraduate students have convinced me to take ocean conservation, digital activism, and plastic pollution more seriously. Likewise, former and current graduate students have provided generative conversations about care, climate, colonialism, disability justice, hashtag activism, incarceration, nonhuman kin, oppression, waste, and water, particularly Lindsey Badger, Ali Branscombe, Warren Cook, E Cram, Catalina de Onís, Jake Dionne, Suz Enck, Logan Rae Gomez, Constance Gordon, Jeremy Gordon, Joe Hatfield, Amani Husain, Yesim Kaptan, Martin Law, Katie Lind, Myles Mason, Cole McGuffey, Norma Musih, Jo Marras Tate, Hunter Thompson, Mikayla Torres, Darrel Wanzer-Serrano, Rachel Vaughn, Isaac West, Bri Wiens, and Kelly Wilz. Gratitude also to Tim Schütz, who has become a rejuvenating interlocutor about the Formosa Plastics Global Archive at the University of California Irvine, research ethics, and digital media.

While finishing the initial manuscript, I enrolled as a student in a six-week online Bennington College class titled "Beyond Plastics," which was taught by the former regional administrator of the US Environmental Protection Agency under the Obama administration, founder and president of the NGO Beyond Plastic, Professor Judith Enck. This experience allowed me to learn from her as well as participate in an extended online intergenerational exchange of over one hundred students. In addition to her content expertise, Enck provides a wonderful model of pedagogical generosity, calling in new waves of advocates and embodying a joyful mix of humor with intersectional awareness and pragmatic political advice. Plastic studies researchers and advocates more broadly continue to teach and inspire me on social media.

My mentors keep retiring but have checked in and taught me over the years. A humble and passionate environmental advocate and mentor, Robbie Cox continues to be generous as a textbook coauthor; his partner, Julia T. Wood, shares honest and insightful interpersonal advice. D. Soyini Madison and Lawrence Grossberg still shape my worldview. Since I moved to Boulder, Jerry Hauser has provided deeply grounding and insightful counsel on professional life and the good life.

Writing should not happen in isolation, even in a time profoundly impacted by quarantines, social distancing, and other labor intensities. Some incredibly smart colleagues whom I also consider trusted friends read early chapter drafts: Jeffrey A. Bennett, Kundai Chirindo, E Cram, Salma Monani, Vincent Pham, and Isaac West. Most generously, Constance Gordon read the entire manuscript early on; Jerry Hauser and Ted Striphas did as well toward the end. Colleagues also shared research: Amanda Carrico on behavior change, Casey Fiesler on digital media ethics, and Patrick Chandler on the US plastic pollution movement. My thinking benefits from episodic conversations with Angela J. Aguayo, Kelly Happe, Stephen J. Hartnett, Gregory J. Seigworth, Anjali Vats, and many more. All errors remain mine, though I am profoundly grateful for the counsel offered, particularly as I was juggling most of this project during a stressful period of care work in hospitals and hospice.

Despite ongoing and acute crises, Boulder is home to an exceptional community of climate research experts. Of late, I particularly have been bolstered by collaborative climate justice scholarship, teaching, and/or service with Karen Bailey, Max Boykoff, Cassandra Brooks, Amanda Carrico, Clint Carroll, Dave Ciplet, Shideh Dashti, Michelle Gabrieloff-Parish, Donna Goldstein, Jill Harrison, Abbie B. Liel, Marianne Moulton, Shelly L. Miller, Dave Newport, Beth Osnes, David Paradis, Shawhin Roudbari, Becca Safran, Manuela Sifuentes, Leah Sprain, Burton St. John, Leaf Van Boven, and Emily Yeh.

My department chair, Tim Kuhn, and my dean, Lori Bergen, have supported my climate leadership on campus. Tim has been a particularly gracious mentor during extraordinary circumstances. CMCI colleagues made memorial donations and shared food, including Karen Ashcraft, Jed Brubaker, Nabil Echchaibi, Steven Frost, Pete Simonson, Jamie Skerski, and Leah. Since they moved away, I miss hanging out more frequently with Tiara R. Na'puti and Lisa A. Flores. Omedi Ochieng has become a welcome interlocutor; he organized a gathering to discuss the introduction, including our colleagues: Leah, Pete, Laurie Gries, Danielle Hodge, and Amber Kelsie.

At the Pulitzer Center's 2022 conference in Washington, D.C., I presented research on climate, labor, and plastics; gratitude to Flora Pereira, Christine Spolar, and Jon Sawyer for their warm support. A version of chapter 4 was presented at Ocean First, and I thank the marketing coordinator, Ingrid Hilbink, a former undergraduate student who has helped inspire me to think more about marine life and digital communication. I presented an earlier version of chapter 3 to the Department of Environmental Studies at CU Boulder, and I am appreciative of their encouragement, especially Emily Beam, Amanda, Cassandra, and Max. From the CU Boulder's Digital Accessibility Office, Laura Hamrick and Allyson Bartley provided training for me, which I found helpful when writing Alt-text for the ebook version of this book.

Grants and funding for the podcast and book include The Waterhouse Family Institute, de Castro Research Award, the CU Boulder Office of Faculty Affairs, Inside the Greenhouse (thanks to Becca, Beth, and Max), and the Department of Communication at CU Boulder. With these funds, financial management (including Laura Burfield, Monica Carroll, Emilia Gaeta, Deborah Schaftlein, and Dawn Williams), and the support of Max as a fellow in CIRES (Cooperative Institute for Research in Environmental Sciences), I compensated talented artists and student workers: Anthony Albidrez, Robert Lino, and Bailey Troutman edited the podcast, and Michael Warren Cook drafted an initial bibliography.

My acquisition editors, Stacy Eisenstark and Chloe Layman, have been unwavering in their support, offering insightful perspectives, commiserating with me about the chaos in the world, and helping me realize my aspiration of becoming a University of California (UC) Press author. The series would not exist without Kim Robinson and Lyn Uhl. Our book series advisory board, including my coeditor, is comprised of many of my favorite scholars—I am grateful for their aptitudes and honored by their support, as well as the UC Press editorial board and two anonymous reviewers. As the hidden abode of books remains complicated, Chad Attenborough and Naja Pulliam Collins thankfully kept track of all the details. Gratitude to my copyeditor, Sharon Langworthy; my indexer, Cathy Hannabach of Ideas on Fire; and the whole UC Press production team for their keen labor.

I benefited from previously unmentioned long-distance gestures from many more people, including but not limited to Balthrop and Carole, Billie and Bryan, Christina, Chuck, Heather, Justin, Krista, Lisa K., Melissa, Michael and Regina, Nina and Dana, Stacy, Stephanie and Kevin, and Sue. Kundai, Carlos, and Emma graciously accommodated me as NCA ECD officers. Nearby, Becca and Sam, Renée, Tracy and Dave, and Val and Noah provided necessary support, from coordinating wildfire evacuations and fallen fences to sharing food and grief. Gratitude to the "workout women" for our network of care throughout the pandemic and a neighborhood mass shooting: Corinna, Donna, Ellie, Jamy, Kat, Kathy, Kelly, Kim, Pam, and Robyn. Catalina, Constance, and E particularly have kept me afloat with thoughtfulness, notifications, and humor. They, along with Isaac, Jeff, Natalie, Suz, and Tiara, generously fed me and my family from afar. Armond, Catalina, Constance, E, Isaac, Kelly, Myles, Natalie, Suz, and Warren helped me feel less adrift by nominating me for the inaugural Daniel C. Brouwer Mentorship Award.

Thanks to my brother Alexis, my fabulous cousins, and the rest of my extended family (through blood, marriage, and friendship) for loving me through it all, despite the miles between us and all the losses we've survived. My parents, Vincent and Carmen, long tolerated, fought, and shared my environmental and social justice values: loving them and being loved in return despite our disagreements probably was my first lesson in impure politics. Both were public school educators, and my dad maintains a bookshelf dedicated to my publications as he is writing the next chapter of his life.

My partner, Ted, has witnessed all of it—my professional heartbreaks and achievements, familial joys and losses, health complications and nourishing meals, mundane juggling acts and unforeseen traumas, and the dozen other book ideas I proposed on our road trips but didn't write over the past decade. Staying together despite our differences and being able to make you laugh with all that is going on professionally and personally feels like a miracle. Thank you for stepping up and still caring.

Niko, my clever, humorous, kind, and creative bambino—who designed the *Communicating Care* podcast cover art—and our four-legged family members (Phoenix, Rogue, and Yoda) remind me to embrace joy, movement, and rest. Niko, I never knew I could love another as I love you.

PREFACE

For decades I have tried to avoid plastics. I take cloth bags to the grocery store, make my daily coffee with a reusable filter, use bamboo toothbrushes, and carry a metal water bottle. But I had not really thought those choices rose to the level of crisis I usually write about. They matter, but as someone who has published about toxic pollution, environmental racism, and climate injustices for decades, avoiding plastics has felt like a privilege I could afford.

And I haven't always avoided plastics. Daily, I write with plastic pens, type on a plastic keyboard, and read with plastic contact lenses. Weekly, I buy fruits and vegetables in plastics. When I travel, using hotel toiletries in single-use plastic has felt like part of the vacation. As the pandemic spread, my cloth masks gave way to N95s, while plastic COVID tests have been taken regularly. When my mom was dying as I was writing this book, I didn't look for cloth diapers or prescription drugs that weren't in plastic bottles, let alone question the IV tubes in her arms or oral care sponge sticks I used in her mouth. Plastics: they haven't felt important enough and yet somehow felt too essential to elude.

The drama on social media about plastics also has felt exhausting, including quick judgments, dismissive comments, and mocking crowds. As I discussed them with students, friends, and colleagues, plastic-related hashtag activism and reactions to it fascinated me, but with all the crises in the world, I wasn't sure—and I'm honestly still not—that I have thick enough skin to seek out engaging that public intensity on purpose.

As I began this project, then, I remained "on the fence" about whether this was a worthy topic and if I was worthy of writing it—until a stranger tagged me on social media to publicize a virtual event, which was amplifying global grassroots efforts to resist a multinational plastics corporation (Formosa Plastics). He found me because he had read my first book, *Toxic Tourism* which included activists I admired who long have been disproportionately impacted by toxic pollution, living and dying with the burden of what it means to be imagined by dominant culture as disposable. The same industries creating toxic petrochemicals in their backyards, they reminded me, have been making and selling plastics. Hesitating to participate publicly in social dramas about plastics, I finally recognized, was my privilege—and one I needed to interrogate: What was holding me back? The answer, I soon realized, was: a lot.

This book has stretched me. I created a podcast to interview people profoundly shaped by where they live as well as to engage visuals circulating in global digital media beyond the locations where they were created. I also have analyzed English-language Global North digital media and tried to account for multilingual trends generated by the Global South. Writing has moved me to reflect on my human existence in a landlocked state as well as marine life communication through oceans and seas, to listen to voices harmed by plastic pollution as well as disability critiques of harmful environmental discourses, and to revisit my own positionalities as well as changing global relations. And much more.

I'm pretty sure I'm still too undone by the events of the past few years to be confident about anything. And yet as I listened, read, reflected on, and began to have more conversations about plastics online and offline, I remembered the anti–toxic pollution adage: there is no safe place. The least, then, I—and maybe any of us—can do is not give up. This book is my attempt to do just that, to once again step into the fray and risk admitting that I care, knowing I'm still unsteady.

Introduction

Care amid Oceans of Trouble

am always pissed off when somebody tells me that we cant do anything about
BAD plastic bags #BanPlasticKE

—JAMES WAKIBIA, PHOTOJOURNALIST AND ACTIVIST (@JAMESWAKIBIA)

last night a guy in a turtle costume grabbed the straw out of my drink, threw it
on the floor, and said 'that's for my homies' #Halloween #StopSucking

—LONELY WHALE, NGO (@LONELYWHALE)

"Festering outrage" #Formosa's #pollution in #Vietnam becomes a tipping point

—INTERNATIONAL POLLUTANTS ELIMINATION NETWORK, NGO
(@TOXICSFREE)

Over the past decade, addressing plastic pollution has felt complicated and overwhelming. Anger, shame, hurt, fear, guilt, and despair abound online and offline.
Like the planet, the reactions are increasingly heated. Such intensity is a sign
something matters.

The most common observation of plastics is their ubiquity. Susan Freinkel
opens *Plastic: A Toxic Love Story* by sharing how she tried not to touch plastic
for a day and quickly failed; instead, she shifted to writing down everything she
touched in a day that was plastic, from toilet seats and glasses to food containers
and computers.[1] Bridging private and public practices, plastics have become integral to what Lauren Berlant more recently called the "intimate public sphere."[2] In
other words, our uses of plastics may feel deeply personal—from objects we place
in our mouths daily to those others may use to assist us in our most precarious
moments—and yet are structured by and structure our collective lives.

And plastic production keeps multiplying globally at remarkable rates since
the mid-twentieth century. In 1950 the industry created approximately 2 million metric tons of plastics. That number rose to 380 million metric tons in 2015.[3]

That's an almost 19,000 percent increase in less than a century. If we don't change course, the World Economic Forum projects, current amounts will more than triple by 2050, resulting in what some alarmingly estimate as a 1:1 ratio of plastics to fish in the ocean by weight.[4]

Further, plastics aren't inert. Stacy Alaimo contends we now can judge that "a study on plastic pollution published in 1973 seems ancient" when it called the "harm 'chiefly aesthetic.'"[5] As the Just Transition Alliance emphasizes, plastics are toxic to public health and broader ecosystems not just as waste but also as petrochemicals throughout their life cycle, including extraction, production, transportation, consumption, and disposal.[6] Our lives are entangled with microplastics, as scientists have detected them in our blood, lungs, and breast milk.[7] On average, people digest about a credit card's worth of microplastics by weight per week.[8] We all are becoming more plastic, even if we haven't fully grasped what that transformation entails.

Unfortunately, recycling won't make plastic pollution disappear. It is estimated that of all the plastic waste generated to date, only 9 percent has ever been recycled.[9] Further, as Roland Geyer, Jenna R. Jambeck, and Kara Lavender Law have documented: "None of the commonly used plastics are biodegradable. As a result, they accumulate."[10]

Plastics, made mostly from fossil fuels, also exacerbate the unfolding climate emergency.[11] Xia Zhu writes: "Plastic is carbon. More specially, almost all plastic is fossil carbon locked up in polymer form."[12] Considering the magnitude of production, Judith Enck emphasizes: "If plastic were a country, it would be the world's fifth largest greenhouse gas emitter, beating out all but China, the U.S., India, and Russia."[13] Estimated to encompass 20 percent of global oil consumption by 2050, multinational corporate producers are looking to plastics to compensate for a decreased demand for oil, gas, and coal as the world transitions away from fossil fuels and toward renewable energy.[14] Greenpeace coexecutive director Annie Leonard stresses: "For the oil and gas industry, plastic is their lifeline."[15] Meanwhile Earth's climate already has begun to "wobble" or "flicker," as we head toward more "tipping points, or critical transitions."[16]

Despite warning signs, this profound proliferation suggests that most of us have embraced plastics—or at least until recently. So who has begun to share "pissed off" reactions of "outrage" about plastics—and why now?

Beyond Straw Men takes hashtag activism seriously by "staying with the trouble" of and beyond the initial hot takes, to try to dwell in and unravel what is being negotiated in the name of plastics.[17] The title is more than a feminist pun on plastic straws and men who promote them.[18] *Beyond Straw Men* attempts to engage plastics-related hashtag activism in ways that don't fall for or recreate straw man fallacies, which set up an imagined opposition for the purpose of showing how easily it can be torn down. My research complicates discourses that conjure false choices through straw man arguments, such as individual or systemic change (spoiler alert: we need both); whether one country is to blame or all; whether environmental advocacy is helpful or harmful; and yes, whether we should stand for or against all plastics. Advocates against plastic pollution consistently accept and even celebrate what I describe as *impure politics*, a contingent array of tactics addressing a complexity of challenges in imperfect yet impactful ways both online and offline.

I came to this understanding by deliberately listening not only to voices where I live in the United States, a country that bears profound responsibility for plastics, but also to advocates in the Global South.[19] Contemporary calls for regulating plastics have not merely served as a distraction led by white, elite environmentalists, despite how they often are portrayed in the United States and the United Kingdom.[20] We are all impacted by—and contribute to—contemporary environmental crises, but not all equally.[21] Addressing global injustices, Raka Shome insists, calls for more research "to theorize through experiences that emerge from the Global South and keep them at the center of our intellectual and political imagination."[22] "Environmental communication from the Global South," Jagadish Thaker emphasizes, "highlights that environmentalism is not just a value reserved for the postmaterialist rich but embedded in everyday struggles of poor communities against land and resource grab by the government and corporations."[23] *Beyond Straw Men* therefore engages voices of the Global South as a way of learning theory, ethics, and politics from, as Mohan J. Dutta and Mahuya Pal describe, "space constituted geographically and communicatively amidst inequalities in the distribution of power."[24]

To clarify, *Global South* often refers geographically to countries primarily located in Africa, Asia, Latin America, and Oceania, despite crossing the equator. Angela Okune argues for *Souths* to underscore the plurality of worlds often obfuscated in dominant discourse.[25] My approach follows David Naguib Pellow's lead in critical environmental justice studies to "include communities of color and poor communities in industrialized nations within the 'South' designation (. . . 'the South of the North') and privileged communities in poor nations within the 'North' designation (or the 'North of the South')."[26] The South of the North includes communities in the US Gulf Coast (sometimes called "the Gulf South"); in contrast, Australia generally is considered part of the Global North (excluding, however, Aboriginal peoples). I invoke *Global South* and *Global North* not to deny or oversimplify these heterogeneities but to reference profoundly uneven historical and ongoing hegemonic global power relations.[27] As D. Soyini Madison observes in her study of water and human rights in Ghana, global neoliberal policies have "increased poverty and broadened economic equalities across the world."[28] Although Global South/North labels shift and are limited, these distinctions remain a pragmatic shorthand.[29]

Beyond Straw Men engages environmental leadership of the Global South and the Global South of the North, to deliberately reflect anti-colonial, deimperialist critiques of plastic pollution as a methodological praxis of *reorienting*.[30] I write across intersectional identities to resist the flattening of global privileges and oppressions, including but not limited to ability, carcerality, coloniality, class, gender, labor, sexuality, race, and species.[31] To situate my own knowledge, I try to position but not center myself.[32] Throughout, I illustrate how the transboundary crisis of plastics is predicated on multiple forms of oppression about who and what is imagined as disposable.[33]

In addition to humans, marine life has been sounding alarms of a plastics crisis, washing up dead with bellies full of indigestible plastics and strangled or otherwise harmed by plastics. Subsequently, *Beyond Straw Men* identifies the ways hashtag activists invoke and are linked to nonhuman systems. Consideration of marine life from an environmental justice perspective does not escape uneven power relations. Subhankar Banerjee argues: "Multispecies justice is not theory or analysis: it is praxis. It brings concerns and conservation of biotic life and habitats into alignment with environmental justice and Indigenous rights."[34] While not everyone addressed in this book aspires to multispecies justice (or "Democracy"), the movement against plastic pollution generally values nonhuman life, as well as water. A tension that regularly resurfaces is how aquatic relations—and biodiversity more broadly—are entangled with plastics in ways that signal threats to ideals of democracy, abolition, justice, and sustainability.

Through attending to hashtag activism from the Global South and about marine life, I have grown to believe that plastics have "*come to serve as the articulator of the crisis.*"[35] That is, while controversies over plastics signify crises about plastics, for reasons noted, they also provide an entry point into a wider range of contemporary contested environmental topics, such as carbon-heavy masculinity, carceral policies, eco-ableism, greenwashing, marine life endangerment, planetary fatalism, pollution colonialism, and waste imperialism. To analyze this complicated conjuncture, we need to consider more than statistics about plastic materials and sciences. "Politics begin with desire," Gerard A. Hauser reminds us, "and desires are tied to our attachments."[36] To better understand the plastics crisis, I believe we should engage attachments—and *detachments*—that arise in public controversies over plastics.[37]

To elaborate: in a founding text of cultural studies, *Policing the Crisis*, Stuart Hall, Chas Critcher, Tony Jefferson, John Clarke, and Brian Roberts set out "to examine why and how the themes of race, crime, and youth—condensed into the image of 'mugging'–*come to serve as the articulator of the crisis*, as its ideological conductor." Discourse about mugging, they argued, was animating a conservative backlash regarding the "British way of life," including perceived threats from welfare, racism, employment, and American culture. Hall and colleagues take this "moral panic" seriously as an opportunity to explore fundamental cultural values of law and order, to unpack what "mugging" was revealing and obscuring in British public discourse at the time.[38]

Today, many invoke the language of "moral panic" to mock dramatic responses they feel are unwarranted (just search "moral panic and" in a web browser). Yet we all have some "morals," and we all "panic" sometimes—at least I do, as the former guides how judgments are made, and the latter appears to be a reasonable reaction to a range of issues today. Dismissive invocations of "moral panic" miss the more complicated questions posed by Hall and colleagues about what studying a conjuncture entails: How do historical and structural conditions enable a particular matter (mugging or plastics) to become an articulator of crisis? Which forces have gone unnamed or underrecognized in public discourse, eclipsed by polarizing frames and dismissive assumptions? And at a fundamental level, how can we understand which discourses of crisis are legitimate so we can attempt to act meaningfully? To engage these complicated questions, it might be helpful to first define some key terms.

THE PLASTICS-INDUSTRIAL COMPLEX AND THE RISE OF RESISTANCE

There are numerous types of *plastics*. "The term *plastic*," as Max Liboiron (Red River Métis) writes, "refers to many types of polymers with many, many associated industrial chemicals. . . . Plastic in the singular misses things that are rather central to plastic activism, plastic science, plastic policy, and other plastic relations."[39] It is important, then, to consider plastics generally, as well as specific plastics in their variety.

Addressing plastics can feel overwhelming, and many environmental advocates dive in by initially focusing on *single-use plastics*. The term *single-use* in the English language has referred to objects "cheap enough to be thrown away" since the late 1800s.[40] Today, single-use plastic generally refers to an item used once—often briefly—before it is discarded and discounted. Think of plastics in the food industry: utensils, bags, beverage bottles, straws, to-go food containers, and individual condiment sachet packets. Single-use plastics often are light, flexible, durable, impermeable, and transparent.

Single-use plastics epitomize *throwaway culture*, which values immediate gratification, convenience, and disposability in contrast to, for example, endurance, reparability, and sentimentality. Throwaway culture is a structure of feeling of dominant culture in the Global North. Borrowing from Raymond Williams, "structure of feeling" references informal social formations that have become so pervasive, they matter profoundly to our lived experiences in felt ways, even if—and perhaps because—we might not always be conscious of them.[41] Single-use plastics have been integral to the social formations of throwaway culture, which emerged from a desire for profit growth in industry.

Consider packaging, which constitutes the majority of US household trash and nearly half of which is single-use.[42] Plastic studies often recall the editor of *Modern Packaging Magazine*, Lloyd Stauffer, declaring in 1956 that "the future of plastics is in the trash can" and "that it was time for the plastics industry to stop thinking about 'reuse' packages and concentrate on *single use*. For the package that is used once and thrown away, like a tin can or a paper carton, represents not a one-shot market for a few thousand units, but an everyday recurring market measured by the *billions* of units."[43] What this anecdote from Stauffer illustrates is how throwaway culture has been manufactured by design—and exported globally through advertising and lobbying for plastics.[44]

This range of actors is why I refer to the *plastics-industrial complex*, which includes industries that extract or manufacture plastics (petrochemical companies), use plastics (including beverage corporations, grocers, packaging companies, and tobacco), and manage plastics (the waste and recycling industry), as well as the institutional apparatuses that enable them, including those that are private (such as advertising firms and industry trade associations) and public (such as governments). Holding this larger system accountable together enables a more accurate understanding of the conditions of possibility of our current conjuncture, even as each facet is complex.

Consider waste management. The United States produces the largest amount of plastic waste per person in the world; while it is applauded for managing plastic waste well, that plan long involved exporting 70 percent of US plastic waste to China, which in turn led to the global mismanagement of plastic waste into the ocean.[45] China, Julie Sze writes of dominant US ecological imaginaries, often is portrayed as "our psychological displacement and doppleganger, our enemy and our salvation."[46] When China announced the "National Sword" policy to ban the import of foreign plastic waste in 2017, therefore, its decision had global ramifications, particularly for the United States.[47]

One response to the National Sword policy was to maintain business as usual by exporting waste elsewhere. Sharon Lerner's reporting during this time illustrates the consistency of the pattern: "In 2019, American exporters shipped almost 1.5 billion pounds of plastic waste to 95 countries, including Malaysia, which received more than 133 million pounds; Thailand, which got sent almost 60 million pounds; and Mexico, which got 81 million pounds.... Ghana, Uganda, Tanzania, South Africa, Ethiopia, Senegal, and Kenya were among the African countries that also received American plastic garbage, most of which was the hardest to recycle and the least-valued plastics.... And that's just what's in the official record."[48]

Interestingly, though predictably, when plastic waste was sent to emergent Global South markets, there was no public uproar in the United States, no related hashtag trending.

As Lerner explains, however, China's policy prompted a turning of the tide of plastic waste throughout the Global South. By September 2020 "Cambodia returned 83 shipping containers full of waste to the U.S. and Canada with a message from Prime Minister Hun Sen: 'Cambodia is not a dustbin.' In January, Malaysia sent more than 8 million pounds of plastic trash back to the U.S. and 12 other rich nations. And in Indonesia, a customs official announced last year that hundreds of shipping containers, many of which had been incorrectly labeled to mask the fact that they contained plastic waste, were being sent back to their 'countries of origin,' including the U.S."[49] These examples illustrate how rejecting plastic waste imports increasingly has become a movement in the Global South— one relatively ignored in discourses about plastics circulating in the Global North.[50] This concerted refusal has mobilized support for the Proximity Principle, which advocates that waste management should occur near the site of consumption rather than exporting it to someone else's backyard.[51]

In turn, more progressive communities have started to recognize the need to manage plastic waste within their own jurisdictions. For example, rallying around ocean conservation, the US city of Seattle's plastic straw ban passed in 2018.[52] The focus on single-use plastics, as advocates repeated, was to start with the seemingly nonessential and most wasteful forms of throwaway culture (an idea complicated later).

Unfortunately, the plastics-industrial complex has lobbied against even such modest steps to reduce consumption. For example, based on a model created by the conservative nongovernmental organization (NGO) American Legislative Exchange Council (ALEC), some US states have preemptively passed legislation banning the banning of single-use plastics, including Arizona, Florida, Indiana, Iowa, Michigan, Mississippi, Missouri, and Wisconsin.[53] ALEC claims: "Many of the concerns about single-use plastic are overstated."[54]

Regardless of barriers, calls to reduce plastic production have continued to make global headlines over the past decade. In 2018 *Collins Dictionary* declared *single-use* the "Word of the Year" given "a four-fold increase in usage of this word since 2013."[55] The United Nations (UN) chose #BeatPlasticPollution as the theme of the 2018 World Environment Day. The host was India, which announced a single-use plastic ban.[56] By 2021 all EU members committed to banning #SingleUsePlastics.[57] And the UN declared a "blue awakening" in Latin America and the Caribbean, launching a #CleanSeas campaign.[58] Despite setbacks during COVID to reduce plastics, members of the United Nations Environmental Programme (UNEP) met in Nairobi, Kenya, in 2022 to negotiate the parameters for a global plastics treaty among almost two hundred countries, to be decided upon by 2024

FIGURE 1. *Giant Plastic Tap #TurnOffThePlasticTap*. Canadian artist Benjamin Von Wong worked alongside locals to collect and arrange plastic waste in the four-story-tall installation, displayed outside the UN headquarters in Nairobi, Kenya, where a global commitment for a global plastics treaty was announced in 2022. *Source:* https://blog .vonwong.com/turnofftheplastictap/. Reprinted with permission of the artist.

(see figure 1).[59] As the BBC reported: "There will be pressure to help countries in the global south dealing with plastic problems created in the global north."[60] This momentum suggests we are living through a conjunctural shift, one we would do well to consider through environmental justice studies.

PATTERNS OF ENVIRONMENTAL INJUSTICE

Before the internet existed, naming *environmental racism* powerfully articulated how People of Color communities have been and continue to be disproportionately polluted (distributive injustice), as well as left out of environmental decision-making (procedural injustice).[61] Robert D. Bullard has emphasized that "environmental racism is an extension of the institutional racism which touches every aspect of our society, including housing, education, employment and law-enforcement."[62]

Perhaps the most fundamental question in environmental justice studies has been one posed by Bullard: "The goal of an environmental justice framework is to make environmental protection more democratic. More important, it brings to the surface the *ethical* and *political* questions of 'who gets what, why, and in what amount.' Who pays for, and who benefits from, technological expansion?"[63] Following Bullard, *Beyond Straw Men* explores this question: Who is paying for, and who is benefiting from, plastics?

Although race/ethnicity is the leading indicator of environmental harms, critical environmental justice studies entails intersectional analysis. For now, consider Jayajit Chakraborty's research finding a pattern in which the US state of Texas disproportionately locates hazardous waste sites near where people with disabilities tend to live.[64] Likewise, the UN has documented how disasters disproportionately harm disabled people: they are more often left behind and turned away during planning and responses, as well as having greater risk because of disruption to services, discrimination, and exclusion.[65]

Understanding who benefits from environmental injustice is telling as well. In his research on Pepsi plastic waste being distributed from Global North nations to India in the mid-1990s, Pellow argues: "Ecological modernization in the United States is made possible through environmental injustice in Asia."[66] Lisa Sun-Hee Park and Pellow coined *environmental privilege* as "the exercise of economic, political, and cultural power that some groups enjoy, which enables them exclusive access to coveted environmental amenities," such as public green spaces, urban trees, and clean drinking water.[67] Environmental privilege also includes the freedom from harm. As someone who grew up near a dump (technically called a "refuse" site), I emphasize that environmental privilege includes the liberty to distance oneself spatially and emotionally from waste infrastructure, as well as what or who is imagined as disposable.

Transnational environmental justice analysis tracks these privileges. Pellow emphasizes that the global waste trade is largely shipped "from Europe, the United States, and Japan to nations in Latin America, the Caribbean, South and Southeast Asia, and Africa," creating a pattern of *garbage imperialism.*[68] More commonly called *waste imperialism* by the plastics movement, this pattern involves both diplomacy and force to secure a global hierarchy. Waste imperialism is predicated on what Leanne Betasamosake Simpson (Anishnaabeg) identifies as *cognitive imperialism* of Indigenous peoples, which has been "aimed at convincing us we were weak and defeated people, and that there was no point in resisting or resurging."[69] The hegemonic norm of convincing oppressed people that resistance is unimaginable or impractical, of course, serves the interests of economic and political elites of the Global North. Before his public execution, poet and anti-Shell activist Ken Saro-Wiwa (Ogoni) named the pattern of fossil fuel violence against his Indigenous community and land "*slow genocide.*"[70]

In 2021 the UN recognized plastics as a source of environmental injustice with a disproportionate impact on Global South communities, especially those reliant on oceans.[71] This pattern reflects the dominant white settler imaginary of *appropriately polluted spaces*, that is, where certain places and people (falsely imagined as "social pollution") are envisioned as expendable, separate, and "relatively invisible to white citizens and centers of power."[72] Given the unequal burden of plastic harms globally, Liboiron succinctly claims: "*Pollution is colonialism.*" "Colonialism," they elaborate, "is more than the intent, identities, heritages, and values of settlers and their ancestors. It's about genocide and access."[73] Whether or not plastic pollution is colonial, Liboiron emphasizes, depends on the specificity of who, where, and why systems of discarding occur.

Leilani Nishime and Kim D. Hester Williams articulate environmentally unjust patterns as *dispossession*: "This inequality of suffering is not only the historical result of colonialism, slavery, and economic dispossession; it is also bound up in the profoundly dialectical relationship of racialized and gendered bodies to the land itself, and in the 'sorrow and suffering' of labor that was, and continues to be, extracted from these bodies for the purposes of mass profit and the propagation of empire."[74] Often linked with dispossession, *extractivism* signals a set of global infrastructures of oppression intertwined with the blood, water, and soil of different ecosystems and cultures.[75] For example, de Onís illustrates how fenceline communities "resist and refuse dead-end relationality rooted in master logics that normalize extractivism and expendability."[76] These racial ecologies extend beyond land to water and beyond humans to nonhuman kin, including oceanic slave trade routes and networks of marine mammals.[77]

Damage-centered research is compelling since environmental injustices are animated by various articulations of crisis. The larger aspiration is not to just document but to contest patterns of environmental injustice. In contrast to systems that thrive off death, the climate justice movement cherishes *regeneration*, prioritizing and honoring cycles that create life.[78] As Craig Gingrich-Philbrook argues in his research on queer intimacy and desire, there is a danger in only focusing on violence and bruises rather than also on wisdom and kisses.[79] Eve Tuck draws on Gingrich-Philbrook to likewise caution Indigenous studies against solely focusing on "damage-centered research" at the expense of "desire-based inquiries."[80] In this spirit, I turn to networked cultures of care as a necessary counterpart of crisis.[81]

PRACTICING CARE THROUGH IMPURE POLITICS

Care is an overused but significant word today, despite disingenuous *carewashing*.[82] Giovanna Di Chiro identifies the importance of care to environmental justice studies through intersectional collectives and solidarity economies.[83] Allison Kenner reminds us that "care is always embedded in broader systems and relations of care—care infrastructures, carescapes, webs of care, nested dependencies, and human-nonhuman relations."[84] Care ethics have long histories and epistemologies in Indigenous, feminist, Black, and queer communities, underscoring the value of care for self, others, and the environment.[85] Shaped by these traditions, I have written about how care work has been cultivated by marginalized communities recognizing the value of a collective sense of dignity and interdependence—not as weakness, but as profound recognition of knowing the interdependent relationships we require to survive, to grow, and to thrive. Care exists in a dialectic with crisis—intertwined and in excess, without which we might only remain reactionary in our politics.[86]

In contrast, as The Care Collective poignantly states: "Neoliberalism is uncaring by design."[87] Likewise, Alexis Pauline Gumbs laments: "Legally and narratively, our society encourages small, isolated family units and an anti-social state reluctant to care."[88] Di Chiro writes: "The neoliberal, green economy is not a climate- or people-caring economy because it ignores or externalizes out of value production, the actual care work that is required to maintain everyday life in all societies: food production, water procurement, childcare, eldercare, healthcare, animal care, housekeeping and waste management, recycling, and water, air, soil, and biodiversity protection."[89] Placing value in a praxis of care, then, is an act of embracing the often racialized, classed, and gendered work required to create and to maintain life while resisting individuals, institutions, and structures that aim to alienate, colonize, estrange, and exploit. Meera Ghani emphasizes: "A #CultureofCare goes beyond human bodies. It operates from a frame of abundance, timefulness, and interconnection instead of scarcity, isolation and indifference."[90]

Impure politics are circumstantial, not universal.[91] Cultures of care, I believe, must reckon with ecological thresholds by abandoning purity and making critical judgments about how much is too much or too little, and for whom. Impure politics are not best judged by if they are the ultimate act, but by their capacity to become consequential within specific contexts. Cultures of care are constituted online and offline through shared values and experiences of what it means to collectively imagine and manifest more viable futures. Despite inevitable setbacks, as the pages that follow show, networked cultures of care can help us navigate the traumatic times we are living in.

Today, ironically, care work often involves plastics. Consider how Freinkel articulates her position: "We want single-use syringes—they're an important guard against spreading infectious diseases. But when plastic gets used for very trivial things, like a bag that carries your groceries from the store to your kitchen, or a Styrofoam cup that keeps your iced coffee cold for half an hour, and then stays on the planet for hundreds of years, that's when you get a problem."[92] Likewise, Rebecca Altman reflects on plastic-mediated relations that keep her kin alive and able to engage the world, as well as the unnecessary plastics that appear unquestioned.[93] And NGO GAIA Asia Pacific tweets: "#BreakFreeFromPlastic-Friday Ultimately, the burden should not fall on the shoulders of the consumers, particularly those living with disability."[94] These negotiated stances—recognizing plastics as both harmful and lifesaving—reflect the impure politics often expressed by anti-plastic pollution activists. This discourse calls for imagining a world where plastics become obsolete—not a denial that some plastics are integral to some of us living today—though exceptions will be addressed in the pages that follow, as well.

While impure politics helps us theorize the lack of perfect choices in public advocacy, it also is useful to consider impurity ecologically. Imagining the environment as a pristine space is a problematic myth caught up in settler fantasies of rigid binaries between nature and culture, as well as human and nonhuman sentience.[95] Articulating an ethic of impurity, María Lugones has written: "Monophilia and purity are cut from the same cloth."[96] Appreciation for impure politics helps us resist alienation from how our own lives exist as part of the environment instead of as somehow outside of ecosystems and solely as threats.

A pluriversal approach to environmental practices and policies cocreates relations through complexity, diversity, and cooperation.[97] Arturo Escobar recalls "the Zapatista's principle of One No (to neoliberal globalization and the patriarchal capitalist hydra) and Many Yeses (multiple transformative alternatives)."[98] As the following pages attest, somewhere between the all-or-nothing of straw man arguments is where those seeking a more just and sustainable future find ourselves negotiating politics—and our (de)attachments with plastics—again and again. *Beyond Straw Men* enters this give and take through hashtags.

AFFECTIVE HASHTAG ACTIVISM

Though imperfect, global information and communication technologies have enabled everyday people to challenge dominant norms of a singular, logocentric public sphere, making possible networks across multiple scales.[99] Advocates tend to use digital media to alert, amplify, and engage broader publics.[100] While social media technologies in general have transformed public advocacy, the hashtag (#) performs a particular role across platforms. Now integral to public visibility and discoverability for those with access to social media, Jeff Scheible observes, "Hashtags have changed the way we think, communicate, process information."[101]

Politically, the hashtag signals both democratic intent to quickly connect strangers using a few keywords and neoliberal appetites for surveillance and soundbites.[102] Although Chris Messina proposed hashtags on Twitter in 2007, for example, they didn't gain traction on that platform until he convinced his friend Nate Ritter to use a hashtag symbol to spark timely crowdsourcing of public safety information and resources for a wildfire in San Diego.[103] While an environmental emergency created the conditions of possibility for hashtags to become popularized on Twitter (and then used to help track people and conversations as new platforms have emerged), hashtags have exceeded these roots, due to public conversations preceding and exceeding online discourses of public safety, marketing, and activism. The hashtag is used today across the political spectrum by a range of institutional and unofficial actors. Messina affirms: "It's not necessarily *due* to the hashtag that we're having these conversations today about race, gender equity, and economic disparity, but it does play a role in leveling the playing field in terms of who can participate without a priori permission or access."[104] Kelly Wilz underscores: "Hashtags should be treated as important contributions to the democratic process."[105]

Sarah L. Jackson, Moya Bailey, and Brooke Foucault Welles define "hashtag activism as a networked activity" offering a "shortcut to make political contentions about identity politics that advocate for social change, identity redefinition, and political inclusion."[106] They emphasize that the "real work" of activism does not only happen offline.[107] Hashtags, Rachel Kuo likewise observes, "allow participants to express their many identities, histories, and experiences to the digital public while also expressing unity with one another."[108] Fatima Zahrae Chrifi Alaoui further stresses "the importance of digital media for articulating multiple, conflicting identities and remaking the self."[109] Hashtag activism may enable networked counterpublics to emerge, rehearse arguments, articulate identities, mobilize action, and publicize shared perspectives.[110]

While not a guarantor of political success, hashtag activism has played a pivotal role in social change. Perhaps most famously, the 1994 rebellion of the Zapatista National Liberation Front (EZLN) signaled "the earliest global social movement of the internet era," which shared firsthand accounts, poetry, storytelling, and more to interpellate transnational support.[111] More recently, Marisa Duarte identifies "repertoires of contention" of three Indigenous movements that "practice a politics of visibility, cultivate solidarity, diffuse an Indigenous consciousness, enforce dominant governments' trust and treaty responsibilities, and remind many of the irrevocable injustice of colonialism."[112]

Successful agonistic hashtag activism of counterhegemonic campaigns, movements, and revolutions resonates through not just the act of sharing information but also *affects*, a term Gregory J. Seigworth observes is "rangy" in its invocations.[113] Agreeing with Escobar, I believe the contemporary conjuncture requires cultural analysis of "a host of formerly unaccented aspects, including ways of being, knowing, and doing (ontology); spirituality; identities; and culture, emotions, and desire."[114] Attuning ourselves to affect exceeds hyper-rational accounts of politics, which fail to account for the collective ways cultures are and are not attached to plastics, to each other, and to broader ecologies. Seigworth and Melissa Gregg emphasize: "Affect arises in the midst of in between-ness: in the capacities to act and be acted upon."[115] For example, E Cram traces "settler whiteness and ecological affect" as integral to shaping settler colonial power in the Rocky Mountain West.[116]

In contrast, drawing on Women of Color feminist thinkers, Claudia Garcia-Rojas argues that affect long has served communities marginalized by white hegemony as "an ethic of survival."[117] Affect then may be constituted through collective repulsions and/or affinities.

Research on digital media of social movements and revolutions underscores affect's significance. Anjali Vats argues that in the United States, 2011 was a turning point in hashtag activism's popular appeal, with the rise of Occupy Wall Street's use of #OWS on Twitter.[118] Likewise, in their study of Egypt in 2011, Zizi Papacharissi and María de Fatima Oliveira have shown how communication through Twitter constituted collective affect during crises, internally shaping how Egyptians imagined themselves and social change through stories, as well as externally charging public reactions.[119] Analyzing the Arab Spring and the Occupy movement, Ronald Walter Greene and Kevin Douglas Kuswa clarify how the affective dimensions of activism, including hashtags, are pivotal to networking people across geographies through (1) "a much wider affective spectrum than simple outrage and indignation" and (2) "a body of protest in communication with other places in protest, thereby composing a common body in the fold between physical locations."[120] Even clicking "like" or retweeting a hashtag enacts concern that may constitute a trend.[121] Engaging climate content on Tik-Tok, Samantha Hautea and colleagues further illustrate how affective publics can help shape a broader "networked atmosphere of concern" through "attracting individuals to mimic" or riff off "similar sentiments" to boost an "orientation towards public affairs" and "popular consciousness."[122] In *Beyond Straw Men*, these affective networked acts of care prove popular and consequential.

OVERVIEW

As noted in the preface, I was tagged in publicity for a virtual toxic tour that highlighted and strengthened solidarity between Global South and Global South of the North by #StopFormosaPlastics activists (see figure 2). My first single-authored book was animated by *toxic tours*: noncommercial expeditions hosted by fenceline communities aimed at mobilizing public support against material and symbolic toxic patterns of pollution. There, I wrote about (face-to-face and face-to-screen) tours designed to reduce the distance between people touring and those hosting. Most commonly on toxic tours, communities bearing a disproportionate burden of environmental harms invite nonresidents (potential allies, journalists, politicians, etc.) to walk or ride through the places where they live, work, play, pray, and bury their dead as a means of creating an embodied sense of presence to ideally transform toxic patterns of pollution and disaster management. Since then, rhetorics of toxic tours have been amplified through hashtag activism and have enabled fenceline communities to network across greater distances.[123]

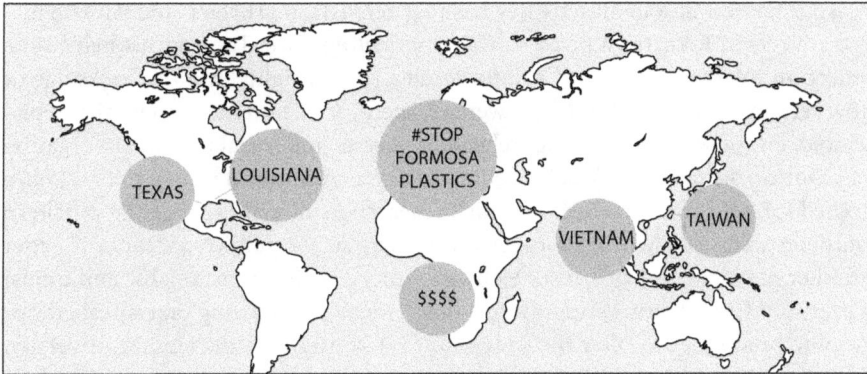

FIGURE 2. Publicity for a virtual #StopFormosaPlastics Toxic Tour, June 30, 2021. The tour featured grassroots speakers from four locations around the world in solidarity, working together to defund, denounce, and divest from Formosa Plastics. *Source:* Adapted from "Formosa Plastics Toxic Tour," *StopFormosaPlastics.org* (blog), July 11, 2021, www.stopformosa.org/post/formosa-plastics-toxic-tour. Reprinted with permission of Tim Schütz.

Beyond Straw Men engages affective, reticulate (multiple and overlapping) publics offline and online, including trending hashtag activism (2015–2022) across platforms, particularly Facebook, Instagram, and Twitter, supplemented by government reports and policies, campaign materials, journalism, scholarship, industry marketing, and interviews.[124] Throughout, I analyze affective counterpublics mobilized by or against plastics-related hashtags to identify how they articulate crisis. As noted, my environmental justice praxis reorients Global North perceptions—akin to what Tiara R. Na'puti (Chamoru) and Joëlle M. Cruz call "flipping the map"—to learn from voices, experiences, and theories from the Global South, Global South of the North, and marine life.[125]

Following best practices of studying social media, all attributed posts of verified public figures (politicians, journalists, celebrities, and advocates) and organizations (NGOs, marketing firms, and governments) were hashtagged and publicly accessible at the time the publisher reviewed this book. Unless noted otherwise, I share hashtagged posts from nonverified, unlocked accounts anonymously, as digital media remains a vernacular space where spontaneous reactions, typos, and grammatical informality can thrive.[126] Even though legally unlocked social media—especially using a hashtag(s)—is searchable by design and circulates publicly, I took precautions because being quoted in a book can lead to increased harassment and surveillance.[127] I revisit social media in each chapter to reflect on the specificity of access and ethics in particular cultural contexts, including illustrative quantitative data of hashtagged trends.[128]

To interview and to amplify key hashtag activists, marketers, and climate justice advocates, I created a podcast, *Communicating Care*. The podcast helped me
reflect on my own biases and comfort zones, gain insights into the backstage of
how certain hashtags unfolded (or were resisted), and learn more about their epistemologies (or ways of knowing) related to plastics and social movement organizing. During the past two decades dedicated to environmental justice praxis, I have
argued for fieldwork practices in which researchers directly engage the subaltern
counterpublics our writing focuses on to help create public spaces for a wider
number of voices and stories, as well as to intervene in unsustainable and unjust
patterns.[129] Overall, my assumption is that—ideally—"drawing on critical ethnographic practices can offer the potential to decentralize—decolonize, diversify,
deanthropomorphize—and to regenerate—rebuild, reimagine, rejuvenate" scholarship.[130] While not traditionally ethnographic, creating a podcast was motivated
by similar commitments to exceed scriptocentric archives.[131]

In the past two decades, podcasts have proliferated within and beyond climate justice networks due to increased access to media technologies and a desire
by some journalists to break silences of major news outlets.[132] Hosting interviews
through podcasts helps create greater transparency about the interviews excerpted
and reaches a wider audience quicker. Podcasts enabled me to speak with international activists and advocates that I could not afford to interview in person, engage
more safely during a pandemic, or meet with less greenhouse gas emissions. As
with any method, the possibilities of podcasts (including visceral, polyvocal, and
accessible) also have limitations (including time and technical expertise).[133] Ideally, podcasts as part of fieldwork repertoires might increase researchers' ability to
amplify voices less heard in mainstream public discourse, bridging epistemological
divides inside and outside of classrooms, as well as between publications and wider
publics. As this book goes to press, my small podcast already has been repeatedly
downloaded, shared, and recirculated; this publicity reflects the ways podcasts may
be used to understand, deliberate, and judge topics of collective interest.[134]

Seeking insights beyond the public transcripts at the time, I invited pivotal
hashtag activists cited on my podcast. One requested a fee I could not afford, and
there are many more I could have asked given additional time and resources.
Despite limitations, I attempted to pay close enough attention to social media
trends to give key voices I did not interview ample credit. Consent to interviews,
reprints, or being quoted does not entail agreement with my arguments or all the
voices in this book.

To make the book more accessible, all hashtags have been converted to camel
case; I have changed, for example, #singleuseplastics to #SingleUsePlastics. If
doing so matters to my analysis, I make note of that.

Turning to this global movement to resist the plastics-industrial complex,
Beyond Straw Men provides two context chapters. Chapter 1 opens in Bangladesh

as a representative anecdote of why plastic pollution has been articulated as a crisis in the Global South and the backlash against those who care. It then summarizes ways plastics harm life, as well as cultural politics that enable—and could undo—their hegemony, including gendered norms and water's entanglements. Chapter 2 addresses barriers to care through three transnational corporate greenwashing strategies of the plastics-industrial complex: the myth of plastic recyclability, Coca-Cola's branding as bright-sided positivity, and how the petrochemical industry (led by British Petroleum) sold the world on carbon footprint calculators.

In the next four chapters, *Beyond Straw Men* turns to the impure politics of specific hashtag activism trends related to plastics and their relevant networked cultures of care. Chapter 3 focuses on Kenya's 2017 ban that resulted from successful, UN-recognized grassroots and government-endorsed hashtag activism (#BanPlasticsKE and #ISupportBanPlasticsKE, respectively). The ban maintains the strictest penalties in the world, both financial and punitive, which appears in tension with an abolitionist politics of care. Chapters 4 and 5 shift attention toward the 2017 US plastic straw drama. First, I consider two Lonely Whale Foundation hashtags that promoted care for marine life: #StopSucking and #StrawlessinSeattle. While they successfully generated social media reach and cultural impact, backlash was swift, including #MakeStrawsGreatAgain. Chapter 5 then foregrounds how disabled networks of care articulated a crisis in response as well, including the #SuckItAbleism trend, as well as how this counterpublic resistance transformed plastic ban conversations and US policies. Following the #StopFormosa coalition, chapter 6 engages hashtag activism articulating care for fish amid the 2016 marine life disaster off the coast of Vietnam: #ToiChonCa/#IChooseFish. Not only did the hashtag trend globally, but it provides an opportunity to engage the plastics-industrial complex beyond bans and single-use plastics. Further, this chapter engages Big Tech's complicity with human rights violations, reminding us not to over-romanticize hashtag activism.

The conclusion reflects across these chapters on the conditions of possibility for plastics to have become a contested articulator of crisis. I also revisit how affective social media may articulate meaningful affinities and galvanize concerted action. Inspired by the impure politics of networked cultures of care, the conclusion then identifies ongoing trends troubling hegemonic plastic relations. Advocates of social change, despite the odds, express anger, humor, joy, reflexivity, and poetic tenderness on the pages that follow. By listening to those advocating for a more viable future, *Beyond Straw Men* underscores compelling ways—despite existing challenges—to become more attuned to international, intersectional, and interspecies interdependence.

1

—

#ThereIsNoAway

Carbon-Heavy Masculinity
and the Life/Death Cycle of Plastics

So that plastic keeps on giving, right?
—JOSÉ TOSCANO BRAVO, EXECUTIVE DIRECTOR,
JUST TRANSITION ALLIANCE

Bangladesh offers a worthwhile touchstone to illustrate why so many Global South nations have come to articulate plastics as a crisis. In 2002 Bangladesh was the first nation to ban disposable plastic bags, motivated in part by the way the bags choked urban drainage infrastructure and created water logging that exacerbated the flooding from monsoon rains. People were literally drowning as discarded plastic bags blocked their ability to drain water.[1]

As leading plastic bag ban advocate in Bangladesh since 1987, environmental journalist, and now NGO director Shahriar Hossian explains: "Flood is the cause of natural calamity but the consequences of flood and waterlog is manmade."[2] Experts estimate that due to its "low elevation, high population density and inadequate infrastructure, . . . by 2050, one in every seven people in Bangladesh will be displaced by climate change."[3] This alarming statistic is a harbinger of how a warming planet increasingly entails living more intimately with disaster, which causes traumas beyond individual weather events. Hossain elaborates: "Changing the erratic weather pattern has also affected our physical and mental health. The climate change in Bangladesh . . . started to impact health and increase the eradication of diseases, particularly the waterborne diseases along with the logged water. And this can lead to the damage of the mental health condition."[4] The way the climate emergency increases a host of risks is why some frame the climate emergency as a "threat multiplier" or what Rosemary DiCarlo described to the UN Security Council as a phenomenon that is "complex and often intersects with political, social, economic and demographic factors."[5] Dipesh Chakrabarty underscores that the climate crisis is therefore a crisis of imagination, one in which we need to think through global human impacts and planetary relations that decenter human agency.[6]

After over a decade of grassroots organizing, many celebrated the plastic ban in Bangladesh; however, like most anti–plastic pollution advocates, Hossain faced intense backlash from the plastics-industrial complex. In the early 1990s, when he began his advocacy, Hossain received threats: "People tried to kill me several times. I requested for protection from the government, but I was denied support. I was forced out of the country instead." Hossain was invited to return from exile in 2001, the year he completed his PhD in urban ecology from Duke University in the United States and the year before the plastic bag ban passed in Bangladesh. Nevertheless, backlash continued. Once the ban had passed, for example, he received an anonymous email: "You hampered our business. You don't know our power. We can purchase minister and secretary any time. So you're a little fly for us. We can kill you anytime. Give up your activity against plastic and us or leave the country forever."[7] This clash between values of survival ("to protect the health and environment") and of elite profit ("you don't know our power") is a common dynamic in environmental politics when subaltern publics challenge hegemonic capitalist industries.

Globally, the false frame of "jobs versus the environment" is a common straw man argument, as green jobs always have been bountiful. While a plastics ban might end some jobs, the creation of all bags or other ways of carrying things requires labor. Bangladesh's plastic bag ban, for example, supports a locally resourced industry: jute.[8] As well as addressing environmental harms, banning plastic bags in Bangladesh enabled a quintessential *just transition* story of how to move from an unsustainable to a more sustainable system without leaving workers behind.

Unfortunately, the impact of the ban in Bangladesh has yet to achieve ideal results. The Bangladesh ban does include incentives for enforcement, both financial (Tk10,000–Tk50,000, approximately US$117–$585) and punitive (one to ten years of incarceration).[9] The country also has directed attention toward the tourist industry and COVID-19 increases in the takeout food industry as incremental steps, maintaining the aspiration of enacting a broader national single-use plastics ban by 2030. Even as laws are vital, a transition away from a plastic-inundated world requires a cultural transformation. As executive director of Bangladesh Environmental Lawyers' Association (BELA) Syeda Rizwana Hasan, emphasizes: "To make it happen, creating mass awareness is the most important thing."[10]

Bangladesh's plastics story is unique but not isolated in the Global South. Ghana, for example, also experienced a flood that led to the deaths of almost two hundred people when plastic trash blocked the drains.[11] In the aftermath of genocide in Rwanda in 1994, the nation became a global leader on plastic bans to address the related waste and to situate "environmental protection as part of the emotional and economic recovery."[12] It is within these life-or-death Global South contexts that momentum for campaigns against the proliferation of plastics should be recognized as a human rights issue, as well as an environmental justice issue. This chapter also explains how the risks exceed disposal or waste.

Plastics, of course, are not the first substance environmental advocates have tried to ban. Previous environmental bans that have been considered successful include, but are not limited to, the banning of dichlorodiphenyltrichloroethane (DDT) sales in the United States to protect environmental and human health, the banning of leaded gasoline to protect public health, and the international agreement to phase out chlorofluorocarbons (CFCs) as part of the 1987 Montreal Protocol to protect the stratospheric ozone.[13] Environmental bans vary in scale and purpose, from cities promoting the banning of cars for a day (to help people reimagine less-polluting transportation) to countries banning plastic microbeads (to protect waterways) to global negotiations to ban mercury (to protect human rights and public health) to no-fishing ocean zones (to help marine life recover).

Bans are imperfect policy choices for imperfect situations, a sign of when a cultural transformation appears infeasible without additional, top-down incentives. Environmental policy bans are a last resort for governments seeking to protect the public good, often to temper the hegemony of multinational corporations. Yet bans are a meaningless regulatory tool if they are declared without funding for the infrastructure to support accountability and transitions. Further, making an object or practice illegal does not stop it from existing or happening. Grassroots campaigns against plastics, then, include bans but also require advocating for changing cultures through which our whole way of life becomes renegotiated.

For those less familiar, this chapter aims to summarize a cultural history of plastic attachments, one fraught with detachment from toxic entanglements and power inequities, beginning with public health and ecological impacts. Drawing on environmental justice studies, the chapter emphasizes who benefits and who pays from the plastics-industrial complex. In conclusion, I return to why connections between water and plastics matter.

PLASTICS 101: THE LIMITS
OF INFINITE TRANSFORMATION

As environmental scientists and engineers emphasize: "The growth of plastics production in the past 65 years has substantially outpaced any other manufactured material."[14] Plastic straws were invented in the 1930s.[15] Tupperware was invented in 1946. Plastic bottles were invented in 1968.[16] If you ever wanted a sign of how fast a culture might be able to change or not, consider that in the United States, two of the largest supermarket chains, Kroger and Safeway, switched to plastic bags just forty years ago, but today plastic bags in marketplaces often feel like common sense or a taken-for-granted habit.[17] "Around the world," the UNEP estimates, "one million plastic drinking bottles are purchased every minute, while up to five trillion plastic bags are used worldwide every year. In total, half of all plastic produced is designed for single-use purposes—used just once and then thrown away."[18] As noted, those numbers are projected to continue to increase dramatically. According to the World Economic Forum, plastics have "increased twenty-fold in the past half-century and is expected to double again in the next 20 years."[19]

The promise of plastics has been the fantasy of living without limits, due to their malleability or "plasticity." This is a foundational myth of plastics that Roland Barthes famously theorized in the mid-twentieth century as "infinite transformation": "More than a substance, plastic is the very idea of its infinite transformation; as its everyday name indicates, it is ubiquity made visible. And it is this, in fact, which makes it a miraculous substance: a miracle is always a sudden transformation of nature. Plastic remains impregnated throughout with this wonder: it is less a thing than the trace of a movement."[20] This tracing of a movement, Barthes warns, shows us the ways that plastics have begun to transform public understanding of human life itself: "The hierarchy of substances is abolished: a single one replaces them all: the whole world *can* be plasticized, and even life itself since, we are told, they are beginning to make plastic aortas."[21] Though plastics were invented in the early 1900s, the fantasy of infinite transformation has helped plastics skyrocket since the mid-twentieth century, canonized in popular culture by the 1968 Hollywood film *The Graduate,* in which a young man (played by Dustin Hoffman) is advised: "There's a great future in plastics."

The American Chemistry Council (ACC) points out that plastics have also saved and improved lives, including in IV tubes, bicycle helmets, automotive air bags, and bulletproof vests. It emphasizes: "Versatile plastics inspire countless innovations that help make life better, healthier and safer every day." The ACC also claims that some plastics can reduce global greenhouse gasses by reducing the weight of vehicles, making them more fuel efficient (which only is relevant while cars are energized by gasoline instead of renewable electricity, but the point remains valid for most vehicles today).[22]

Once they were synthesized, the uses for these polymers appeared endless. Recall that plastics were invented, in part, out of a desire to solve environmental crises: slaughtering elephants for ivory used in billiards and killing tortoises for their shells, which were used in a wide range of everyday objects, from dollhouses to hairbrushes.[23] Since then the US military-industrial complex has used plastics in body armor, helmets, parachutes, ropes, and more.[24] Early film was created on celluloid, and mass music is distributed on plastics, including vinyl records, tape cassettes, CDs, computers, and cell phones. Today, plastics are used for everything from pipes, light switches, and flooring to clothing, acrylic nails, and boats. Plastic culture, as it is called, has fostered the fallacy that our more affordable, more convenient, more pliable futures are limited not by planetary limits but only our imagination.

Popular culture reflects collective affinities and anxieties about plastics. While grassroots movements attempt to "flip the script," mainstream media tends to normalize and promote single-use plastics in the fabric of storylines.[25] Consider the scene in the 1999 film *American Beauty* claiming the most sublime object the young man ever filmed was a plastic bag drifting in the wind above a sidewalk for fifteen minutes. Romanticizing plastic vibes has not fully disappeared, as heard in the 2020 global K-pop hit "Wrap Me In Plastic" by Momoland and Chromance. Identification with plastics, however, sometimes reflects a deeper ambivalence, as pop superstar and influencer Katy Perry's 2010 breakout song of pride and reinvention kicks off with the question: "Do you ever feel like a plastic bag?" There also have long been bad popular cultural aftertastes. Consider popular cultural expressions of living in "a world gone plastic" (as pop group MKTO sings) or of people who "act plastic" (as in movies such as *Mean Girls* or songs like Unghetto Mathieu's "Plastic"). In these contexts, plastics signify the artificial, superficial, and fake facets of throwaway culture.

This indecisive legacy of plastics prompted an entry point into anti-plastics advocacy. A 1990 Environmental Challenge Fund ad in *Time* magazine, for example, juxtaposed the Egyptian pyramids with a polystyrene clamshell box and the message: "Your cheeseburger box will be around even longer."[26] Likewise, a 2022 campaign led by the Egyptian environmental NGO veryNILE built the world's largest plastic pyramid to raise awareness of the magnitude of the problem.[27] Such campaigns continue to goad public reflection: Is plastic waste a cultural accomplishment worthy of pride, like the Wonders of the World? And, if so, what —and who—is this plastic civilization leaving behind?

Like our love/hate relationship with toxics more broadly, most of us today seem to admit our romance with plastics is problematic, but we are less ready to break up with plastics' durable affordances—and sometimes, as chapter 5 addresses, some people literally cannot live without them.[28] To assess the ethical and political dimensions of plastics from an environmental justice perspective, it remains worth recalling Bullard's question: "Who pays for, and who benefits from, technological expansion?"[29] The following pages engage that inquiry in reverse order.

Who Benefits? Carbon-Heavy Masculinity in the Global North

Plastics are a commodity that creates astounding profits for a few economic and political elites of the Global North. As the beneficiary of corporate welfare that uses regulations to manipulate market success, the fossil fuel industry is provided $584 billion on average every year in subsidies from G20 governments through "direct budgetary transfers and tax expenditure, price support, public finance, and investments."[30] The Australian-based Minderoo Foundation published a report in

2019 identifying the twenty polymer producers that create more than half of all single-use plastic waste globally, and the top one hundred accounted for 90 percent. Primarily based in the Global North, transnational petrochemical companies are the main beneficiaries. The top three producers of single-use plastic waste reported are ExxonMobil, Dow, and Sinopec.[31] Each of these companies has made extraordinary profits and was founded by people portrayed by the industry as extraordinary men in their organizational narratives.[32]

According to ExxonMobil, for example, approximately US$4.7 billion in earnings was estimated in the second quarter of 2021, with investments in "key projects, including Guyana, Brazil, [and] Permian [a basin in Texas and New Mexico, US]."[33] Exxon is a US-based multinational petroleum and petrochemical enterprise, a descendant of Standard Oil Company, founded by John D. Rockefeller. Exxon's own timeline of its history begins in 1859, when "Colonel Edwin Drake and Uncle Billy Smith drill the first successful oil well in Titusville, Pennsylvania."[34]

Meanwhile Dow, which was established in 1897 in the United States, claims US$39 billion in 2020 net sales, with fewer than forty thousand employees in a little over thirty countries.[35] Dow's founder, Herbert Henry (H. H.) Dow, a Canadian-born US citizen, had "indomitable optimism [that] helped him persevere against those who nicknamed him 'Crazy Dow.'"[36] In the beginning of the plastics revolution, it was on-brand for ads by Dow and other producers of plastics to regularly place normative men and boys at the center of their campaigns, solidifying them as defining figures by which the public should measure US progress (see figure 3).

Not to be forgotten, founded in 2002, Sinopec is a Taiwanese-based company that "traces its history back to May 4th, 1948 when the Taipei Mutual Savings and Loan Company was established."[37] The founder, Ho Show-chung, was indicted in 2017 for illegal financial dealings, with headlines announcing: "Taiwanese tycoon needs 90 minutes to find NT$400 million (approximately usd$14.5 million) for bail."[38] Nevertheless, Sinopec reported profits in August 2021 alone of US$1.416 million, as well as profits from Bank SinoPec of US$421 million.[39] Although Taiwan has a long history as an occupied, colonized island (by the Dutch, Spanish, Japanese, and Chinese), Sinopec might be what is categorized as an industry of the Global North in the Global South.

The stories of origin of each of these three companies reflect a structural pattern: the co-constitutive relationship between plastics and what Stacy Alaimo calls "carbon-heavy masculinity" or "the hegemonic masculinity of impenetrable, aggressive consumption."[40] In the cast of the masters of the largest plastic producers—Colonel Edwin Drake, "Uncle" Billy Smith, "Crazy Dow," and now-indicted "Taiwanese tycoon" Ho Show-chung—we find militaristic, outlandish, and filthy rich Great Men romanticized for and profiting from the objectification

FIGURE 3. Dow Chemical advertisement featuring a European American boy eating apple pie, surrounded by plastics, linking plastics with dominant US culture, 1950. *Source:* Dow Chemical Historical Collection. Reprinted courtesy of Science History Institute.

and subordination of nature. Today, their corporately packaged stories constitute a master narrative of aggressive masculinity overcoming odds or hardships to rationalize, promote, and normalize their unsustainable carbon relations. These compulsive stories of origin, ads, and other corporate propaganda reveal the often taken-for-granted master narrative of the plastics-industrial complex—the carbon-heavy masculinity that powers innovation and wealth—therefore, any challenge to the maverick captains of capitalism or the plastics-industrial complex they have built risks cultural and financial impotency.

This master narrative also deflects a larger pattern of privilege. As noted, my starting assumption from critical environmental justice studies is intersectional, because, as Julie Sze writes, "unjust environments are rooted in racism, capitalism, militarism, colonialism, land theft from Native peoples, and gender violence."[41] Given the preponderance of literature on the "Anthropocene," which universalizes humanity's role in this geologic age, it seems important to underscore that every company noted in this book was founded by and every lead advertiser hired appears to embody carbon-heavy masculinity. While marking gendered discourses might feel dated, given the broader gender spectrum that has become more common (thankfully) to recognize, privilege long has given those in dominant roles what Berlant called the "freedom to feel unmarked."[42] Feminist analysis of toxic politics, then, seems significant to studying the uneven intersectional power relations of plastics—not to rationalize Global North women as exempt from responsibility, but to identify hegemonic masculinity of the Global North as a critical discourse that has enabled the hegemony of plastics (and the climate emergency). As the charity-driven NGO Oxfam reports, to address carbon requires "challenging stereotypes that promote growth and individual consumerism as normal, desirable, 'powerful,' and 'masculine.'"[43] A "feminist response to global climate change must challenge," as Alaimo emphasizes, "the tendency to reinforce gendered polarities and heteronormativity."[44]

To consider the patterns of who pays for carbon-heavy masculinity and coconstituted patterns of environmental injustices, I turn to the harms of a life cycle analysis (LCA) of plastics or what I call "the life/death cycle" to denaturalize the necropolitical appropriation of biological and ecological "life cycle" stages for toxic commodities such as plastics.[45] The following data-driven analysis may trigger some readers; if statistics about ecological or human collapse are not productive for you, skip this next section, knowing what the evidence shows: while plastics can save lives, plastics also can kill. Further, their negative health impacts follow systems of cultural hierarchy.

Who Pays? The Life/Death Cycle of Plastics

As noted, *plastics* covers a wide variety of materials. Liboiron makes the stakes of the variety in relation to ecologies of harm clear: "The singular term 'plastic' is horribly misleading, given the hundreds of polymer types and blends and the many more chemical additives they harbor and leak. Each poses a different threat (or not) in different environments. Large plastics from fishing gear entangle marine mammals, while tiny plastics that make their way into gills and bloodstreams can cause inflammation, but not in all species. Plastics might be a global problem, but they are not universal."[46] While this book primarily focuses on single-use plastics, as Liboiron reminds us, the broader production of plastics implicates a much more diverse set of commodities with a range of impacts that reflect uneven power relations. Further, addressing waste is not linear but multidirectional and carries on in ways that are both generative and destructive.[47]

Throughout their life/death cycle, plastics are toxic. "A recent review by the United States Environmental Protection Agency (US EPA) revealed that out of 3,377 chemicals potentially associated with plastic packaging and 906 likely associated with it, 68 were ranked . . . as 'highest for human health hazards' and 68 as 'highest for environmental hazards.'"[48] The toxic chemicals involved throughout the creation, use, and disposal of plastics have negative health risks from the placenta to epigenetically lasting multiple generations.[49]

Like the climate emergency, it feels worth repeating: we all are impacted by the unsustainable patterns of the plastics-industrial complex, but we do not contribute nor are we impacted equally. International researchers and policy experts have identified multiple human rights violations in relation to plastics. According to the Geneva Environment Network:

- Children suffer a silent assault on their right to health, and often on their right to life, where plastic toys, utensils and other products contain toxic substances that leach and enter their bodies;
- Coastal and fishing communities suffer in their ability to enjoy the right to food, as a result of dwindling fish resources from plastic pollution, not to mention contamination from oil extraction and tankers disasters;
- Workers in the oil and gas industry, and communities around production facilities, often suffer violations of their right to health as a result of exposure to hazardous substances emitted in the production process of plastics;
- Users and consumers often experience obstacles to the enjoyment of their right to environmental and health information about the volumes and hazards of plastics;
- Indigenous peoples and local communities endure violations of their right to water from massive plastic pollution; and,
- Countless individuals experience an assault on their right to personal integrity from the exposure to hazardous substances that leach from plastics and find their way to their bodies (as demonstrated by the recent study finding microplastic in all placental portion, maternal, fetal and amniochorial membranes).[50]

As expected from reviewing who benefits most from the plastics-industrial complex, women also disproportionately pay for carbon-heavy masculinity. In terms of everyday material impacts, beyond jobs as waste pickers and domestic laborers, the UN identifies disproportionate toxic exposure of plastics on women through everyday lived experiences, including makeup and childbearing risks.[51] In a case study of Ghana, a World Economic Forum report claims: "The plastics value chain is a near-perfect example of how gender norms and roles lead to inequalities when they are not addressed head-on."[52]

Commodity cycles generally involve five phases: extraction, production and manufacturing, transportation (to be sold or transported as waste), use or consumption, and post-use (landfills, incineration, litter, etc.). Each stage of plastics harms the environment, including public health risks, and disproportionately impacts some communities more than others. The health and climate impacts of "burying and burning" plastics again disproportionately land on Global South communities, who more often live near and/or work at landfills and incinerators, becoming exposed to toxic chemicals that leach into water supplies or off-gas into the air.[53] Further, studies of zero waste systems have found that jobs in this less-sustainable plastics economy provide lower wages and worse working conditions than careers in zero waste systems.[54]

To focus just on extraction, all plastics are polymers, larger molecules constituted by smaller ones. Most plastics mentioned in this book are single-use and used in the food industry, referring to synthetic plastics, which require drilling for crude oil, fracking gas, and/or mining for coal. Fossil fuel extractivism can harm the health or life of workers, pollute waterways, diminish water supplies, deplete soil, increase disasters in nearby communities, harm the health of those living nearby, and create global greenhouse gases.[55]

I do not have space in this book to summarize all the environmental injustices related to the extraction of fossil fuels, but part of who pays the "first and worst," as climate justice advocates say, are fenceline communities that long have been oppressed. According to the Just Transition Alliance, for example: "Indigenous Peoples bear a disproportionate impact given that Indigenous communities are often targeted for oil extraction, waste dumping and pipelines that carry the oil from the extraction site to the refineries and ultimately the chemical plants."[56] Stories of this colonial pattern of extractivism may be found among the Ogoni in Nigeria's Niger Delta, the Adivasi in India, the Achuar in the Peruvian and Ecuadorian Amazon, the Standing Rock Sioux Tribe in the United States, and many more, globally for generations.[57]

The US fracking boom particularly "has reversed the fortunes of the plastics industry," creating a plastics industry "renaissance," according to the ACC.[58] "The expected result of all this investment," Zoë Carpenter reports, "is a spike in the amount of plastic produced globally, as manufacturers in Asia and the Middle East ramp up their own production—with capacity increasing by more than a third in the next six years alone, according to an estimate from the Center for International Environmental Law (CIEL)."[59] Extractivist logics, then, continue to proliferate plastics globally, even as they accelerate our trajectory toward more climate chaos.

The remaining four phases of plastics follow this pattern: harming the environment, including workers and local communities—with a disproportionate impact on those with less power. Here, I highlight only one life/death findings for the next phases (production, transportation, and consumption):

The dangerous chemicals cracker plants release are associated with a host of serious health issues [for workers and communities], including: increased rates of asthma, lung and respiratory infections, heart problems, fatigue and nausea, poor birth outcomes, neurological issues such as memory impairment. In large enough doses over time, many of these chemicals can cause cancer.[60]

An international fleet of cargo ships carries 5 to 6 million containers across the oceans at any given time, transporting about 90 percent of the world's consumer goods. These containers—each of which can contain upwards of 60,000 pounds of cargo—are filled with all types of items . . . made at least partly of plastic, or are wrapped in plastic packaging. . . . [A]t least hundreds, and as many as ten thousand, shipping containers are lost at sea annually. . . . Large plastic items can lethally entangle marine wildlife. And over time plastic in the oceans breaks up into tiny pieces and absorbs toxins from the water; it's often ingested by animals, eventually sickening and killing them. And the ocean's plastic pollution problems get transferred back to land when debris washes up on beaches.[61]

People of Color and people that live in low-income neighborhoods more often only have discount retailers (Dollar Stores) as their only choice to purchase housewares, self-care products, and pre-packaged food. Plastics are abundant and there is no alternative. These same consumers cannot just buy their way out of plastics goods and the danger associated from chemicals in plastic.[62]

Once again, these are just representative findings of the impacts related to the life/ death cycle of plastics.

The last stage, the afterlife of plastics, is contested, as it varies greatly depending on which plastic and where it ends up; some plastics in the depths of the ocean and soils remain intact.[63] Some companies do reuse plastics in clothing, including outdoor apparel, athletic wear, and firefighting gear, reintroducing toxic chemicals into human contact.[64] If plastic waste does photodegrade (it cannot biodegrade), however, it breaks down into microplastics, which is why they remain pervasive today. The blessings of the durability, transparency, and flexibility of plastics also become curses, especially once entangled with water.

WATER DEMOCRACY IN A SEA OF PLASTICS

Perhaps the greatest number of headlines reflecting the pervasiveness of a growing concern over plastic waste has been dedicated to polluted waterways, which is why much of this book—like anti–plastic pollution advocacy itself—has ended up engaging waterways and marine life, even though plastics pollute terrestrial and aerial spheres as well.[65] Plastic pollution, however, is not just gathering in one patch in the middle of one ocean.[66] Plastics have saturated the entire water cycle. Plastic waste is found trickling through groundwater in the United States and United Kingdom, raining in Rocky Mountain National Park, blowing in the wind atop the Pyrenees, and being digested by deep sea creatures in the Mariana

FIGURE 4. Video still from *Greenpeace Philippines—Dead Whale*, September 25, 2018. Although this was a staged event, dead whales washing ashore with stomachs full of plastic signals a crisis, including on social media. *Source:* www.greenpeace.org/philippines/.

Trench (the lowest place on earth).[67] Researchers discovered that microplastics in oceans may not only float back to the shore, but also may travel through sea spray mist.[68]

Plastics are ubiquitous in part due to their aquatic entanglements, which also means that plastics disrupt vital life-supporting systems provided by water (see figure 4). As David Helvarg writes: "Although the tropical rain forests have been called the lungs of the world, the oceans actually absorb far greater amounts of carbon dioxide. Microscope phytoplankton in the top layer of the sea act as a biological pump. . . . Scientists recently have come to recognize ocean currents as key to the creation of climate, clouds, and weather."[69]

To counter alienation from the vital role of water to life, as many Indigenous cultures do, Vandana Shiva encourages the value of remembering and cherishing water culture. Consider her distinction between imagining *water as a commodity* (something to be bought and sold) and *water culture* (part of the balance of all life):

To me, water culture is the consciousness of water, the consciousness of being immersed in a water cycle, the consciousness of knowing that we are 70% water, and that the planet is 70% water, and to tread extremely lightly to ensure that water balance is not destroyed. Heightened water awareness creates water culture and water cultures build into them cultures like the sacredness of India's rivers. . . . The culture that creates is extremely different from the culture which sees water running into the sea as wasted and sees rivers as wild women to be tamed and creates the most violent technologies for rerouting rivers, imprisoning rivers and drying out rivers. That idea of control that develops technologies that disrupt the water cycle and impair the water culture goes hand in hand and [is] leading to the current thinking that water is just another *commodity* on the planet, you don't have to give it any special respect.[70]

Water culture, then, prioritizes the relationality of water. Attuning ourselves to water's significance as a public good can help protect human and nonhuman life. In contrast, Shiva argues, water as a commodity reduces water to a resource to control, like women. Commodifying water allows us to imagine that it is acceptable to deny some people access to clean water and to pollute water in a way that would harm the health of water, marine life, or people.[71] It requires that we forget how, as Stephanie LeMenager reminds us, the labor of water remakes worlds.[72]

Moved toward and through "the ocean, islands, atolls, and archipelagos," Na'puti calls for an "oceanic orientation" to research to reimagine "enduring empire and colonialism" by attending "to peoples' experiences as interconnected exchanges and kinships that belong to these places."[73] While *Beyond Straw Men* doesn't focus on Indigenous communities, Na'puti's call resonates with the methodological reorientation of the following pages, particularly communities attuned to aquatic kinships, resisting imperialism, and the Global South movement to challenge pollution colonialism.

To protect water, environmental advocates often emphasize the ideals and practices of democracy, which Shiva links to *vasudhaiva kutumbakam*, the Earth family of Indian cosmology that imagines human and nonhuman life in a continuum.[74] This long-standing value of water and nonhuman life in many cultures articulated through several belief systems exceeds the financial benefit of a few by prioritizing biodiversity and listening to grassroots voices. Most people today are more familiar with the common Indigenous belief system across cultures that "water is life."[75] Antiprivatization, procommons Italian water movement advocates also use a slogan that resonates: "*Si scrive acqua si legge democrazia*" ("We write water, we read democracy"). As Chiara Carrozza and Emanuele Fantini argue, this motto has "articulated its struggle in terms of water as human right and commons, framing the fight against water privatisation as a paradigmatic battle for democracy,"[76] Likewise, in "The Story of Bottled Water," Leonard argues that instead of subsidizing plastics, we should prioritize water infrastructure as a public good, including improving tap water and increasing access to public drinking fountains.[77] Perhaps unsurprisingly, then, water democracy returns throughout this book, like a tide reminding us to not abandon the pull of the moon.

CONCLUSION

Opening with Bangladesh, this chapter reorients attention from popular Global North campaigns often based on consumption and run by foundations toward the stakes of the anti–plastic pollution movement as articulated by the Global South. The upswell of advocacy from the Global South is not one motivated by lifestyle choices, but of the choice to prioritize life. In this context, plastics, climate disasters, water, and conservative policing of dissent intertwine in ways that threaten individual survival, as well as global and planetary relations.

Despite the fantasy of plastics' infinite transformation, plastics can be limiting. The extraction, manufacturing, transportation, consumption, and disposal of plastics cause ecological harms, including to public health. Plastics are problematic for all life on Earth, though disproportionately impacting everyone with less power.

To critically interrupt the hegemony of the plastics-industrial complex, anti–plastics pollution advocates have begun to articulate a different story, one that identifies how plastics—while sometimes life-saving—also already have begun destroying life itself, burdening those without the privilege of proxemics or ability to distance themselves from what and who dominant culture imagines as disposable. Bullard reminds us that an environmental justice analysis should examine who benefits and who pays. The corporations that have benefited the most from the technological expansion of plastics promote stories of origin that romanticize carbon-heavy masculinity, illustrating how—in addition to patterns of environmental injustice identified already—reifying toxic gender norms is central to the ways the plastics-industrial complex imagines and promotes its master narrative.

Including and exceeding Bangladesh, entanglements of plastics and water cycles are predicated on alienation from water culture, which democratic movements to increase regulation of plastics resist. These relations speak to the pervasiveness of plastics and water, as well as to the ways our lives are profoundly shaped by both. The next chapter turns to greenwashing strategies of the plastics-industrial complex, which perpetuate the normalization and naturalization of these inequitable and unhealthy patterns.

2

Have a Coke and a
#FootprintCalculator

The Myth of Recycling and
Transnational Greenwashing

Destroy life on Earth. plant a tree. tweet about it.
—SHARON LERNER, JOURNALIST (@FASTLERNER)

Weekly, signs of climate chaos appear: stronger storms, hotter fires, longer droughts, and more infectious diseases. Writing about the petrochemical industry's response to a hurricane that landed in the US Gulf Coast, Lerner's brilliantly distilled tweet (in the epigraph) succinctly captures the arc of this chapter.[1] Generally, of course, planting trees is a worthwhile act, and one many of us support in a wide range of contexts. Lerner, however, is referencing a longer news story in which she writes about the pattern of how fossil fuel industries claim to care about the environment but continue to cause more harm than good. This public relations strategy of *greenwashing* "refers not only to 'greening' the appearances of products and commodity consumption, but also to the deliberate disavowal of environment effects."[2] Greenwashing signals when environmental harms are perpetuated by those claiming environmental values. Social media has provided the latest platforms for this age-old trick.

Although greenwashing has been receiving more attention lately, it has long enabled environmental injustices. In 1991, for example, Michael R. Reich conducted a study of toxic disasters in three countries (Italy, Japan, and the United States) and found consistent communication patterns: everyday people tend to try to spread or publicize the disaster by sharing experiences, gaining recognition, and creating the grounds for redress, while economic and political elites of the Global North tend to want to contain the story by minimizing and mitigating publicity. Reich identifies three institutional strategies of privatizing toxic crises:

dissociation (stonewalling by avoiding public engagement), *confrontation* (coercively and symbolically discrediting or blaming the victims/survivors or media, as well as portraying the institution as a victim), and *diversion* (passing the buck to another institution and political theatrics of public hearings when decisions are made).[3]

In 1999, in dialogue with the International Climate Justice Network, the San Francisco–based Transnational Resource and Action Center (TRAC) identified a broader set of patterns: "Climate Justice provides an alternative to the 'solutions' corporations have proposed to the climate problem—false solutions which are divisive, inequitable and unjust. Their response . . . is not different from past corporate responses to environmental problems—to DENY the problem, DELAY solutions, DIVIDE the opposition, DUMP their technologies on the developing world and DUPE the public through massive PR campaigns."[4] The five Ds of industrial privatization or containment of public discourse are why advertising and public relations have been vital to the rise of plastics. Recognizing the importance of cultural acceptance to its success, the plastics industry has invested billions of dollars and countless hours controlling the public agenda through paying attention to public opinion, social norms, reputation, and controversies.

Lerner's, Reich's and TRAC's examples are not recognized solely by those resisting the hegemony of the plastics-industrial complex. As noted in the introduction, 2017 was a banner year for turning the tide of plastics. That year, *Plastics News*, which reports to plastic product manufacturers, as well as their suppliers and customers, emphasized that the industry was on alert: "Social media has the plastics industry on its toes. . . . Don't believe me? . . . [D]o a search on 'plastic straws.' You'll get hundreds (if not thousands) of recent results, including many with hashtags like #StopSingleUse, #EndOceanPlastics, #RefuseTheStraw and #StrawsAreForSuckers. A year ago, plastic straws were barely part of debates about single-use plastics and ocean pollution. Today, it's a movement that's growing every day. And that's just straws. You'll find the same results if you search on plastic bags, polystyrene cups, plastic water bottles or balloons."[5] While one might imagine this story ending with a manifesto in response to the crisis of plastics production, this is not that Hollywood tale. Instead, the author suggests: "Still, there's plenty that the industry can do to help. Participate in Operation Clean Sweep. Volunteer in community beach and river cleanups. Show people outside the industry that being in plastics doesn't mean you're in favor of litter."[6] Sound familiar? It should.

The plastics-industrial complex has long promoted the reductive frame that the only problem with plastics is litter. As a public relations strategy, in the United States: "Beginning in the 1950s, big beverage companies like Coca-Cola and Anheuser-Busch, along with Philip Morris and others, formed a non-profit called Keep America Beautiful."[7] The stated mission was to promote beautifying the nation, including individual actions, such as deterring littering. The NGO partnered with the Ad Council to create some of the most iconic environmental public service announcements (PSAs), like the 1971 "Crying Indian" PSA and the 2013 "I Want to Be Recycled," which reinforced this anti–"litter bug," pro-recycling message.[8] "Keep America Beautiful" carries on today.[9]

Granted, many of us have participated in a litter cleanup, and doing so can create a sense of civic pride and/or accomplishment. In the United States, consider the popularity of "Adopt a Highway" signs. Likewise, the social media challenge #TrashTag, created by outdoor company UCO Gear in 2015, asked people to clean up litter in wilderness areas and then post a picture of before and after. The hashtag gained a second life in 2019 when US-based Byron A. Román created a Facebook "#challenge for all you bored teens," which his friend based in Guatemala reposted in Spanish. More than three hundred thousand shares on Facebook and twenty-five thousand posts on Instagram responded with #TrashTag, #TrashTagChallenge, #TrashChallenge, and #BasuraChallenge.[10] Even more popular was YouTuber MrBeast's #TeamSeas campaign to crowdfund US$30 million to remove 30 million pounds of trash from beaches and oceans, which was accomplished in three months.[11]

Perhaps no one in the world has been more dedicated to cleanups than Afroz Shah, a lawyer in Mumbai, India, who began picking up litter from beaches in 2015 every weekend and, as this book goes to press, has been doing so *for over four hundred weeks*. Shah—clearly more dedicated to cleanups than most, if not all people—made his aspirations explicit in an interview: "Cleaning is one part, but it's not the solution. We are drowning in plastic. The bottles, packets, wrappers, packaging to preserve the food is what travels and lands (in the ocean). You have to reduce garbage in this world and change the way our packaging is made. So, it's about what you can do as a person and as a system."[12] Linking his weekly practice and his call for systemic change, Shah is one of many global anti–plastics waste advocates who refuses to bifurcate individual actions and structural transformation, as well as litter from production. When the plastics-industrial complex sponsors a cleanup day, then, it is important to ask if the companies are making the same connections or if they are using the focus on individual actions to deflect corporate and government accountability.

CORPORATE TRASH TALK

The plastics-industrial complex is one of the most successful advertising stories ever. Plastics have been cocreated with literal trash talk—we didn't come to this conjuncture naturally. Rather than explicitly claiming not to care about the environment, the transnational corporate greenwashing of plastics has pivoted on the appearance of green values. The following pages summarize three of the more popular greenwashing corporate campaigns: the myth of plastic recycling codified by the chasing arrows icon, Coke's bright-sided branding, and BP's carbon foot-print calculator as an exemplar of the hypocrite's trap. Each has helped plastics flourish by deflecting corporate responsibility through fostering self-blame, creating challenges for the activists featured in the rest of this book.

Wishcycling and the Myth of Plastics Recycling

One of the most popular green slogans has been the three Rs: Reduce, Reuse, Recycle.[13] It ends up, however, that plastics recycling itself entails greenwashing. As of 2015, approximately sixty-three hundred megatons of plastic waste had been generated, around 9 percent of which had been recycled, 12 percent incinerated, and 79 percent accumulated in landfills or the environment.[14] As noted, analysts predict plastic production will triple by 2050. These numbers are not due to a lack of green messaging and inaction on the part of everyday people; they reflect what happens to plastics once the recycling bin is collected.

Let me start this story a different way: recycling—I still do it, and I believe it matters. Every day, I sort my waste between trash, compost, and recycling. Every week, we put bins on the curb for pick up. Composting and recycling entail no charge; trash costs money to be collected. Despite identifying as an environmental advocate, however, I inevitably stand above waste bins with a piece of plastic in my hand sometimes wondering: "Can I recycle this?"

Aspirational recycling, or *wishcycling*, is the name given by waste and recycling workers to objects people hope are recyclable, even when they are not.[15] Part of the confusion lies in the proliferation of plastics that are hard or impossible to recycle: polystyrene or foam, plastic cups, toiletry tubes, plastic bags, six-pack rings, frozen food boxes, plastic lids, that plastic "pizza saver" (as it's called by the industry) found in pizza boxes, and so on. To decide whether plastic waste is recyclable, we often turn the object around and look for a number in three chasing arrows that form a triangle. Sometimes we throw the questionable plastic in the recycling bin anyway because—consciously or not—we wish it were recyclable—and that is not an accident.

Roger Bernstein of the ACC has argued that concerns about plastics have been grounded in either fear or guilt. On fear of what he called "environmental self-protection," Bernstein claims industry's response primarily has been funding research by experts to reassure the public that environmental harms are minimal or nonexistent. On guilt about participating in throwaway culture, he recalls: "Seven of the major resin makers, including DuPont, Dow, Exxon, and Mobil, launched a special initiative . . . to ramp up plastics recycling."[16] This alliance led to the so-called recycling symbol.

As Finis Dunaway recalls, this now popularly recognized icon was created in 1970 by Gary Anderson, who then was a student at the University of

Southern California, as part of an Earth Day youth art contest sponsored by the Container Corporation of America. Anderson's iconic design was inspired by M. C. Escher, a popular Dutch graphic artist, and an inspiration of Escher's, the Möbius strip, a surface with no boundaries, appearing to create an infinite loop.[17] Dunaway describes the winning design: "Out of all the images that [first] Earth Day would bequeath to the environmental imagination, none would be more significant than the recycling logo . . . [which] attempts to hold the apparently contradictory trends of individual responsibility and environmental regulation perfectly in balance. . . . The logo presented a new aesthetic of environmental hope . . . a reassuring vision of a constantly regenerating future."[18] The three chasing arrows, then, created a circular logic, an endless cycle of possibilities, a promise that people need not worry about consumption or the broader proliferation of plastics—if we can just make the effort to find a recycling bin. Recycling makes many of us feel good because it gives meaning to everyday life through a moralized behavior.[19] And we might even gain a sense of redemption about this newly imagined virtuous consumption that contributes to the myth of plastics' infinite transformation.[20]

To be clear: for metals, glass, and cardboard, the promise of recycling largely is realized in places with adequate facilities. Old aluminum cans may be upcycled as a new can or rain gutters or bicycles. Used glass may be reborn as jars and bottles. Likewise, cardboard may be transformed into more cardboard, a cereal box, tissues, or writing paper. There are many ways recycling works. But plastics are a different story.

In 1988 the Society of the Plastics Industry (SPI) developed a system of codes for sorting plastics—*which include plastics that are not recyclable materially and/ or economically*. Generally, for example, most communities in the United States can recycle #1 PET or PETE (polyethylene terephthalate), including soda, water, and condiment bottles, as well as polyester (and mylar balloons, which recycling centers often do not accept), and #2 HDPE (high-density polyethylene), including thicker plastic bottles, like detergent containers, but also sometimes snowboards, boat parts, or sewer piping. Some communities have the capacity to recycle #5 (polypropylene), which includes some food containers, single-use straws, medical parts, and some automotive parts. Most communities cannot recycle #3 PVC (polyvinyl chloride), including hoses, medical tubes, and plumbing, or #6 PS (polystyrene), including foam food containers, many plastic hangers, and single-use utensils. Further, the remaining two categories are not recyclable: #4 PDPE (low-density polyethylene), including shopping or carrier bags and packaging, and #7, which stands for "other" plastics that fall outside the previous codes, such as many baby bottles, sippy cups, and water cooler bottles. Similar systems were set up globally, including recycling in Japan and the European Union's Green Dot.[21] The plastics industry then spent millions on advertising.[22]

Remarkably, the plastics-industrial complex persuaded governments to place these numbers in the chasing arrows logo—*even though the numbers include nonrecyclables and do not even indicate numbers from most to least recyclable.* Linking the code with the chasing arrows added to the manufactured confusion. The government paid—and continues to pay through taxpayer dollars—for the infrastructure to allegedly recycle plastics. As Dunaway persuasively argues, "This neoliberal solution depended upon massive government support, from new recycling laws to the funding of recycling programs. Between 1989 and 1992, for example, curbside pickup programs expanded almost sevenfold across the United States."[23] This strategy set the stage for the plastics-industrial complex to improve its public image by drawing attention to disposal as the primary plastics challenge and by linking the management of plastic waste to the public good, which deferred producer responsibility not only for disposal but also for the rest of the plastic life/death cycle.

Perpetuating the myth that recycling is relatively simple in the United States, plastics recycling became ubiquitous in labeling and curricula for school-aged children. These early and often repeated educational lessons that reinforce wishcycling are important because trying to recycle nonrecyclables makes it less cost efficient to recycle and more challenging to recycle the plastics that can be processed. "Rather than something which represents a continuous, unbroken loop," as Rudy Sanchez emphasizes of the linkage between the recycling icon and these industry codes, "the very promise of a circular economy—became a series of arrows chasing after themselves and going nowhere."[24] While some communities have begun regulating the cascading arrows to only indicate what is recyclable there, these regulations remain uneven and have far to go globally. (If you want to know how confusing and futile single-use recycling can be, try to correctly classify three to-go coffee cups.) And yet plastics recycling persists as a pervasive greenwashing disinformation campaign, as a myth.[25]

A myth is a foundational story of a culture.[26] Myths shape what we believe to be true and therefore our actions.[27] The myth of plastics recycling has enabled the profitable plastics-industrial complex to convince governments that everyday people should pay for the collection of their waste, recyclable or not. Through taxpayer-funded PSAs, public schools, and more, governments have taught a fundamental lesson of so-called environmental education: focus guilt, and therefore blame, on individuals, not corporations.

Again, environmentalists agree: we should recycle the recyclable. The myth of plastics recycling, however, perpetuates two falsehoods: (1) the cascading arrows indicate that a plastic might be recyclable, when they currently are placed on industry agreed-upon nonrecyclables; and (2) recycling is the responsibility of individuals and governments, not the corporations that market and profit from them. As Max Liboiron and Josh Lepawsky emphasize: "Recycling infrastructure creates a framework where disposables become naturalized commodities instead of foregrounding waste redesign, reduction, or most importantly, elimination."[28]

Coke's "Living Positively" as Bright-Sided Branding

Recall that Bernstein of the ACC argued public concerns about plastics are grounded in guilt or fear. The plastics recycling myth addresses guilt, but that leaves fear: What if more people begin to worry that plastics will cause harm? The Coca-Cola Company offers a successful exemplar of branding aimed at assuaging fears that plastics are unhealthy and unsustainable—or at least at creating a cultural norm to not share such fears. While Coca-Cola does not make plastics, activists often critique its role in increasing the demand for plastics, including its bottles (see figure 5). "Coca-Cola makes about 3 million tons of plastic packaging a year. That's roughly 200,000 bottles a minute."[29]

Today, thanks to its "ad men," as advertising creators have long been called, the Coca-Cola Company is one of the most recognizable brands in the world—from the use of red with white lettering to the bottle shape and slogans. Again, it does not claim to be anti-environmental. Consider the 2007 Coca-Cola branding campaign called Living Positively, which covered these areas for social responsibility: active healthy living, the community, energy and climate, sustainable packaging, water stewardship, and the workplace. Each area was represented with two white arrows to represent the company's partnerships (and seemingly to play off the recycling icon).[30] The advertisers claimed the new "sustainability platform" and "brand icon" for Coca-Cola were emblematic of designer Bruce Mau's philosophy that "the power of markets, brought to bear on the world's real problems, is the power to change the world."[31] "Think Positively" was not the first attempt to link Coca-Cola with uniting or saving the world; have you heard the 1971 "I'd like to buy the world a Coke" jingle? Or the current tagline: "Refresh the World. Make a Difference." "Living Positively" is emblematic of Coca-Cola's broader lifestyle branding strategy that resonates with the tenets of positive psychology.

Positive psychology aspires to help people heal and thrive through optimistic thought ("Visualize success"; "Manifest your dreams"). Popularized by self-help books, its promise is that one can even defy illnesses (cancer, cardiovascular disease, depression, etc.) through attitudes of "gratitude," "hope," and "personal growth."[32] There appears to be some validity in this approach. Consider, for example, the ways laughter can improve one's mood or how some goals are achieved after imagining their possibility.

When taken up by a transnational corporation, however, positive psychology can have a chilling effect. As a branding approach to assuaging health and environmental fears, positive thinking can become a neoliberal tactic to focus on individual consumption instead of collective action, personal feelings instead

FIGURE 5. Video still culture jamming Coca-Cola from Greenpeace International, "Story of a Plastic Bottle—Greenpeace," April 8, 2021. Globally, many environmental advocates have linked the brand Coca-Cola with the ongoing single-use plastics crisis, calling for greater corporate accountability. *Source:* www.youtube.com/watch?v=CLeccbkBZzs.

of material changes, and promises of the future instead of problems in the present. Daniela Ripoll argues in her analysis of the Coca-Cola *Viva positivamente* (Live Positively) campaign in Brazil: "These appeals translate themselves into the construction of a positively healthy body . . . and into the institution of positive thoughts and attitudes . . . via consumption."[33]

As many feminists have argued, in addition to problematically reducing social change to shopping, there are limitations to only expressing positive moods, especially when life entails loss, oppression, and precarity.[34] Barbara Ehrenreich names this hegemony of positive thinking in her aptly titled book, *Bright-Sided*. Her analysis explores the cultural roots of American optimism and how corporations have used positive thinking to manufacture self-blame to deflect systemic change. She argues that in the case of cancer, relentless positivity can imply that everyone who dies just didn't wish hard enough.[35] As someone who has lost kin to cancer, that extreme message lands on me as what Berlant calls "cruel optimism."[36]

Sad someone died? Don't like your job? Feel ill? Worried about the planet? Instead of mourning, organizing with a labor union, lobbying for improved health-care benefits, and taking climate action, Coca-Cola's enduring slogan since 1979 has been: "Have a Coke and a Smile."[37]

When the company sponsors events like COP27 (the UN Conference of the Parties 2022), why might it be worthwhile to speak up about Coca-Cola's branding? While Coca-Cola is part of the global push to privatize water and sell an unhealthy sugary drink, let us focus on its role in the plastics-industrial complex.[38] Consider, for example, a Coca-Cola ad using the recycling icon with the words "refresh recycle repeat" and the tagline: "When you're done, your bottle's not. Please recycle." The press release announced the company had made three hundred bottles from ocean plastic waste as an indicator of the company's intention to reduce its plastic use.[39] This story, then, encourages the repetition of consumption, assuring potential consumers that if we just recycle, we can leave other worries behind. Likewise, another ad ran with "Refresh, Recycle, Reuse" to feature how a chair was made of upcycled Coca-Cola plastic waste; as the advertiser explains, "We took an empowering—not a scolding—approach to upcycling."[40] This approach emphasized the value of positive appeals in creating voluntary corporate change, as opposed to reducing consumption or scolding companies for promoting plastics. Again, Coca-Cola's branding has been coupled not only with a focus on littering and recycling in the United States (Keep America Beautiful) at the expense of producer responsibility, but also perpetuates waste imperialism lobbying against glass and refilling programs globally, including in Fiji, Samoa, and other Te Moananui (Pacific Island) nations.[41]

For years, Coca-Cola has been named the world's worst plastic brand polluter by the coalition #BreakFreeFromPlastic, which advocates for structural change through, for example, its annual global brand audit campaign to collect, calculate, and publicize plastic waste littered on beaches and streets across forty-five countries (see figure 6).[42] In 2019 Coca-Cola produced 2,981,421 metric tons of plastic and emitted 14,907,105 metric tons of carbon dioxide, a major greenhouse gas contributing to the climate emergency.[43] The Break Free From Plastic Movement also has identified Coca-Cola as one of the worst offenders in false narratives about plastics.[44]

Further, as Lerner observes regarding Kenya, "where some 18 million people live on less than $1.90 per day, the responsibility off-loaded by some of the most profitable companies in the world falls to some of the poorest individuals in the

FIGURE 6. Video still from #BreakFreeFromPlastic, "Youth Leaders Reveal the #BrandAudit2021Top Corporate Plastic Polluters," October 24, 2021. *Source:* www.break freefromplastic.org/brandaudit2021/.

world."[45] In 2018, for example, even after sending a delegation in person to a dump in Kenya, where residents were looking for funding to help their youth-driven plastic collection bank, Lerner reports: "Charles Lukania, programs manager of Dandora HipHop City, says that afterward, he sent a proposal and budget to some of the Coca-Cola marketing staff who had visited outlining how the company could support its bank project. But the visit—and the proposal—didn't lead to any funding. Instead, 'they offered to give us a fridge full of Coke the kids could buy.'"[46] Likewise, when Coca-Cola sponsors youth cleanup days, its main contribution often is to donate free beverages—all packaged in plastic—for volunteers as a reward for spending the day cleaning up plastic pollution.[47] Talk about being bright-sided.

Of course, environmental advocates from #BreakFreeFromPlastic, reporters like Lerner, and even I could be told that we are not contributing "positively" when writing about such shortcomings. And so we can be portrayed as scolding or not believing enough in the goodwill of this industry. By framing criticism, dissent, or discord as always-already negative, Coca-Cola reduces the ability of everyday people to call out greenwashing.[48] In this context, those who complain often risk becoming labeled as the problem for pointing out problems. As Sara Ahmed argues of feminist complaints, any criticism or dissent from the norms of positive therapeutic consumer discourse may be framed as coming from willful subjects who are "killjoys," creating hostile conditions for honest conversations about negative impacts. Given this atmosphere, she suggests reclaiming the feminist killjoy as a posture of resistance.[49]

BP's #CalculateYourFootprint as the Hypocrite's Trap

Before jumping into the specifics of the carbon calculator that BP (formerly known as British Petroleum) popularized in 2004, some historical context might be illuminating. BP was preceded by a deal in 1901 commonly called the "D'Arcy Oil Concession." This agreement was made between a British man, William Knox D'Arcy, and the shah of Persia (now Iran), Mozzafar al-Din, in which D'Arcy was granted the exclusive rights for sixty years to prospect for oil in Persia.[50] D'Arcy hired another British man, George Reynolds, as a geologist to drill. BP begins its origin story with these lines: "To find oil in Persia, George Reynolds and his caravan of explorers had lived through seven years of harsh heat, gastric illnesses and disappointments. The next seven years would be no less difficult for the Anglo-Persian Oil Company, which would one day become BP."[51] The language of grit, hardship, the "harsh" Persian environment, and physical ailments is rife throughout BP's narrative of its history. This frame reflects longer patterns of colonial travelogues that manage to denigrate the very places it is profiting from to rationalize extractivism. BP's narrative elements also portray a common myth of meritocracy, which propagates the fallacy that those who work the hardest will achieve the most success.[52]

Throughout history, however, this so-called success has been predicated on the industry being granted explicitly or implicitly the rights of people without the same responsibilities. An example is how Winston Churchill convinced the British government to become a major stakeholder in the company to help it avoid bankruptcy in turn for providing oil to Britain's naval fleet.[53] Corporate subsidies from governments are a recurrent theme of the fossil fuel industry, which would have died off long ago if the industry were run as a truly capitalist enterprise without corporate welfare.

Today, BP is highly profitable.[54] Perhaps the most notable organizational shift in the company historically occurred in 1998, when BP joined with Amoco (American Oil Company), "marking the largest industrial merger at the time and the largest foreign takeover in the US." As BP describes it: "The $48.2 billion deal was the precursor for further consolidation of the oil industry in the early 2000s, including mergers between Exxon and Mobil, Chevron and Texaco, Total, Fina and Elf, and Conoco and Phillips."[55]

In 2000, with the help of the advertising firm Ogilvy & Mather, BP rebranded itself.[56] According to Adweek, in 2004 Patrick Collins became the "worldwide client service director—the No. 2 post—on the agency network's $300 million BP account."[57] Ogilvy and BP's partnership led to their winning the Brand Development Campaign of the Year for "Taking BP Beyond."[58] The brand included not only a new visual identity but also a story about the company's future as "innovative, progressive, environmentally responsible and performance-driven."[59] BP promised to reinvent energy by diversifying investments in renewable sources and go "beyond petroleum."

The company, however, quickly fell short of its advertised aspirations: "In March 2006, a BP oil pipeline caused one of the largest oil spills in Alaska's history. In 2010, another BP explosion on its Deepwater Horizon oil rig unleashed the largest marine oil spill in history. Under financial pressure, BP eventually sold off many of its solar and wind assets, quietly abandoning the rebrand."[60] It was amid these controversies that BP launched the now iconic carbon calculator.

The human ecological footprint metaphor initially was developed in the early 1990s by Canadian researcher William Rees and his PhD student, Swiss-born Matthis Wackernagel, at the University of British Columbia's School for Community and Regional Planning. The goal was to help humans realize how practices of consumption environmentally impact the planet. The tool would allow someone to input basic data to calculate how much area on Earth would be needed to support one's consumption; ultimately, such calculations aimed to illustrate Earth's finite resources, as well as global disparities. For example, some of these calculators and related exercises point out that everyone living in the United States is estimated to have twice the carbon footprint as the global average because of government services, such as roads and police/military.[61]

In 2007, another Ogilvy-BP collaboration led to the award-winning "carbon footprint calculator."[62] Ogilvy's and BP's version focused on carbon, which reflected growing public concerns at the time about the fossil fuel industry's global greenhouse gases.[63] The campaign involved print and television advertisements, as well as an interactive digital tool. A 2003 thirty-second ad, for example, began with the question: "What size is your carbon footprint?" Then it starts a visual montage of people on the street guessing what "carbon footprint" even means. The ad next states: "We all can do more to emit less. Over the next 4 years, we're planning to implement projects to reduce emissions by another 4 million tonnes." And then it directs the audience: "Learn how to lower your carbon footprint at bp/com/CarbonFootprint."[64] If people visited the website, it might declare: "It's time to go on a low-carbon diet" and poll individuals about consumption habits (regarding food, shopping, and travel).[65] Within the first six months, the calculator had almost eight hundred thousand hits, which was considered incredibly successful at the time.[66]

The BP calculator once again epitomizes neoliberal individualism, reinforcing self-interest as a primary motivator and deflecting collective structural change or corporate accountability with a distracting new tool focused on every individual needing to "diet." "The company's branding and advertising activities," as Julie Doyle persuasively argues, "constitute a way of creating a new discourse of energy that draws upon existing discourses of the environment and sustainability to deflect attention away from its dominant ethos, in the context of a common social world increasingly aware of global climate change."[67] Rees himself now argues: "It may sound cynical, but mainstream governments and the corporate sector really have no interest in making the fundamental structural changes needed for the economy to become 'sustainable.'"[68]

Like the recycling icon, as Mark Kaufman reports, "The genius of the 'carbon footprint' is that it gives us something to ostensibly do about the climate problem. No ordinary person can slash 1 billion tons of carbon dioxide emissions. But we can toss a plastic bottle into a recycling bin, carpool to work, or eat fewer cheeseburgers."[69] The calculator, in other words, offers a sense of agency—despite a lack of structural transformation.

While many had begun to point fingers at the petrochemical industry for planetary destruction, the company's response was to ask everyday people what *we* were doing, as if we have the same impact as the largest petrochemical merger in world history: How can we expect them to change if we haven't? It's the common trope of the "gotcha" moment from the environmental backlash playbook, which encourages a feeling of self-blame if one's actions have any negative ecological impact. Relatedly, in writing about coal industry rhetoric, Jen Schneider, Steve Schwarze, Peter K. Bsumek, and Jennifer Peoples identify the "hypocrite's trap" as "a set of interrelated arguments that attempts to disarm critics of industries that provide particular goods or technologies based on the critics' own consumption of or reliance on those goods." The trap is set through three moves: (1) claiming activists don't know "how the world *really* works"; (2) emphasizing that activists are "complicit" through individual consumption of fossil fuels; and (3) shaming activists as hypocrites void of moral power, "reinforcing the neoliberal bromide that There Is No Alternative."[70]

In addition to flattening accountability and deflecting corporate accountability, the footprint calculator focuses on a narrow neoliberal understanding of consumption that does not include diet, food waste, or a range of topics that do have climate consequences. Consider how a version of the tool still exists on the BP website today under "Travel" and has participants make choices between driving cars or taking planes.[71] There are no options for mass transit or bicycling, let alone walking, and arguably the topic of "travel" over "transportation" makes carbon appear to be a lifestyle choice rather than a whole way of life. In a more recent reboot, the campaign has been tweeted; includes an app; and has the hashtag #CalculateYourFootprint, with the following three steps: (1) "Calculate your emissions"; (2) "Pay to offset"; and (3) "Fund Carbon reduction projects." Each step invites everyday people to use the calculator and fall into the hypocrite's trap—from which the solution appears to become donating money back to the industries primarily responsible for creating climate chaos.

Carbon calculators grapple with the complicated crisis of climate by inviting individuals to answer a few questions that require no research so that an algorithm can quantify impact almost instantaneously for users to adopt. As a digital platform simplifying and providing quantifiable comparisons in just minutes, the calculator resonates with a broader trend of gamification in environmental communication, which also resonates with positive psychology to incentivize behavior change through more carrots (rewards) and fewer sticks (punishment). As I explain with J. Robert Cox: "Gamification is the encouragement of something through play, often involving point scoring, competition, and rules. These apps can associate environmental awareness with fun (instead of sacrifice) and motivate people who enjoy competition."[72] The carbon calculator is designed to encourage a sense of complicity in the climate emergency and then agency to address individual impacts. In response to individual factors inputted, the message sent is that adjusting lifestyle choices might have a measurable impact on the incalculable threat many of us have begun to sense. In this context, reducing carbon is framed as a friendly website or app-based competition, where points can be shifted depending on actions chosen and users can compare their scores with others.

Unfortunately, there is little evidence that suggests the corporate gamification of carbon footprints has created lasting, long-term, sustainable environmental behavior changes.[73] Consider, for example, how competitions to reduce water use can end with people taking longer showers afterward as their reward to themselves for their temporary sacrifice. Further, by design, the choices tend to be limited to quantifiable logics and individual choices. As Jeremy Gilbert observes, "the key mechanism of neo-liberal governance" might be "to offer more 'choice' but less democracy" so that we need to rethink how to critique the "hegemony of consumerism and competitive individualism."[74] In the case of corporate gamification of individual climate impacts, the calculator does not necessarily amplify the voices of everyday people in decision-making processes about how we are governed nor encourage significant structural transformations. The equation never suggests protest, dissent, ending subsidies, or otherwise trying to hold BP or other multinational corporations accountable for the climate emergency, including the ways they have lobbied to stop responsible climate regulations.[75] The footprint calculator might raise consciousness about First World privilege, but does it challenge global dynamics of what or who has not been counted as worthwhile?

The debate is long on individual versus institutional change, particularly in climate studies, one that is revisited throughout this book, from littering to refusing single-use plastics. For now, Doyle's analysis of BP's carbon calculator further cautions against the equation of environmentalism with advertising of multinational corporations in general: "Addressing climate change requires changes in practices, as well as beliefs and values. Analyzing BP's branding strategies urges us to think about the broader consequences of equating environmental (and climate change) politics with corporate capitalism, and the role of green branding within this."[76] As a result of dissatisfaction with the carbon calculator paradigm, for example, Emma Pattee has suggested the idea of a *climate shadow* instead, which includes consumption but also choices (such as the number of children and use of money) and attention, which exceed purchasing as our sole gauge of impact in the world.[77] Many have pointed to voting and conversations as actions exceeding capitalist consumption but mattering to collective life. When calculating impact, Bill McKibben emphasizes that it is "multiplication, not addition": "Putting up a clothesline is a fine idea: 1,016 pounds of carbon, remember. But if you join Project Laundry List to fight for the idea of clotheslines, you become, in essence, an Amway salesman for positive change."[78] The footprint or shadow of the average person in the Global North certainly weighs more in these equations.[79] Beyond institutional change through governments and corporations, another option is to choose policies to hold the world's wealthiest individuals accountable and have a significant impact in reducing greenhouse gas emissions overnight through, for example, banning private jets, superyachts, and space tourism.[80]

CONCLUSION

The plastics crisis is about much more than litter. Although cleanups are helpful to local communities, they are bandages that don't stop the fact that we are bleeding every day from the impacts of the plastics-industrial complex. We also need systemic responses to the massive amount of plastic waste being created daily. And remember: more—not less—plastics are being produced daily because, as Greenpeace executive director Annie Leonard emphasizes, "For the oil and gas industry, the stakes are higher too, because single-use plastic is their Plan B."[81]

Overall, plastics manufacturers try to minimize negative publicity through a range of greenwashing strategies to mitigate financial or reputational costs. Applied to plastics, Reich's previously noted pattern holds: *dissociation* ("we create life-saving bike helmets"; "we sell refreshing soda"), *confrontation* ("the problem is litter, not the creation of the product"; "calculate your own footprint"), and *diversion* ("we planted a tree" or "our company is sponsoring a beach cleanup" or "buy our positive brand to feel good"). The plastics-industrial complex's linking of the recycling icon with its codes to sort plastics promotes a myth of recycling as part of the illusion of infinite transformation. Coca-Cola's positive branding deflects fear and dissent. BP's #CalculateYourFootprint deflects accountability through inviting a guilty sense of hypocrisy. These campaigns set up a straw man argument of a choice between focusing on individuals or on systemic transformation, a neoliberal fallacy that encourages the former and not the latter.

The more complicated and honest response is that all politics are impure. The idea of purity is a powerful myth for many, shaping religious and secular cultural practices about how to avoid danger throughout the world and moralize behavior.[82] The fascist history of creating a hierarchy contrasting "the clean and the impure" further signals the danger of insisting on a rigid articulation of purity politics.[83] As Lawrence Grossberg contends, "impure politics" might be all we can do.[84] When writing about boycotts and buycotts, I have agreed, suggesting "consumer-based advocacy is not most productively engaged as the sole or perfect tactic of resistance. Instead, the more compelling question becomes, given contexts in which no tactical choice is pure, how does consumer-based advocacy change the world, if and when it does?"[85] Further, I argue that "the term 'impure politics' not only implicates the contingent array of tactics from which an advocate can choose, but it also provides a way to underscore the complexity of the adversaries one might face during contemporary global ecological crises."[86] Likewise, considering the impure politics of creative appropriations of the law, Isaac West emphasizes that we should "think of political advocacy within the frame of context and contingency" and "attend to dominant logics as opportunities."[87] "Contradiction," as Valerie P. Renegar and Stacey K. Sowards argue, "also provides an important element of flexibility that is necessary for facing new and complex social circumstances."[88]

As a result of greenwashing, part of the challenge environmental advocates face is how good the plastics-industrial complex is at making many of us feel guilty or fearful in order to avoid accountability while making record profits. On the surface, there's nothing terribly wrong with any of these campaign goals: recycling, living positively, or self-reflection on one's own consumption. Challenging unsustainable patterns and policies, however, requires that we recognize the impure politics involved in all environmental decision-making and remember structural transformation is necessary to address crises bigger than one person's capacity. Despite the odds, it is this imperfect advocacy against plastic pollution that is gaining momentum.

3
———

From #BanPlasticsKE
to #ISupportBanPlasticsKE

Pissed Off Online, Picturing Participation,
and Policing Pollution in Kenya

Plastics are politics.

—DR. JUDI WAKHUNGU, FORMER CABINET SECRETARY
FOR ENVIRONMENT AND NATURAL RESOURCES, KENYA

The first ethnographic lesson I was taught happened during a semester abroad in Kenya in the 1990s, hosted by the School for International Training.[1] The assignment was to observe and record our day to the best of our ability without using cultural assumptions. I included an account of traveling in a small, decorated bus playing hip-hop music called a *matatu*. My teacher circled the word "crowded" in my fieldnotes and asked what it meant. She was right: "crowded" could mean many things, depending on one's experiences—did I just mean all the seats were taken? That there was standing room only? Or that bodies of passengers became completely pressed up against each other? Just as one's sense of crowded is deeply shaped by our lived experiences, so are our understandings of plastics. This chapter focuses on plastic bags as contested cultural commodities in Kenya.

In addition to ongoing negotiations for a global response to the plastics crisis, over one hundred countries are committed to legislation regulating plastic bags. Africa leads the world in single-use plastic bans.[2] Although facing unique contexts, each of these policies was prompted by grassroots movements and government reports on the public health impacts of waste, livestock health, water pollution, and more green harms that complicate the oversimplified narrative that single-use plastics are merely a littering problem.[3]

This chapter focuses on Kenya's ban on single-use plastic carrier bags (with or without handles or gussets), which stands out due to the prominent role hashtag activism played in its passing and the inclusion of the strictest penalties in the world to date. Kenya's ban involves enforcement of up to approximately US$40,000 in fines or four years in prison for manufacturing, importing, or selling plastic bags and up to approximately US$1,500 or one year in prison for using them; exceptions include garbage bin liners, medical waste, construction, and food packaging.[4] Then Kenyan cabinet secretary for environment and natural resources Dr. Judi Wakhungu, the first woman to be appointed to that position, announced the ban on March 15, 2017.

This was not the first ban during Wakhungu's tenure.[5] The year before, Wakhungu had led the effort to recommit Kenya to stopping poaching of elephant tusks and rhino horns by helping organize an international media event, which began with a twenty-one-day amnesty for surrendering wildlife "trophies." On the day of the press conference, April 30, 2016, President Kenyatta sparked the largest fire to date created with ivory, igniting over 105 tons of ivory (equivalent to approximately 6,700 elephant tusks) and 1.35 tons of rhino horn at Nairobi National Park.[6] Hoping to protect Kenya's biodiversity through these keystone species, Wakhungu shared: "Although the destruction of ivory and rhino horn will not in itself put an end to the illegal trade in these items, it demonstrates Kenya's commitment. . . . Poaching is facilitated by international criminal syndicates and fuels corruption."[7] The fines, then, had to be larger than the fees poachers were being paid by international syndicates, which also are intertwined with human and drug trafficking.[8] Beyond indicating the significance of Wakhungu's role as a global environmental leader, the burning event also signals her keen awareness going into the plastics controversy that while something may be made illegal (such as poaching or single-use plastic bags), as environmental advocates also often are acutely aware of, there is a need to transform attachments to objects some people have come to fetishize (ivory and horns) or take for granted (plastic bags).

The 2017 ban also wasn't the first effort to ban plastic bags in Kenya, with two notable attempts occurring in 2007 and 2011.[9] Environmental movements and leaders widely acknowledge that "Kenya has historically been a bastion of environmental leadership across the continent."[10] The first Kenyan many heard advocate for a plastic bag ban was Professor Wangari Maathai. She was the founder of the Kenyan Greenbelt Movement, which inspired the planting of more than 50 million trees globally, as well as the first African woman to win the Nobel Peace Prize, the first female to earn a PhD in Kenya, and a successful author.[11] Globally celebrated, within Kenya Maathai was both adored and feared for her intersectional politics of "feminism, environmental sustainability, and democratic governance."[12] Maathai was, as Kundai Chirindo has observed, "without a doubt one of the most influential environmentalists the world has ever known."[13]

Maathai, then, was a significant voice historically when she articulated plastic pollution as a crisis in Kenya—first as an activist and eventually as a member of Parliament and assistant minister of the environment: "The *mottainai* (waste reduction) campaign—both in Kenya in particular, but, generally, in Africa—has been a message of reuse, reduce, and recycle. And it has been around plastic . . . which are often used just once, and then they are thrown into the environment. And they end up in the dumpsites. They end up on the ground. They end up on trees. They end up in rivers. They end up in stomachs of domestic animals and, sometimes, I'm told, they kill the animals."[14] Maathai used her international platform to advocate for a plastic bag ban and called the reversal of the 2007 ban due to plastics-industrial complex pressure "very, very painful."[15]

Prior to Maathai's passing in 2011, the petrochemical industry had too much sway with the national Kenyan government and lobbied heavily to prevent any limits on their businesses. The Kenyan Association of Manufacturers (KAM) claimed that a "ban would cost 60,000 jobs and force 176 manufacturers to close while also contributing a loss in exports."[16] In Pritish Behuria's comparative study of plastic bans in Rwanda, Uganda, and Kenya, an interviewee shared that KAM was known as "a big boys club"; nevertheless, the governmental leadership of Wakhungu KAM's lobbying eventually was outweighed by the relatively larger economic scale of the tourist industry and the growing public pressure to maintain Kenya's reputation as an environmental leader.[17] The intersectional legacy of colonialism has remained palpable as part of Kenya's environmental status, as Omedi Ochieng writes: "This idea of Africa was articulated with the interest of capitalism that conceived of the continent as one rich in natural and human resources for exploitation. The discourses of racialism and capitalism dovetailed with longstanding Christian teachings . . . . These notions often spurred early engagements with Africa as a space for colonial conquest and, then later, . . . as a space for tourism."[18] White settler conservationists in Kenya often have continued the paternalistic rationalization of controlling lands in the name of wildlife or national parks to continue to profit from Kenya.

The success of the 2017 ban in reducing plastic pollution has been attributed to the efforts of hashtag activist, journalist, and photographer James Wakibia. Representative headlines declared: "Meet James Wakibia, the campaigner behind Kenya's plastic bag ban"; "Fed Up With Plastic, This Man Got Kenya to Ban It"; "How this man is helping to solve Kenya's waste problem."[19] Wakibia's hashtag activism began with #BanPlasticsKE, which he describes as less of a campaign and more of him sharing how he was feeling "pissed off." Once he garnered Wakhungu's attention publicly on social media, however, he quickly shifted his hashtag to launch a broader participatory campaign, using her #ISupportBanPlasticsKE. Although no one person is solely responsible for social change, and there is a tendency in history to overshadow the labor of women by spotlighting men as the protagonists of most tales, it is clear Wakibia's hashtag activism, as the *New York Times* reported, "helped inspire Kenya to enact one of the world's toughest plastic bag bans."[20] Further, the campaign "put the issue on the international stage" and the ban helped "re-assert" Kenya's "environmental leadership credentials in the region."[21]

This chapter revisits the touchstones I introduced in the opening of *Beyond Straw Men* within the context of Kenya. First, I elaborate on how Maathai's single-use plastic bag ban advocacy set the stage for the subsequent ban through impure politics. Then, I provide cultural context for how plastic bags were articulated as a crisis. Next, I analyze the hashtag activism and campaign that imagined a broader affective counterpublic, including insights from podcast interviews with Wakibia and Her Excellency Professor Wakhungu. I engage networked cultures of care through research on participatory campaigns and environmental advocacy, particularly attuned to the ways African national politics have been navigated on social media to enable affective resistance. In closing, the impacts of the ban are assessed, including the environmental policy's use of financial (fines) and punitive (incarceration) leverages. Enforcement in carceral systems, as environmental justice and civil rights leaders have argued, often reinforces hierarchies of who can and who cannot pay to escape accountability. While Kenya's single-use plastic bag ban policy is significant to consider given the nation's role as a global environmental leader and recognition of Wakibia's hashtag activism as pivotal to its passage, it also provides an opportunity to question aspects of the celebration of "the strictest ban in the world."

MAATHAI'S PLASTIC BAN ADVOCACY REVISITED

Given her significance, it is not surprising that Wangari Maathai was recognized as a major influence on single-use plastic bag ban advocacy. As Wakibia shared with me: "Indeed, efforts or cause to ban plastic bags did not start with me, you know, me James Wakibia. From 2005 to 2007 to 2011, Kenya had tried banning plastic bags with no success. Wangari Maathai was at the frontline advocating for the plastic bans for a cleaner environment. She was an icon who I greatly admired. I loved her enthusiasm for environmental and human rights. . . . If I want to point out one person, one specific person that really inspired me, was Wangari Maathai."[22] Likewise, Wakhungu credits Maathai as the first person she heard advocate for a plastic bag ban in Kenya, noting: "Well, Professor Wangari Maathai, you know, for those of us that . . . consider ourselves environmentalists, . . . we were all mentored by her."[23]

Notably, Maathai's environmental advocacy emphasized not only the principle of collectively caring for nature, such as trees, but also the intertwined struggles for human rights, particularly peace, democracy, women's rights, and the rights of future generations. When she began to turn to plastics, then, her discourse often illustrated the ways single-use plastic bags had become ubiquitous, negatively impacting many facets of everyday life. In Maathai's memoir, *Unbowed*, for example, she addresses plastic pollution:

> When I look at Nyeri today, I am reminded that when I was a child, people carried beautiful, colorful baskets of different sizes and types made from sisal and other natural fibers to and from the markets to transport goods. These baskets were part of the local handicrafts industry. Today, these baskets are hardly used and instead are made for tourists. The people meanwhile use flimsy plastic bags to carry their goods. These plastics litter the parks and streets, blow into the trees and bushes, kill domestic animals (when they swallow them inadvertently), and provide breeding grounds for mosquitoes. They leave the town so dirty it is almost impossible to find a place to sit and rest away from their plastic bags.[24]

Here, Maathai establishes her disdain for plastic bags as "flimsy" and "dirty," littering and harming animals (both nonhuman and human health). She also expresses a nostalgia for more traditional and more sustainable ways of carrying goods: baskets made of sisal that were locally grown and crafted. In the conclusion of her memoir, she notes her book is published on recycled paper and her *mottaini* (waste reduction) campaign.[25]

Ellen W. Gorsevski argues that this passage in Maathai's memoir suggests a politics of purity in relation to plastics: "The use of action verbs, 'litter,' 'blow,' and 'kill' emphasizes the unsettling movement of the plastic across town in everyday life. Plastics make it nearly 'impossible' to 'rest,' thus the plastics seem to appear in this passage like an invading force. She [Maathai] implies there seems no proper place in nature for plastics."[26] Gorsevski's reading of an appeal to nature within Maathai's reflections is understandable insofar as that one passage might suggest. Further, plastics may appear to be a quintessential colonial substance insofar as they are produced by transnational corporations with little benefit to local people. Similarly, despite his skepticism of "panics over plastics," George Paul Meiu concludes: "Produced elsewhere, proliferating against the collective will, and difficult to control, for postcolonial subjects, plastic is dubious at best—a reminder that globalization, like colonialism, works through multiple, nested forms of alienation and marginalization. It is then not as much *through* plastic, but *against* it—through its repudiation—that postcolonial subjects imagine order, autonomy, and autochthony."[27] Further, I provide examples throughout this book of anti–plastic pollution advocates who do argue that plastics are a form of colonialism and waste imperialism.

Yet I would argue that Maathai's critique of single-use plastic bags was not a universal condemnation of plastics. In her memoir, Maathai recalls "older children had inflatable plastic water wings (readily available and fairly cheap) that allowed them to swim safely in the big pool."[28] In the *New York Times* in 2005, she is quoted making this explicit statement: "I'm not saying don't use plastic at all."[29] Further, Maathai explains that her nonprofit organization, the Greenbelt

Movement, which she dedicated much time and effort to launching and maintaining, is dependent on plastic bags to grow trees: "The other thing of course, is that in the Greenbelt Movement, we use a lot of plastic bags because we grow trees in plastic bags and we nurture them for several months before we transplant them into the ground. And, here, in the Greenbelt Movement, we encourage people to reuse the plastic bags." For me, this statement clarifies that Maathai was not anti-plastics or even anti–plastic bags. And she was honest about her plastic attachments, from swimming safely in the water to retaining water for trees. Maathai's critique, therefore, was not pure—she was not calling for the eradication of plastic completely in all forms, but rather in a very specific form. It is important, then, not to overgeneralize her carefully weighed position as an individual (as opposed to part of the plastics-industrial complex) to avoid setting up her—or the many following in her footsteps—as a straw man.

"AFRICAN FLOWERS" AND "FLYING TOILETS"

Prior to the 2017 ban, plastic bags in the Kenyan landscape became so commonplace that some began ironically to call them "African flowers."[30] Kenyan economist Clive Mutunga states: "You can't miss these bags. It's gotten to the point where it's almost become our national flower."[31] *New York Times* reporter Marc Lacey paints the following picture:

> All over Nairobi, and all over Africa, are ugly artificial blooms that mar the landscape and that environmentalists want plucked up and removed. These flowers are cheap, thin plastic bags that are tossed to the ground by consumers. This kind of litter has reached a critical mass in Kenya—clogging streams, choking animals and piling up into little mountains of disease. . . . The bags are so pervasive in this part of the world that many have taken to calling them 'African flowers,' as if they were local varieties of roses or bougainvillea.[32]

Here, a foreign correspondent identifies not only the ironic neologism "African flowers" but also, as Maathai emphasized, the ways these eyesores created more than aesthetic problems, including blocking drainage infrastructure, harming animals, and spreading disease.

As the "African flower" discourse reflects, when the rhetoric of a "war on plastic" emerged in Kenya in the 2000s, people felt a sense of being surrounded by single-use plastic bag waste in the waterways, beside roads, in the trees, and inside livestock, as well as creating disease. This new plasticscape was not the picturesque landscape for which Kenya has been well known and upon much of which the local economy depends through tourist and outdoor recreation industries.[33]

Another neologism arose during this time called "flying toilets." Due to a lack of convenient sewage or wastewater infrastructure, some living in poverty in the capital of Nairobi, Kenya, were defecating into plastic bags, tying them closed, and throwing them as far as possible. Although this practice allowed "single-use" plastics to receive a second use, this growing trend created concern about the spread of disease, in addition to unsightliness and other problems.[34] For example, Johnson Kaunange, a resident and wheelchair user, told the *Guardian* that they made mobility and accessibility more challenging: "I don't know when the flying toilets started, but they are not good. You never know where they are going to land or where they will fall when it rains. My wheelchair often rolls over the bags and splits them, and then the stink on the wheels is disgusting."[35] Plastic bags, then, became unsightly for residents and tourists, as well as hazardous.

UNEP has been headquartered in Nairobi, the capital of Kenya, since 1972. It is perhaps unsurprising, then, that when UNEP decided to study plastic bags as a regional environmental challenge in 2005, it chose Kenya as a case study for solid waste management. The focus on plastic bags, according to the report, was due to the "magnitude" of the crisis and the attention it was receiving in the public sphere from politicians, environmental advocates, and everyday people: "In Nairobi and indeed all other urban centres in Kenya, plastic bags of all sizes and colours are found dotting the landscape. Besides this visual pollution, plastic bag wastes contribute to the blockage of drains, are consumed by livestock at great danger, and take many years to degrade. Furthermore, Wangari Maathai, the assistant environment minister in Kenya and 2004 Nobel Peace Prize winner, has linked plastic bag litter with malaria. The bags, when discarded, can fill with rainwater offering ideal and new breeding grounds for the malaria-carrying mosquitoes."[36] Public aesthetics, degradation of public services (such as storm water drains), public grazing, and public health all are identified once again as part of the plastic bag crisis in Kenya. UNEP also emphasizes that many of the characteristics that make plastic bags popular are why they have ended up creating problems: affordability leads to "excessive consumption and a tendency for misuse," and easiness to store or transport (due to "thinness and lightness") results in bags being "too thin and fragile to be reused."[37] In addition to disposal and consumption problems, UNEP also emphasizes the petroleum and natural gas used in the production.[38]

In Kenya, specifically, plastic bags had proliferated in just a couple of decades, replacing traditional baskets of banana fibers, sisal, and other local, more sustainable, often biodegradable materials. UNEP documented at the time of this 2005 report that "approximately eight million bags (are) given out every month by supermarkets and two times as many in the informal sector"; in addition, plastic bags were sold in the informal sector and as "designer" plastic bags.[39] In Kenya, when collected, plastic bags often have been taken to a dumpsite to be burned in the open with no controls.[40] The Greenbelt Movement, which Maathai founded, emphasizes that "the burning of plastic bags produces dangerous chemicals (furans and dioxins) which are not only harmful to the environment but also human health," including "lung and respiratory disorders," cancers, "hormonal imbalance, skin diseases, child birth defects, decreasing fertility and suppressed immunity."[41] In summary, UNEP recommended a ban on bags of less than thirty microns and a levy for thicker plastic bags, concluding: "The process of selection, design and implementation of economic instruments to manage a selected solid waste management problem—plastic shopping bags in Nairobi—has reached a critical stage."[42]

Bans were not the initial solution, of course. As noted, the plastics-industrial complex encourages cleanups and anti-littering campaigns, which unsurprisingly were common first steps to clean up plastics in Kenya as well. Yet, as Wakhungu recalls, they did not solve the plastics problem:

> My focus was on single-use simply because, unfortunately, in Kenya, we do have a littering culture, you know, where people just litter wantonly without thinking about who was going to clean up after them. We had several cleanups.... [W]e've had movements, where the last Saturday of every month, all people in the community come and clean up. But it wasn't effective at all, because as soon as you cleaned up on Saturday, by Sunday, there was litter thrown all over the place. Also, our collection of . . . waste has not been the best, and that's why I also initiated the waste management bill, which was just gazetted.[43]

Recognizing the futility of cleanups and focusing solely on littering, Wakhungu began a multistep process to pass a single-use plastic ban in stages, which included working with KAM, the public, the National Assembly, and the Senate for support beyond plastic bags. Although Wakhungu did not achieve all her aspirations, she led an effort to do more about single-use plastics than had ever been done before in Kenya's history. The social media campaign of Wakibia, then, became an opportunity for her to encourage and to showcase public participation in this multiyear legislative process.

ON WAKIBIA'S HASHTAG ACTIVISM

Born in 1983 in the town of Nakuru (approximately 93 miles or 150 kilometers from Nairobi), James Wakibia long has resented environmental destruction. When asked about the beginnings of his environmental advocacy, he recalls deforestation during his primary school years and a dumpsite in his hometown. Wakibia focused on plastic pollution because it was "the most visible problem at the dumpsite."[44] He is a professional photographer and journalist who studied communication and media at Edgerton University, since it was the program in Kenya most aligned with his love of photography.[45]

The use of hashtag activism by social movements and revolutions in the Middle East and Africa has been well established for over a decade, with the 2011 Arab

Spring in North Africa as the quintessential case study. Many new media scholars identify the rise of specific uprisings with digital platforms to mark their interrelated significance, including "the 2009 Iranian 'Green Movement', the 2011 Egyptian social movement and the 2011 Arab Spring movement," emphasizing "the importance of social media such as Twitter and Facebook to organize, communicate and raise awareness."[46] As Tebogo B. Sebeelo emphasizes in his study of the hashtag activism of #ThisFlag (Zimbabwe) and #RhodesMustFall (South Africa): "The African political landscape, in particular, has become a site of contest where new digital technologies, especially social media, have significantly made inroads into an arena that was traditionally the preserve for traditional media."[47] Increasingly, then, policymakers pay attention to media platforms to learn about and to shape public opinion.

These studies, of course, only include a segment of populations, since not all have access to the internet. Sub-Saharan Africa, in particular, "has a lower level of internet use than any other geographic region." In Kenya, as of 2018, according to a Pew Research study, 30 percent owned a smartphone, 50 percent owned a basic phone, and 20 percent owned no phone.[48] From 2015 to 2017, internet use leveled out in Kenya at about 39 percent.[49] This time period maps an era of increased democratization.[50] According to some, Kenya is "the most digitally advanced country in sub-Saharan Africa," where politicians now regularly use social media to campaign and engage publics more broadly.[51] Sometimes called the "Silicon Savannah," Kenya also serves as regional headquarters for "multinational technology companies like IBM Research, Google, and Microsoft," and media reports of the area often hold "ties to market research performed by corporate firms working for the likes of multinationals such as Coca-Cola and Unilever."[52]

When I asked Wakhungu about the role of social media in relation to Kenya's government, she emphasized that government officials are encouraged to have active social media accounts to be open to public feedback and to share with the public what government officials do. She cautions, of course: "We were encouraged to use that as a way of communicating with the public, [although] the public also use it to abuse." Nevertheless, she finds "it's a very effective tool for conveying a message and in Kenya, it's super, super effective."[53]

ARCHIVAL ACCOUNTING AND AFFECTIVE PUBLICS

One way to account for hashtag activism is through quantitative analysis of data. In this case, the numbers verify that these hashtags did trend significantly in Kenya and to a lesser extent abroad. #BanPlasticsKE (with and without an "s" after "plastic") was used by Wakibia starting in 2015. His use of #BanPlasticKE included 378 tweets by 174 contributors with 9,246,237 potential impressions and a potential reach of 4,365,218. The tweets overwhelmingly were in English (341 of 378). The top influencers include investigative journalist and human rights defender Joyce Kimani (with 3.09 million impressions); photojournalist, politician, and activist Boniface Mwangi (with 1.74 million impressions); a Kenyan mobile, web, and SMS platform designed to share information about city access and transportation, Ma3Route (with 1.23 million impressions); and Kenyan Traffic (with 1.13 million impressions). Wakibia follows them with 893.52 thousand impressions.

From 2015 until 2021, a subsequent hashtag of #ISupportBanPlasticsKE (with an "s" on plastics) was used in 8,760 tweets by 1,464 contributors with 112,234,355 potential impressions and a potential reach of 16,432,020.[54] The tweets overwhelmingly were in English (6,521). The top influencer was Wakibia with 28.58 million impressions, the next closest being 12.43 million by Kenyan Traffic. The most popular contributors included NTV Kenya (3 million), Boniface Mwangi (1.74 million), and Ma3Route (1.23 million), as well as Wakhungu (299.96 thousand followers) and Senator Susan Kihika (366.99 thousand followers).

From 2015 until 2021, #ISupportBanPlasticKE was used in 244 tweets by 118 contributors with 3,194,335 potential impressions and a potential reach of 799,966. The tweets overwhelmingly were in English (200 of 244). The most popular contributors included, once again, Joyce Kimani, with 712.5 thousand impressions, and Wakibia as the fourth top influencer with 345.21 thousand impressions. The new second and third top influencers include the advocacy organization WanjikūRevolution Kenya (with 490.61 thousand impressions) and Maskani, the self-defined "Online-Offline collaborative space. Social and civic issues. An informed + An activated citizenry = Transformation" (with 443.26 thousand impressions).

Quantitative analysis of hashtag networks provides evidence of frequency and circulation. Yet as Carrie A. Rentschler argues, "Techniques of web scraping and large-scale data mining of Twitter feeds" may "reproduce an extractive resource-based model of social media research."[55] The following pages, therefore, turn toward qualitative accounts based on my analysis of this archive, secondary news accounts, and podcast interviews with Wakhungu and Wakibia. Sharing social media in this context has minimal risk for those I quote, though I did request permission from nonverified accounts and chose an image of people arrested without an identifiable photograph.

As noted, hashtag activism's appeal—and what makes anything trend on social media—is its capacity to riff and boost affective affinities. Studying movements such as the events that led to the resignation of Hosni Mubarak in Egypt and the US Occupy movement, Zizi Papacharissi argues these less traditional spaces for civic engagement are important to study not only for the information shared (and I would add quantifiable "likes"), but also for their affective dimensions, intensities "that mix fact with opinion, and with emotion, in a manner that simulates the way that we politically react."[56] These online interactions then create what she calls "affective publics," defined as "networked publics that are mobilized and connected, identified, and potentially disconnected through expressions of sentiment."[57] In her work with Maria de Fatima Oliveira, Papacharissi emphasizes the importance of studying how crisis is imagined as a story and "the interplay between social networking sites like Twitter, journalism, and political engagement."[58] Papacharissi argues that these are "networked structures of feeling that help us tell stories about who we are, who we imagine we might be, and how we might get there."[59] Likewise, Wakibia's following story has moved people within Kenya and globally to reimagine who "we" are, potential policies, and collective practices in relation to plastics.

From Everyday Life to Pissed Off Online

Before the hashtags, Wakibia was documenting everyday life in Nakuru. Grounded in a love for where he lived, he made his hometown the focus of his photography starting in 2013. As he recounts: "I had this idea to document the streets. So, every day for about a month, I would wake up very early in the morning and take specific kind of pictures of Nakuru town—they could be streets, structures, people, environment." Wakibia initially had multiple social media accounts, to keep his professional work distinct from his activism, at least initially, including one he began as early as 2009 called Streets of Nakuru, with the handle @StreetsofNakuru, where he first used the hashtag #BanPlasticsKE (and eventually #ISupportBanPlasticsKE).[60]

By 2015, however, the visible pollution and Wakibia's growing frustration goaded him out of his anonymity. Wakibia emphasizes that anger was his motivation: "My campaigns to call for ban on plastic bags was fueled by anger. Seeing so much plastic in the environment, something that I really wanted changed."[61] He was not alone, as was apparent from the eventual popularity of his efforts online. Arguably, then, Wakibia tapped into anger not only as an individual emotion, but also as an affect of Kenyan culture at the time.

Wakhungu, a very elegant and knowledgeable diplomat, surprised me when I asked if anger was how others felt because she too identified with anger:

> He was *angry* that the regulators were not doing anything about it and that's what motivated him. Other people were *angry* because of Kenyans' wanton littering, you know, the fact that we just have a littering culture and people didn't seem to care about the fact that . . . their space that they valued so much was *so* filthy. So, some people were *angry* about other people's lack of responsibility. I was just, I was *angry*, simply because—I would say all of the above. Because I could not—I was sitting there in a very powerful position and yet—I could not *marshall* the entire country to rally behind us and say that this was needed. Other people were *angry*, because they didn't want to lose the convenience of using . . . the plastic bag.[62]

James Wakibia ✔
@JamesWakibia
...

why #BanPlasticKE picture says it all ,

9:37 PM · May 20, 2015 · Mobile Web (M2)

James Wakibia ✔
@JamesWakibia
...

am always pissed off when somebody tells me that we cant do anything about BAD plastics bags #BanPlasticKE

10:21 AM · May 20, 2015 · Twitter Web Client

FIGURE 7. First two tweets by James Wakibia (@JamesWakibia) using the hashtag #BanPlasticKE to promote a plastic ban in Kenya, as posted originally on Twitter, May 20, 2015. Reprinted with permission.

Wakhungu's account of the affective reaction of so many for such a wide range of reasons suggests that the hostility and frustration that plastics evoked in Kenya were of a broader intensity that I previously had not realized. The collective frustration with a *lack of care* was palpable.

Reflecting his keen interest in documenting everyday life and expressing his anger, Wakibia's first two tweets using hashtags about banning plastic link a broad goal ("ban plastic") and a commitment to a specific location ("KE" for Kenya). Following a usual tweet storm pattern, the posts appeared within ten minutes of each other, making them worthy of considering together (see figure 7). What's initially striking to me about these two photographs is their banality. Wakibia is a professional photographer. Looking at his broader portfolio online, he has a photographer's eye for light and shadow, composition, and other professional photography aesthetics. These two photos, however, appear less stylized, more like scenes Kenyans may have witnessed walking, biking, or busing to work countless times in their everyday lives. The landscape picture shows a farmer in a scene with a grassy hillside, ditch, and farm polluted by plastics—yet it also appears fuzzy and relatively flat in its brightness. The picture of the bag appears to be one anyone could create. The photographs, then, reflect an aesthetic of identification, seemingly common, relatable scenes.

Wakibia's first tweet includes the text "why #BanPlasticKE picture says it all," following a logic that visuals may speak for themselves. As an outsider to Nakuru, however, I just assumed this picture was of a ditch on a hillside. It turns out, however, that the green hill is his "hometown's biggest landfill." Neverthe-less, the specificity of the site was less important to even his own choice to tweet and retweet that photograph often. As Wakibia shared: "I think what really made me share that picture is the message that was in the picture, you know, it has a green surrounding, if you look at, and there is these contaminants, the plastic, so basically trying to say 'plastic is destroying our green environment' and which is something that I was really against."[63] As I revisit shortly, juxtaposing people and green landscapes with plastic pollution is a theme carried throughout most of Wakibia's pictures in the campaign.

The second tweet of a bag highlights the hashtag and a live action embodiment of thumbs down. Wakibia told me he doesn't remember whose thumb it was, but that this tweet captured his feelings at the time. The intensity of the tweet isn't subtle, between the iconic thumbs down, the word "BAD," and his tweet that he is "always pissed off" when people tell him that nothing can be done about plas-tic bags. As he recalls: "You know, anger makes people do things, you know, good and bad. And, to me, and especially for my campaigns, anger gives courage to speak out to express it and act better. Most of the time, I acted on pure emotions, and my heart is soft and easy to break. So, when I see an environment polluted, I see blood, my heart bleeds, and I wish that can change."[64] This revelation of how a feeling like anger can be related to its seeming opposite, vulnerability ("soft and easy to break"; "my heart bleeds"), speaks to the intertwined relationship between crisis and care often felt on a personal level.

The "courage to speak out" manifested in this moment as Wakibia tagging government officials online. Wakibia shared how tagging worked to lobby the state on behalf of the public good:

> Using a handle or a hashtag, you know, first of all, you're able to track it. You're able to know how far it went, how many people are reached, you know. And, also, if you wanted to reach a specific person in government, you just tag them, you know, there's no red tape of information. If I wanted to talk to the President and he's on Twitter, I'll just tag the President. If I wanted to talk to the Minister for Environment, Madame Judi Wakhungu, I just tag her with a message and, immediately, when she opens her phone, the first thing she sees is my message.

While this approach of tagging the government does not work for everyone, it did in Wakibia's case, which follows the broader success stories of African social movements using social media to constitute networked cultures of care in ways that include and exceed state governance.

On September 29, 2015, Wakhungu noticed Wakibia's pictures on Twitter, which reflected her own desire for a plastic ban, and riffed off his hashtag to tweet #ISupportBanPlasticsKE.[65] Given her significant role in Kenya, Wakibia believed that Wakhungu's tweet could change everything and tweeted or retweeted the new hashtag seventy-two times that day alone. The task for Wakibia soon became how to shift from his individual opinion to a collective affirmation of the counterpublic advocacy goal. As he recalls: "Just the fact that she tweeted with my hashtag saying that she supported the ban on plastic bags meant everything. That was everything for my campaign. It was trending and then the government came on Twitter, to say that, you know, they supported the ban on plastic bags. So the hashtag changed, inspired by her, by Cabinet Secretary of the Environment because she said, 'I support.' [And] . . . I thought, why not make it personal? Why not personalize the campaign?"[66] The evolution of Wakibia's hashtag illustrates his willingness to give up his recognizable trend for a related one with the potential for a larger following and more significant impact.

That moment is when Wakibia also shifted from solely tagging strangers online to approaching strangers in the streets and encouraging a viral social media campaign. On Twitter, Instagram, WeChat, and Facebook, he encouraged people to personalize the hashtag by posting their own pictures of pollution or self-portraits, providing support for a ban. As he recalls: "Streets of Nakuru started as a Twitter handle, then a Whatsapp group (streetnakuru), then it became a grassroots movement."[67] Of course, there were other actions in between, including demonstrations, cleanups, and petitions, but social media was vital to Wakibia's publicity and the finally successful ban on single-use plastic carrier bags in Kenya two years later, as it made visible growing public support.[68] Since public participation in Kenya's government is influential, the task for those supporting the ban was to grow the movement.

PICTURING PARTICIPATION

How does one person start approaching strangers to build a movement? Wakibia narrates his approach to offline grassroots advocacy to grow hashtag activism as follows: "I would carry a banner. . . [and] I would meet random people in the street to tell them what is happening, about my campaign, about plastic pollution. And if they said that indeed, plastic was a big problem and . . . the plastic bags need to be phased out, they will take a picture holding the placard: ISupportBanPlasticsKE as a sign of commitment . . . it's as if they are . . . signing a petition."[69] After this grassroots organizing in the streets, Wakibia then would post people's photographs and statements at night, tagging them if they had social media accounts and using the hashtag to publicize the campaign (see figure 8).

Sometimes Wakibia enlisted the help of friends to approach strangers, since not everyone was eager to talk to a stranger in the street:

> When you are doing it online, it's different from when you're doing it on the ground in the streets. You know, you're meeting random people you're trying to tell them

FIGURE 8. Post making the link with public health from one of Wakibia's accounts (@Isupportbanplasticske), May 5, 2017. Wakibia encouraged people in Kenya to hold his campaign sign, "#ISupportBanPlasticsKE," and took pictures of them, which he then posted on social media and invited them to share with a quote about why they support a plastics ban. Reprinted with permission.

something unless they just tell you, "[H]ey, I'm busy. . . . I have no time, I have no time for you." Others be like: "Oh, what is what is it? what are you talking about?" And so it was very interesting some time because the way I would try to bring them closer I would . . . go with a friend in the street, he will take a picture, and then people will be curious, and they'll be like: "Oh, what's happening?," and then I will now talk to them: "Oh, this is what I'm doing. I'm running a campaign if you, you know, if you agree with me, you can take this picture and then share it online on Twitter. If you have a Twitter handle, I going to tag you and also follow you." So, it was all: "Oh, this is cool! Let me take a picture."[70]

Wakibia's reflection on how the #ISupportBanPlasticsKE campaign worked beyond digital technologies speaks to how a network of care may be constituted through modeling behaviors and sparking conversations to peak the curiosity of strangers.

As noted, Wakibia's approach worked to publicize the campaign. His repetitive common message resonates with research on how hashtag activism can be used to build a public mood and create a sense of identification and solidarity through public discourse and create common actions.[71] Posts that began trending on Wakibia's social media feeds showed portraits of people (appearing to be from a range of genders, occupations, and ages) with the hashtag message in a variety of settings both smiling and not, as well as pictures of plastic pollution.[72] Support was shown from everyday people in the streets to CEOs, and from park rangers

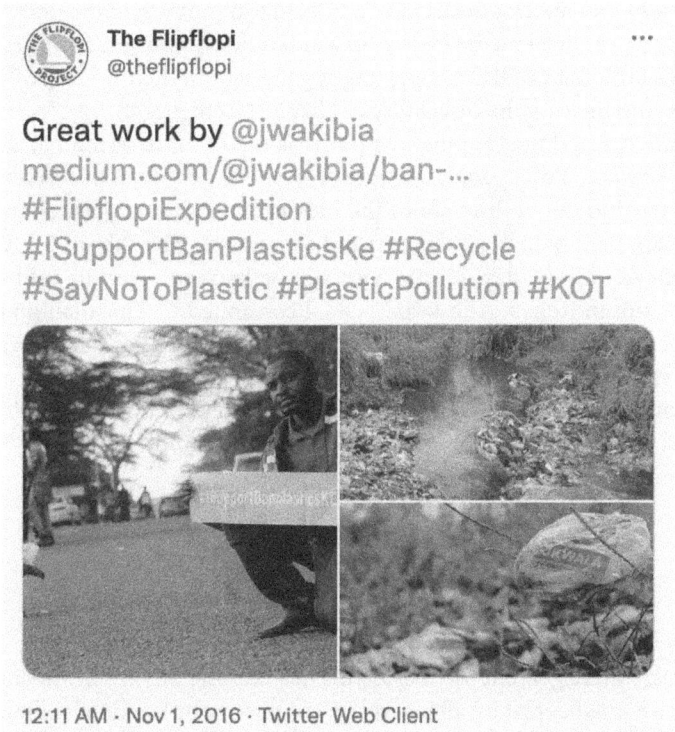

The Flipflopi
@theflipflopi
...

Great work by @jwakibia
medium.com/@jwakibia/ban-...
#FlipflopiExpedition
#ISupportBanPlasticsKe #Recycle
#SayNoToPlastic #PlasticPollution #KOT

12:11 AM · Nov 1, 2016 · Twitter Web Client

FIGURE 9. The Flipflopi (@theflipflopi), "Great Work by @jwakibia"; and Wakibia, "Ban Plastic Bags in Kenya Campaign." The East African NGO Flipflopi launched in 2016 to promote a circular economy for plastics to protect marine habitat and oceans. On social media, it promoted Wakibia's campaign, sharing his photographs. Reprinted with permission.

in uniform to public figures, such as models and politicians. He also continued to circulate photographs of plastic waste, featuring impacts on waterways, on streets, and near parks, and harming wildlife.

Wakibia's social media presence complemented his offline advocacy and vice versa, illustrating how both were personal and embodied for him. His hashtag printed on a sign provided the in-person campaign a consistent message with his online hashtag activism, providing new social media content, broader networks, and a wish that a ban was possible. The offline interactions sparked a desire for public participation and then nurtured that curiosity in the streets with a growing online network that continued to center pictures to imagine a growing public investment in collective action. This campaign underscores Seigworth and Gregg's critical insight that affect is anticipatory, a "bridge of not yet, to the next."[73] Photographs shared on social media by Wakibia and other supporters included pictures of plastic pollution, as well as people holding the sign in their homes, offices, or in the streets, anticipating the possibility of change (see figure 9).

Some still resisted the idea that the future is not set or determined, as Wakibia recalls: "Of course, I met people who are critics. I met a guy in the Streets of Nakuru and I tried to talk to him about the message of banning plastics and he told me, 'Ah, nah, nah, nah. Plastic is part of life, we can't do without it.' And I told him: 'I believe, . . . one day, plastic bags will be banned.' And he told me that's impossible, not in Kenya. And I said, 'Ok.' I continued."[74] This man on the street story echoes the chemical industry's longtime fatalistic public relations campaigns noted previously, arguing that plastics are necessary and unable to be restricted. It also shows Wakibia's resilience in imagining the ban despite previous failed efforts and dominant cultural beliefs, as well as the importance of his role as an effective organizer of an alternative future.

POLICY IMPACTS AND POLICING
POLLUTION BACKLASH

In 2017 Wakhungu declared: "Plastic bags now constitute the biggest challenge to solid waste management in Kenya. This has become our environmental nightmare that we must defeat by all means."[75] The victory of Wakibia's campaign was declared fairly universally when Wakhungu announced the ban in Gazette Notice No. 2334 and 2356, citing the crisis once again: "Many Kenyans have applauded the environmental watch dog (NEMA) for taking the bold step to save the country from the plastic bags menace which has resulted in major consequences to our environment and other sectors of our economy including livestock, fisheries, tourism and the built environment. This is compounded by the fact that plastic bags take over 100 years to degrade. It is known that 100 million plastic bags are handed out in Kenya by supermarkets alone."[76] Some, like John Kariuki Njuguna, worried that loopholes would allow manufacturer use for industrial zones and disproportionately penalize small vendors, calling it "discriminatory as the effect on the environment does not discriminate the source of the pollution."[77] Yet Wakhungu reassured Reuters—and has repeated in many venues—that enforcement would begin with companies: "Ordinary *wananchi* (Kiswahili for everyday people) will not be harmed."[78] The goal was to stop manufacturing in Kenya and to incentivize manufacturers to create more sustainable products.

Despite some of its members being in the business of producing more sustainable products, KAM predictably sued the government over the ban, claiming potential financial losses and job losses, upholding a long tradition of global industrial framing of environmental policy in a false dichotomy of jobs versus environment (as if environmentally sustainable labor does not exist). After multiple legal cases were filed, the highest Kenyan court upheld the single-use plastic bag ban on February 28, 2017. The judges claimed the law was reasonable, constitutional, and justified, and "although some ordinary Kenyans could suffer social and economic losses as a result of the ban, the plastic ban was for the common good of the general public and as such lawful."[79] This legal stance against KAM provides legal precedent for plastic pollution to be interpreted as antithetical to the public good. It is not clear, however, that Kenya will be able to expand the ban to plastic bottles or other single-use plastics.

Wakibia himself declared the campaign over on August 27, 2017, announcing the end of his hashtags to make way for new articulations, like "less plastic is fantastic." He does lament that the plastics ban became limited to such a specific set of uses of a specific subset of single-use plastics, hoping that in the future Kenya might "completely phase out all plastic bags," as well as "disposable plastic utensils and straws."[80] In our podcast interview, he was prioritizing plastic bottles more as a necessary next step in reducing plastic pollution, and he clarified that he does agree some plastics, including personal protective equipment (PPE), will continue to be needed to protect public health; his "hope is that there will be some innovation" so that "in the future, we will have more sustainable products." Overall, however, Wakibia has celebrated the campaign as a success: "The ban has been successful 80%. I'm not a scientist, but, in my own opinion, and it has been quite successful. And for that, I applaud the government."[81]

Two years since the ban's passing, NEMA declared it a success, emphasizing that, beyond punitive top-down measures, the ban had changed Kenyan throwaway culture already: "Most importantly, the Authority has reported increased levels of compliance with ban. Notably, the public has changed attitude towards their view on plastic bags and their usage in relation to the environment. Moreover, there been visible cleanliness in most towns that previously used to be dotted with plastic carriers bags hanging loosely on buildings and trees. 80 percent compliance with the plastic ban has been recorded."[82] Shifting attitudes and visible results felt like palpable victories, even as NEMA admits that some still resist change. The conversation has not ended, but impacts have been immediate on Kenya's whole way of life, as Wakhungu had desired. Even the flying toilets have decreased in frequency.[83]

Yet no solution is uncomplicated. Using bans as an environmental policy tool often turns on financial (fines) and punitive (incarceration) consequences. Fines are a justice issue insofar as they can exacerbate who can and cannot pay, as well as bringing into relief who is or is not fined at all. Unequal burdens of financial penalties that are not tied to income make the impacts disproportionate. Consider how approximately 16 percent of Kenyans live on less than US$694 a year, whereas ExxonMobil pays CEO Darren Woods approximately US$23.6 million annually.[84] A fine, then, of US$455–$1,365 is almost all or more than one person's salary, as opposed to being relatively insignificant to another.

Incarceration also has environmental justice implications. Since 2017 Pellow and his students have published annual reports on the intersections of prisons and environmental justice considerations, including but not limited to toxic pollution, the use of prison labor for fighting fires, and impacts of the climate emergency on disaster risks at prisons.[85] In their 2020 report, Pellow's research collective identified poor prison conditions globally as environmentally unjust, including reports of Kenyan prisons having "bad ventilation, bucket toilets" and being "filthy," leading to disease outbreaks, including "scabies, tuberculosis, and diarrhea."[86] Prison abolition, according to Angela Davis, is a shift from "a justice system based on reparation and reconciliation rather than retribution and vengeance."[87] Likewise, Ruth Wilson Gilmore argues for invoking a sense of presence for a broader commitment to the life-affirming principles of environmental justice: "Abolition is about a presence, not absence. It's about building life-affirming institutions."[88] In a study of feminist hashtag activism and anticarceral politics of collective care, Rentschler contends that transformative justice requires "re-directing the desire to punish toward structures of community accountability and away from solutions based in police action and criminalisation."[89] Given these arguments, it is worthwhile to ask the significance of celebrating—or questioning— "the strictest policy."

While plastic bag use and littering have declined in Kenya since the ban began, the decision to police the crisis remains controversial. The National Environment Management Authority (NEMA) Kenya is somewhat equivalent to the US EPA insofar as it is the government agency charged with creating and implementing environmental protections (though it appears to enforce more and research less). Two years after the plastic bag ban was initiated, NEMA Kenya reported "over 500 arrests and 300 prosecutions. Those found culpable have been fined between Ksh 50,000 and Ksh 150,000 (which is approximately usd$455–$1,365) with some jailed."[90] With the average salary in Kenya at approximately 3,000 less than 150,000 ksh, again, the fine is heavy for most. Although many have expressed gratitude for the reduction in pollution in a relatively short period of time, some have begun to stir a backlash by asking equity questions about who is being held accountable and who is not.

A few news sources picked up on a Twitter controversy in February 2020, for example. NEMA Kenya tweeted that it had arrested three vendors with approximately five hundred bags. Even though that amount does suggest a midlevel trader in banned plastic bags, celebrity Boniface Mwangi retweeted the post with the comment: "Poor traders will be jailed for upto [sic] 4 years for using plastic bags or pay millions in fine. Unfortunately @NemaKenya will never arrest the rich people who dump industrial waste on our rivers or the ones who run

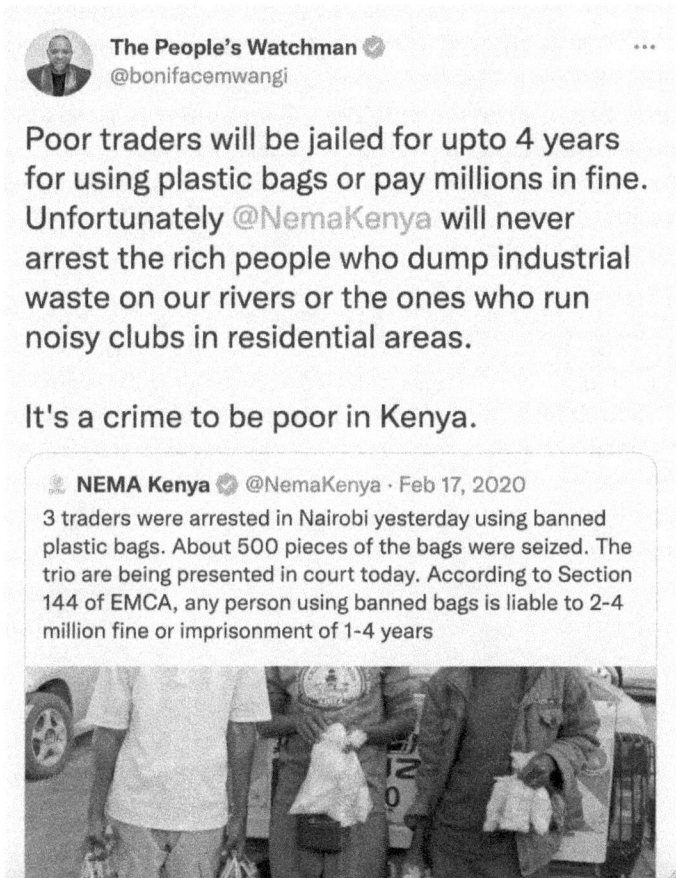

The People's Watchman ✓
@bonifacemwangi

Poor traders will be jailed for upto 4 years for using plastic bags or pay millions in fine. Unfortunately @NemaKenya will never arrest the rich people who dump industrial waste on our rivers or the ones who run noisy clubs in residential areas.

It's a crime to be poor in Kenya.

NEMA Kenya ✓ @NemaKenya · Feb 17, 2020

3 traders were arrested in Nairobi yesterday using banned plastic bags. About 500 pieces of the bags were seized. The trio are being presented in court today. According to Section 144 of EMCA, any person using banned bags is liable to 2-4 million fine or imprisonment of 1-4 years

FIGURE 10. Post by NEMA Kenya announcing plastic bag ban arrests, February 17, 2020. Boniface Mwangi (@bonifacemwangi) retweeted the post to criticize who is held accountable and who is not.

noisy clubs in residential areas. It's a crime to be poor in Kenya."[91] Mwangi received 1.8 thousand likes and 1,000 retweets for his post (see figure 10). While many agreed that plastic waste was a problem, there was public controversy in response to NEMA's post about whether the poor were targeted more than the wealthy manufacturers for enforcement of the ban. The head of campaigns for the Wangari Maathai Foundation and Green Generation Initiative, Elizabeth Wathuti (@lizwathuti), commented on Mwangi's post: "The poor have a voice too and the rich can also be arrested! No one is above the law. We must hold everyone accountable!"[92]

One way to deter the disproportionate burden on everyday people is to focus only on those who are in the business of plastic bags, but where does one draw the line? If five hundred bags are not enough, what is? Another approach is to focus only on bag manufacturers. NEMA Kenya also has tweeted about fining effluent waste and larger distributers.[93] One of its larger arrests made headlines on May 29, 2019, when NEMA coordinated with Kenyan DCI (Department of Criminal Investigation; somewhat like the US Federal Bureau of Investigation) officers to arrest a major distributor, with 3.445 million pieces of plastic confiscated.[94]

Yet what else could be in place to deter single-use plastic bags without financial and punitive consequences? Encouraging alternative bag industries and subsidizing their distribution with the fines collected from larger manufacturers might be one approach to shifting the burden of change, in addition to salary-based fines. Improved sanitation for all to reduce the need for bags as defecation carriers also would help, which some in Kenya are working on. Tracking incarceration rates for producers versus distributors might also provide accountability for those who profit the most from single-use plastics.

CONCLUSION

Wanuri Kahiu's Afrofuturist film short *Pumzi* (2009) is set decades in the future, when a postapocalyptic child emerges after a water war into a scene of toxic waste, with the camera notably pausing on plastic bottles and plastic bags.[95] The protagonist is a curator at a virtual natural history museum and follows her decision to plant a seed in soil, despite being told not to do so. After she plants the seed, she waters it with a plastic bottle (see figure 11). James Wachira persuasively argues the Afrofuturist film was shaped by impacts of the climate emergency and Maathai's environmental advocacy.[96] *Pumzi* circulated globally and was critically acclaimed. It also could have become part of the conditions of possibility of a successful plastic bag ban in 2017, reflecting the recognition of the cost of unregulated waste and how living without plastic still feels impossible to imagine, even in science fiction. Pointing to any one person or action is too reductionist to explain how social change works.

The nuance of contemporary environmental advocacy exceeds rigid, oversimplified binary debates. Revisiting Wangari Maathai's impure politics about plastics, then, is significant not just due to the importance of her unbowed voice within Kenya and globally as an intersectional environmental leader, but also because of the ways her call for a ban were never meant to be totalizing. Likewise, Kenya's current plastic bag ban is not a complete ban, even of plastic bags. With this quite modest target of a specific type of single-use plastic product, the ban is an impure step in a longer transition.

FIGURE 11. Still frame from *Pumzi*, a Kenyan short film about the period after a water war. Plastic waste often persists in postapocalyptic science fiction imaginaries. *Source:* Wanuri Kahiu, dir., *Pumzi* (Awali Entertainment, 2009), https://vimeo.com/46891859.

An ethic of care can be expressed in many modalities. A report on plastic ban campaigns by the Sustainable Lifestyles and Education Programme of the One Planet network, which is a partnership between Japan and Sweden, assessed Wakibia's campaign as successfully avoiding common pitfalls and "grounded in a positive social norm and includes a statement of commitment, tapping into a positive appeal ('I support')."[97] While the personalization of "I Support" as a gesture of calling in more allies did appear promising despite the odds, Wakibia clearly and repeatedly argues that anger motivated him because a place he loved (the streets of Nakuru) was being polluted. As Wakhungu observed, however, anger was expressed by many at the time as more than an individual emotion: anger with the "litter culture," anger with the lack of government response, anger with the broader public for littering, and so on speak to a collective affect in Kenya at the time. The discourses of "African flowers" and "flying toilets" also reflected intimate, lived experiences animated by disgust and frustration, not joy or pride. It feels important, then, to not reduce accounts of emotion or affect during ecological crises and instead to recognize how anger may become a felt, collective experience that may move people to articulate a crisis, to network cultures of care, and to mobilize policy. While practices of care may be linked to feelings of healing, honoring, and hoping as ideals, embodied performances of care as a praxis might also be stirred by anger—and disgust, frustration, outrage, and so on—at a lack of care.

Since Kenya's ban was passed, single-use plastic carrier bag pollution has been reduced, though some report enforcement as uneven and that bags are brought in from Uganda and Tanzania. Each country in the continent of Africa faces unique contexts that overlap and differ from Kenya's, like the assumed differences between the United States and Canada and different nations in North America or Sweden and Greece as nations in the European Union. Further, ethnic diversity within Kenya is articulated across a range of political positions. This chapter, then, isn't meant to be an overgeneralization about the continent or nation or broader movement, but rather an engagement with a few of the key voices that have mattered most to plastic hashtag activism.

Unfortunately, plastic bottles and other unnecessary sources of plastic pollution persist in Kenya, as does industry pressure to keep normalizing single-use plastic. As discussions continue in Kenya about this ban and more plastic waste politics to come, Wakibia, who considers himself a human rights and environmental advocate, has been emphasizing the message of corporate accountability to counter the recycling myth: "Plastic pollution is not a consumer problem; it is a producer problem. The Plastic Industry has been very efficient at perpetuating a flawed narrative blaming the consumers for the plastic pollution problem and running away from responsibility. If we are to effectively address the problem of plastic pollution, we must first acknowledge that it's those that produce it that must take full blame. We cannot continue beating around the bush and pretending we do not know the source of our biggest environmental problem. It is a product of the petrochemical industry."[98] While Wakibia continues to "demand better for our planet," he emphasizes the need to challenge flawed arguments and narratives. In his statement, he calls for redesigning of products, expanding efforts for a circular economy, and enacting a bottle deposit and return scheme for extended producer responsibility (EPR).[99] EPR would shift the financial and planning burden back to the industries selling plastics to collect, sort, and recycle. If we excluded plastics that cannot be recycled, the recycling symbol might become more helpful.

In 2020 Kenya passed a ban on all single-use plastics from public beaches, national parks, forests, and conservation areas, a policy trend revisited in chapter 5. Najib Balala, Kenyan cabinet secretary for tourism and wildlife, suggested: "This ban is yet another first in addressing the plastic pollution catastrophe facing Kenya and the world, and we hope that it catalyses similar policies and actions from the East African community."[100] Indeed, each country that creates a plastic ban begins to turn the tide on global attitudes toward plastic consumption and pollution. As noted, Bangladesh was the first, and many more countries have followed (including Australia, China, Italy, Mexico City, Rwanda, and Thailand), as well as US states and cities (including California and New York City, with more charging fees for bags). Efforts such as the largely industry-supported Kenya Plastics Pact also continue to bring new actors into the conversation about solutions to the changing economy and culture.[101] Leveraging national bans and forming transnational alliances to counter the hegemony of transnational corporations are vital tactics of the Global South—imperfect but meaningful attempts to recognize, as Ochieng writes, "the contingency and fragility of the emergence of creaturely life on the planet."[102]

While every culture faces related industry pressure, the ways digital media is used and the messages that resonate—or not—vary. This chapter focused on Kenya due to the role of hashtag activism in mobilizing publics and publicizing public support for plastic waste reduction and its strict response to policing crisis. Wakibia's hashtag activism amplifying his photography does suggest how a care network may be constituted through creating visuals, leveraging social media, lobbying government officials, and organizing with others to bring about collective action. Wakibia's shift from #BanPlasticsKE to #ISupportBanPlasticsKE illustrates the dynamic collaborative nature of hashtag advocacy, and his trends attracted more attention and lasted longer than most.

The government also played a vital role in mobilizing this public support online to justify environmental policy. This chapter, then, extends Bram Büscher's study of digital platforms and South Africa by addressing the following question: "How do all these unprecedented platform developments influence environmental and conservation praxis, and vice versa?"[103] In this case, social media empowered new forms of public participation and publicity that have expanded more traditional ones. Wakhungu's amplification of Wakibia's efforts provided motivation for him and greater publicity for public opinion to support her goal of banning plastics, allowing their mutual advocacy goals to benefit by complementing each other.

Some have complained about the ban. Bans themselves are a top-down solution. Policing (through financial or physical punishment) is also a sign of public failure to transform a culture. For environmental justice, if liberation is an aspirational goal, then we need to critically reflect on when financial (fines) and punitive (incarceration) leverage is used as a disincentive for environmental degradation. In Kenya, there has been some backlash about inequitable enforcement, which reflects broader concerns in the environmental justice movement that policing is prejudiced and abolition should guide our care ethics. To date, policing this crisis has played an important role in limiting the environmental and public health impacts of multinational industries on an African nation. Both Wakibia and Wakhungu claim they did not believe industry would have been motivated to find alternatives or that people would have transformed their practices without a ban. Both also appear acutely aware of the need for broader collective action and global policies. Both have emphasized the importance of focusing enforcement on the industry and distributors, not everyday people. Environmental advocates need further discussion about whether policing provides the justice we are striving for—and if not, how else does one sway multinational corporations in an age of planetary crisis? Further, in the United States, where prison labor is sometimes used to manufacture plastics, the plastics-industrial complex intersects with the prison-industrial complex in ways that suggest troubling one implicates troubling the other.

On the *Nairobi Ideas Podcast*, Wakibia offers this humorous way of thinking about ongoing global plastics advocacy: "We have to work as a global community to protect the environment. Kenya has to do its own part to its side of the environment. America has to do its part to protect its side of the environment. Because we live in one environment, you know. . . . Because you cannot urinate in one corner of the swimming pool and expect the other corner to be clean."[104] As the old environmental justice idiom goes: "The air blows and the water flows." So, too, do plastics, it seems. No country alone can create a just transition for all.

The next chapter then turns to a US city that partnered with a nonprofit marketing organization to nudge residents and industries to reduce single-use plastic consumption. In contrast to Wakibia's grassroots campaign that ended up connecting with government officials, Seattle's campaign began by coordinating with the municipal government to popularize an already decided upon ban. Foregrounding care for the oceans and marine life, the next campaign marketed a playful take on why the ocean too is angry.

4

Engaging #StrawlessInSeattle and #StopSucking

The Loneliest Whale, Sporting Fun, and American Exceptionalism

We Americans are reluctant to learn a foreign language of our own species, let alone another species. But imagine the possibilities. . . . We don't have to figure out everything by ourselves: there are intelligences other than our own, teachers all around us. Imagine how much less lonely the world would be.

—ROBIN WALL KIMMERER, POTAWATOMI SCIENTIST AND AUTHOR

Listening to nonhuman kin, particularly marine life, and caring for the ocean are entangled with global public advocacy to reduce plastics. For example, some trace the current wave of the US marine conservation movement to 2015, after biologist Christine Figginer shared a dramatic video of pulling a plastic straw out of a sea turtle's nose.[1] Others appear haunted by Captain Charles Moore's 1997 naming of the "Great Pacific Garbage Patch" or photographer Chris Jordan's 2009 photograph series, *Midway: Message from the Gyre.*[2] Some recall cutting plastic six-pack rings in the 1980s due to images of sea life strangled by plastics. Each of these examples provides opportunities to "imagine the possibilities," as Kimmerer suggests (in the epigraph), to learn from another species and to feel less lonely.

This chapter turns to the Lonely Whale Foundation (Lonely Whale), founded in 2015 by actor Adrian Grenier and film producer Lucy Sumner, which "strives to inspire empathy towards marine species and develop life-long advocates for ocean health."[3] Based in New York City, Lonely Whale describes itself as "an incubator for courageous ideas that drive impactful, market-based change."[4] As a nonprofit marketing organization, it connects companies, policymakers, NGOs, and celebrities to incubate ideas, as well as to promote goods and services "to prevent plastic waste from entering the ocean."[5]

73

Lonely Whale made international headlines in 2017 with two hashtags: #Stop-Sucking, a month-long August launch of a national push to reduce single-use plastic straw consumption, and #StrawlessInSeattle (riffing off the 1993 Hollywood film *Sleepless in Seattle*), a month-long September launch that preceded and coordinated with the city of Seattle, Washington, announcing a ban on single-use plastic straws and utensils. Despite previous efforts, such as Ocean Conservancy's #SkipTheStraw, Lonely Whale's campaigns became perhaps the highest profile plastic-related hashtag trends to date.[6]

As in Bangladesh and Kenya, Seattle's 2018 ban did not occur overnight. It was the result of at least a decade of lobbying in the face of aggressive corporate opposition. Whereas the trending Kenyan hashtags called for a broad plastic ban and ended up focusing on a narrowly defined single-use plastic item, Lonely Whale's hashtag focused on straws, even though utensils were part of the broader ban. Seattle's first related ordinance in 2008 required one-time-use food service items be recyclable or compostable. In 2009 polystyrene packaging (often recognized by the brand name "Styrofoam") became banned in food service. The next year, businesses were required "to use compostable or recyclable food serviceware," "provide recycling and compost bins," and "sign up for collection service." Plastic utensils and straws were exempted until the 2018 ban, which included a fine of up to US$250 for noncompliance and an accommodation for medical reasons (which I revisit next chapter).[7]

Lonely Whale's success pivoted on the role of influencers (actors, athletes, chefs, musicians, and pundits) in catapulting plastics into popular culture.[8] Stuart Hall argued: "Popular culture is one of the sites where this struggle for and against a culture of the powerful is engaged: it is also the stake to be won or lost in that struggle. It is the arena of consent and resistance. It is partly where hegemony arises, and where it is secured."[9] Rather than providing a straw man or dismissively cynical reading of Lonely Whale's celebrity-funded, government-driven, industrial-partnered initiatives, this chapter is written with the belief that it is worth considering the role of Lonely Whale's hashtags in the struggle over articulating plastics as crisis.

To do so, I consider behavioral science informing Lonely Whale's campaigns, including insights from a podcast interview with Lonely Whale's director of digital strategy. Next, the chapter focuses on the #StopSucking and #StrawlessinSeattle campaigns. Although they were widely lauded as successful, backlash was swift from conservatives and progressives alike. To give the disability critique ample attention and to keep it distinct from anti-environmental sentiments, I suspend addressing that backlash until the next chapter. This chapter attends to the #MakeStrawsGreatAgain backlash motivated by American exceptionalism. First, however, I want to provide context for Lonely Whale's namesake.

WHALE SONGS AND BLUE ECOLOGY

Lonely Whale was named after "the loneliest whale" or "52," in the early 1990s by a navy engineer who heard a frequency of 52 hertz from one whale but no other.[10] To understand how this whale became imagined as lonely and why many people care, it is worth considering the whale conservation movement.

"Save the whales" often is articulated by North American and European environmentalists as necessary to save ourselves.[11] For example, in 1961 the environmental NGO World Wildlife Fund (WWF) was established, in part, to save whales. Its founder, Sir Peter Scott, is often quoted as saying: "If we cannot save the whales from extinction, we have little hope of saving mankind and the life-supplying biosphere." WWF traces the expansion of this movement globally: "By 1925, the League of Nations recognised that whales were over-exploited and that there was a need to regulate whaling activities. In 1930, the Bureau of International Whaling Statistics was set up in order to keep track of catches. This was followed by the first international regulatory agreement, the Convention for the Regulation of Whaling, which was signed by 22 nations in 1931. However, some of the major whaling nations, including Germany and Japan, did not join and 43,000 whales were killed that same year."[12] This momentum led to the signing of the International Convention for the Regulation of Whaling in 1946.[13] The relative lack of attention to and denigration of the ocean depths and scales of marine kinship eventually gave rise to what some call "blue ecology."[14]

US popular culture began listening to whale communication in the mid-twentieth century using military technology. In 1964 the US navy shared recordings of whale sounds with music scholar and biologist Kay Payne and her husband, Roger (also a biologist, often solely credited with the discovery); they were moved to tears. Recognizing a beginning sound, a pattern, a capacity to change and more, the Paynes argued the recording shared key characteristics of music.[15] Added to the US National Registry in 2010, the recording titled *Songs of the Humpback Whale* was released in 1970, a watershed year for the institutionalization of environmental movements in the United States. The record quickly earned the status of "the most famous nature album in American history."[16] As reported in *Wire* magazine, the "greatest single pressing of any album of recorded music was *Songs of the Humpback Whale*. In 1979 *National Geographic* magazine inserted a flexible sound page inside the back cover of all of its editions in 25 languages."[17]

Although some countries continued whaling for decades, by 2010, based on overwhelming evidence of "the complexities of cetacean minds, societies and cultures," an international organization of scientists signed the "Declaration of Rights for Cetaceans," calling for their "right to life, liberty and wellbeing."[18] This statement has led to a broader international discussion of ethics and whale rights, especially as whale tourism has increased globally.[19] Whales not only "regularly interact with close kin, including close maternal kin, they also frequently interact with more distantly related and unrelated" whales.[20]

The complicated communication networks of whales, involving both intelligence and emotions, relatives and strangers, is part of why 52's unique frequency generated attention. "The call, possibly a mating signal," Andrew Revkin reported, "suggests that the animal lives in total, and undesired, isolation."[21] A sense that one might die alone, unable to communicate with, and therefore be unheard by, anyone else struck an international chord. For example, 52's story inspired a Taiwanese musical drama film, *52Hz, I Love You,* based on multiple love stories in Tapei on Valentine's Day. The human discovery of 52 also instigated a children's novel, *Song for a Whale,* about a twelve-year old Deaf girl who learns of a lonely whale, identifies with the isolation, and decides to create connections on her own terms.[22] There also have been many songs, including "Whalien 52" by K-pop superstar boy band group BTS, about loneliness.[23] An irony of our times, then, is that while we are surrounded by billions of people in person and online, as well as by nonhuman kin, a sense of loneliness—or lack of attachment—appears to be a globally felt structure of feeling.[24]

In 2021 filmmaker Joshua Zeman created a documentary, *The Loneliest Whale: The Search for 52,* as part of his self-declared obsession.[25] Its executive producers include Grenier, cofounder of Lonely Whale, and longtime celebrity environmental advocate Leonardo DiCaprio. *The Loneliest Whale* recalls not only how Zeman and others have been moved by 52, but also a brief, bloody history of how whales were killed as a main source of energy in the nineteenth century, which led to the endangerment of many whale species.[26] From whales being overhunted to achieving beloved iconic status, the film underscores the affective affinity many sense listening to whale songs. Moreover, Roger Payne recalls the growing attachment to whales as a motivation for politics: "We probably never would have saved whales unless we heard them sing. It made people care. And when people care, they can change the world."[27]

On a *Lonely Whale Podcast,* Grenier emphasized the importance of cultivating care between people and ocean life to its approach: "We can't care unless we connect. The first and foremost necessity is to try to connect the ocean to our everyday experience. . . . Initially, it was just a very simple communications challenge and opportunity: how do we get people to connect with the ocean?"[28] This quest to (re)connect humans with our oceanic relations posed an even greater challenge for Lonely Whale: How could it not only foster a sense of wonder or desire to learn from marine multispecies and oceanic connections, but also challenge throwaway culture? To do this, the group turned to straws.

"THE GATEWAY PLASTIC"

Most accounts date the invention of straws back to the Sumerians of Mesopotamia in 3000 BC, who used grass ones to drink beer.[29] It was not until 1888 in the United States that someone decided to file a patent for straws made of manila paper.[30] In the 1930s the first bendable straw was invented and patented in the United States by Joseph B. Friedman, because "sitting in his brother's fountain parlor," he noticed his young daughter struggling to use a paper straw when drinking a milkshake.[31] When Friedman decided to market them, he advertised the "Flex-Straw" as an "adaptive technology in hospitals . . . to help reclined patients drink from a cup . . . [because] they were sanitary, cheap, sturdy, temperature-resistant and suited for children and epilepsy patients." Since then the bendable plastic straw has been "championed as an icon of 'universal design'—a principle that ensures that all designed objects and systems are accessible and pleasing to everyone regardless of age, economic status or physical abilities."[32]

Lonely Whale claims that "straws are among the top 10 items found during beach clean ups and can do so much harm to seabirds, turtles and other marine creatures."[33] This insight comes from Ocean Conservancy, which hosts an annual international coastal cleanup of roughly eighteen thousand miles with approximately half a million volunteers; consistently, the top ten items found are cigarettes (which have plastic in their filters), plastic food wrappers, plastic beverage bottles, plastic bags, plastic caps/lids, plastic utensils/plates/cups, plastic straws/stirrers, glass beverage bottles, metal beverage cans, and paper bags.[34] Lonely Whale also shares a commonly cited statistic by Milo Cress, founder of Be Straw Free, that the United States uses more than 500 million plastic straws daily, which are too lightweight to be recycled.[35] Dune Ives, former Lonely Whale executive director, emphasizes that the "campaign isn't just about plastic straws, it's about building a connected and motivated global audience focused on the issue of ocean health."[36]

Notably, the ACC won a victory against a 20 cent plastic bag fee in 2008 in Seattle by playing on fear: "spent $180,000 to gather enough signatures to put the issue on the ballot, then devoted another $1.4 million to overturn the fee— the most spent on any Seattle referendum. The industry campaign relied largely on scare tactics, falsely claiming that the fee would cost the average consumer an extra $300 a year."[37] Bags, then, already had failed to sway public opinion.

Ives, who also has a PhD in psychology and was a cofounder of the Green Sports Alliance, explains Lonely Whale's choice to target straws as a process of elimination and due to their relatability:

As our team explored the myriad options of plastic pollution to find the perfect entry point to incent behavior change, we found plastic water bottles too endemic, plastic bags already somewhat politicized, and no viable alternative for the plastic cup in ALL markets. That's why we chose the plastic straw. To us, it was the "gateway plastic" to the larger, more serious plastic pollution conversation. Plus, plastic straws are social tools and props, the perfect conversation starter. In starting the conversation by pairing something playful alongside our gross human over-consumption ("500 million consumed daily in the United States alone") we aimed to nudge people toward understanding the issue.[38]

Plastic Pollution Coalition CEO Dianna Cohen likewise told *Business Insider*: "We look at straws as one of the gateway issues to help people start thinking about the global plastic pollution problem. They've been designed to be used for a very short amount of time, and then be tossed away."[39] Similarly, at the launch in Seattle, Grenier repeated: "We see straws as a 'gateway plastic' in understanding the pollution problem and a simple way to motivate consumers and industry leaders to take greater action against all single-use plastics."[40]

Most of us might be more familiar with the idea of a "gateway drug" than a "gateway plastic." Behavioral scientists have studied the gateway hypothesis since the mid-1970s, researching whether there is sufficient evidence that early use of seemingly more minor drugs (such as alcohol, tobacco, or cannabis) might lead to more addictive drugs later in life.[41] The gateway model of behavior change research has expanded to other realms, including reproductive health, diet, and exercise. Research shows that facilitating a successful behavior change, however seemingly small, can help one imagine greater self-efficacy or agency to change behavior with more substantial impacts.[42]

While behavioral scientists of public health tend to find evidence that gateway actions matter, those who focus on environmental advocacy express reservations. For example, a professor of civil and environmental engineering and codirector of the Stanford Center for Ocean solution cautions: "The risk is that banning straws may confer 'moral license'—allowing companies and their customers to feel they have done their part. The crucial challenge is to ensure that these bans are just a first step, offering a natural place to start with 'low-hanging fruit' so long as it's part of a much more fundamental shift away from single-use plastics across the value chains of these companies and our economy."[43] A concern for moral license by behavioral science is called *negative spillover,* in which an act might lead to undesirable unintended consequences. One study, for example, found that people who brought their own reusable bags to grocery stores were more likely to reward themselves with unplanned affordable and unhealthy items like candy or chips.[44] Another found that if policymakers leverage a nudge, they—and their constituents—might be less likely to support implementing more substantive approaches.[45] Research on human behavior, however, doesn't seem to suggest that one environmental behavior *necessarily* will lead to negative spillover. Reviewing research on spillovers likewise tends to find that consistency (the desire to want to be and want to appear to be void of contradictions) and environmental identity (a sense of social identity that shifts behaviors) can weigh the scales toward sustainable impacts.[46]

To return to Lonely Whale, Ives elaborates that once a gateway plastic was identified, the organization's theory of change was nudging: "So how did we get here? We nudged. We nudged sports fans to care a little bit more about their home turf, nudged foodies to think twice about the filet of fish on their plate, nudged leading chefs to rethink their restaurants' plastic footprint, and we even nudged an entire city to forever forego single-use plastic straws and cutlery."[47] In psychology, nudges are alterations to the architecture of choices that encourage individuals to voluntarily change behavior. This approach, according to the literature, modifies social norms through peers (they didn't use a straw, so I will follow) or even everyday institutional messaging (like a hotel sign suggesting reusing towels instead of changing them daily).[48] For larger impacts, nudges can add up. For example, one study found presenting renewable energy as the standard option (instead of as an alternative) led to 80 percent adoption in households.[49] Nudges therefore provide a way to rethink popularizing adoption of requirements like bans and stand in contrast to negative frames.

Lonely Whale's approach attempted to link an affect of loneliness to nudges: "Let's Make Ocean Conservation Personal. We live in a lonely, plastic world. But together we can change that. The more we fill our world with disposable stuff, the more it disconnects us from each other. Lonely Whale . . . [is] encouraging behavior change away from single-use plastic and toward a healthy, thriving ocean."[50] Again, since the ban in Seattle was already planned to move forward, the decision to focus on individual behavior was to complement structural change by promoting a cultural conversation to rethink throwaway attitudes and practices.

It also is important to remember that during this time in the United States, the deregulation of environmental protections felt overwhelming to many. As part of its rollback of environmental regulations, then US President Donald J. Trump's administration had given way to a plastics boom.[51] This environmental backlash perhaps contributed to an even greater sense of exigency for a gateway campaign that offered actions an individual could feel good about accomplishing, despite the lack of national environmental leadership.

ON LONELY WHALE'S SOCIAL MEDIA

According to the Pew Research Center in 2021, statistics on social media in the United States remain relatively stable, with seven out of ten people using You-Tube and Facebook and young adults favoring Instagram, Snapchat, and Tik-Tok.[52] Seattle ranks in the top ten "most social media savvy cities" in the United States, "building a reputation as a place where social media geeks like to hang out."[53] Seattle also is home to many multinational corporations, such as Amazon (a retailer and technology company), Starbucks (a coffee chain), Nordstrom (a luxury department store chain), and Alaska Airlines. Of these, Starbucks and Alaska Airlines, as well as Live Nation (a music concert, ticket, and brand marketer) and Dell (a computer company) pledged to ditch the single-use plastic straw.[54]

Market-based advocacy has become popularized as transnational corporations increase their power and impact over many nation-states and as some governments have failed as environmental leaders.[55] In turn, environmental research in communication and cultural studies increasingly has focused on market-related advocacy, including but not limited to neoliberal approaches to public goods and services, "consumer-based advocacy," "commodity activism," "market-based metaphors," corporate and trade group "marketplace advocacy campaigns," "radical consumption," and "consumer activism."[56] With the conflation between consumers and citizens promoted by neoliberal politics, there are possibilities for cultural shifts through markets, as well as limitations to deprioritizing government-based initiatives. Beyond green marketing and consumer behavior, "market advocacy aims to harness financial power by, for example, consumers withholding or spending money through boycotts or buycotts or shareholder activism, such as voting to divest and reinvest stocks of a company to reflect values of a campaign."[57] These campaigns leverage consumer and investment power to shape markets and rearticulate human-environmental relations by tapping into vernacular expressions of public opinion.

Lonely Whale's approach exemplifies contemporary participatory social media culture, which blurs lines between privately funded public relations and viral public engagement. For example, Bridget Tombleson and Katharina Wolf argue that "public relations professionals need to become cultural curators, who are equipped to construct meaning from audiences, who have now become content creators in their own right, and encourage a true participatory environment that sees cultural values shared as an organic exchange, rather than a manufactured one."[58] Top-down public relations are not bottom-up social movements, but they shape public controversies over the collective good.

There is no doubt that celebrity social media challenges for participatory campaigns matter. A quintessential exemplar is the 2014 "Ice Bucket Challenge," which dared people to post a video online pledging to donate to the ALS (amytrophic lateral sclerosis) Association, then pour a bucket of ice water on their heads and challenge someone else to do the same.[59] More than just a popular online trend, the Ice Bucket Challenge raised US$115 million in donations, which enabled seventeen ALS-related clinical trials and the identification of five new genes that contribute to ALS.[60] Likewise, celebrities "offer novel engagements with climate change that move beyond scientific data and facilitate more emotional and visceral connections with climate change in the public's everyday lives."[61]

As with all impure politics, however, celebrity culture is not above criticism. I have emphasized: "There is a risk in articulating environmental politics to what is 'sexy': one easily can eclipse structural change and what Hall called 'socialist politics' with bright lights, loud noises, and big egos."[62] Further, celebrities tend to live carbon-intensive lifestyles.[63] Nevertheless, celebrity influence contributed to Lonely Whale's success.

IMPACT

Lonely Whale's website and engagement initiatives were recognized as cutting edge. In 2019, Fast Company described Lonely Whale as one of the Most Innovative Companies: "an important voice in the movement to stop the use of single-use plastic straws, pushing for shifts in corporate policy (by partnering with airlines, tech companies, and MLB and NFL teams) to youth engagement (it trained 300 global youth on plastic pollution and how to create their own campaign to reduce plastic straws) to city- and country-level policy change under the banner of, 'For a Strawless Ocean.'"[64] Innovating multiple approaches to market-based advocacy, including branding, public relations, and event planning, Lonely Whale won Silver Distinction in the "Social Good Campaign" category from the Shorty Awards and a Bronze Effie award in the "Positive Change" category.

Claiming it "influenced true policy change," the Lonely Whale's "organic #StopSucking content saw a social reach of 74 million" and gained widespread news attention.[65] The campaign involved a launch (or "provocative activation") in Seattle and partnerships throughout the city with sports stadiums and more than one hundred restaurants to remove single-use plastic straws for the month of September. Lonely Whale claims to have "stopped 5 billion plastic straws from entering the ocean" and "spurred Seattle to ban straws completely."[66]

Lonely Whale's data shows a range of audiences for both the broader organization and its popular campaigns. From sports and music fans to environmentalists and marine conservationists, the organization's media outreach covers similar demographics.[67] In addition to the Stop Sucking campaign, Lonely Whale summarized its media impact: "The campaign reached 103 global markets with over 334 million media impressions across 270 unique stories. The Strawless In Seattle campaign demonstrated that ocean health can penetrate a variety of markets and audiences—from sports fans to business leaders."[68] Its company partner, Dittoe PR, assessed that the month-long Strawless in Seattle campaign in September 2017 alone received "248 media hits, 290,946,957 media impressions and a total publicity value of $2,896,547."[69]

Emy Kane, then Lonely Whale director of digital strategy, offers insights into the backstage process, which involved competing in a sea of hashtags:

> At that stage, the hashtag was important because it was a way for not only us to be able to track the campaign from a who was talking about it perspective, but also the cumulative reach as well . . . for us having a hashtag like #StopSucking was essentially kind of our secret sauce because we were able to leverage in so many different moments to really kind of generate that buzz and that word of mouth. And I also think that was a huge reason why we had so much success with our influencers initially as well: because they thought it was fun and it was funny. And it was a great piece of content for them to create around and co-create with us for every challenge that they would launch on their channels.[70]

Kane is clear that the "fun and funny" hashtag "secret sauce" of Lonely Whale generated social media buzz and enabled it to track its impact.

#SuckerPunch: OCEANIC FEELINGS

To pilot the #StopSucking campaign, Lonely Whale launched a #SuckerPunch popup event in Austin, Texas, at SXSW (South by Southwest, an annual film, interactive media, and film festival) in March 2017, followed by a PSA.[71] Lonely Whale hosted a photo booth with a slow-motion camera (often used for comic effect) to pose with a plastic straw (to be upcycled into skateboards). Outside the booth was a screen sharing the videos. The SXSW event "was lauded by the Ad Council as part of a new model in the most effective way to execute successful social good movements."[72] Many people shared videos of themselves being sucker punched online with a sense of self-deprecating humor and pledging to stop using plastic straws. Particularly resonating with younger adults, UpWorthy's posting of the SXSW sucker punch highlights reel had almost 700,000 views and a reach of 1.9 million.[73]

Drawing on this content, a one-minute #SuckerPunch PSA was created, which opens with sounds of ocean waves crashing on a show and the words: "In the U.S. alone, we use **500 million** plastic straws every day. Many end up in the ocean. And the ocean is **angry**."[74] Then, showing the words "Sucker Punch," the soundtrack starts to play the upbeat musical opening to "Go Off" (performed by The Skins). Images in the compilation video shift from a montage of people drinking from plastic cups with plastic straws to them being hit in slow motion by a muppet-like wet orange and red octopus tentacle, knocking the plastic straws and cups out of their mouths (see figure 12). Featured people include a range of ethnicities, genders, and occupations, including celebrity cameos by Boots Riley (director, entertainer, and activist), Brooklyn Decker (*Sports*

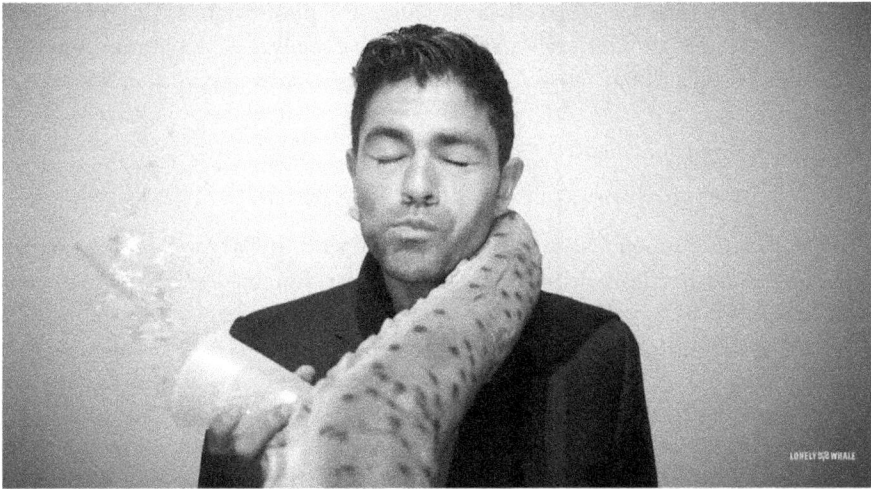

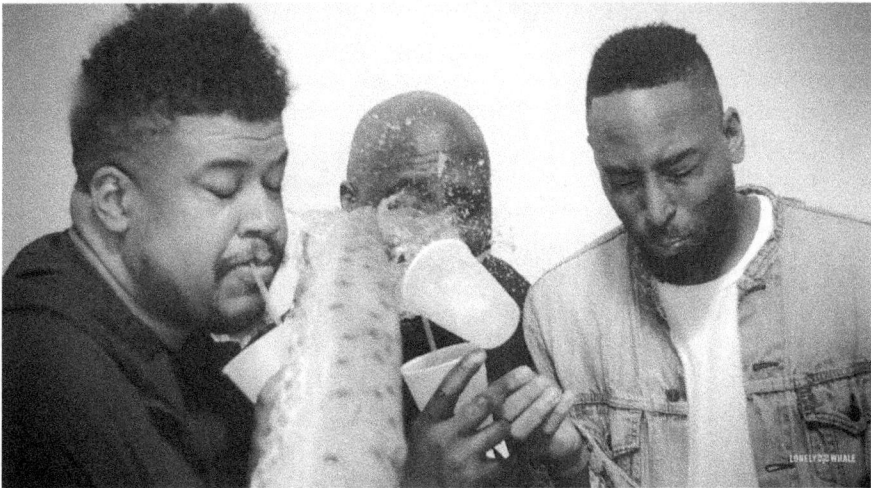

FIGURE 12. Video stills of Adrian Grenier (top) and De La Soul (bottom) from Lonely Whale, "Sucker Punch," March 31, 2017. To illustrate the ocean's anger using levity, Lonely Whale's #SuckerPunch campaign featured celebrities and everyday people being hit by a fake octopus tentacle in slow motion while holding single-use plastic cups and straws, which were all upcycled afterward. *Source:* www.youtube.com/watch?v=rfFpz8KM-9E.

Illustrated swimsuit model), De La Soul (hip-hop trio), The Skins (musicians), and Neil deGrasse Tyson (scientist).

In a sense, this concept offers a playful invocation of multispecies "justice," inviting participants and audiences to imagine what marine life might feel about human plastic consumption. Despite the humorous mood, this slap-in-your-face

PSA also embodies a wake-up call about single-use plastic habits. The popularity of this #SuckerPunch pilot event gave proof that Lonely Whale's humorous concept would prompt many to generate social media conversations about plastic pollution.[75]

#StopSucking: SELF-DEPRECATING HUMOR

The pop-up led to the creation of another PSA with the larger campaign name #StopSucking, featuring celebrities such as Grenier, Emanuelle Chriqui (actress), Van Jones (political commentator), Xiutezcatl Martinez (environmental justice advocate and hip-hop artist), Kendrick Samson (actor), and Amy Smart (actress) (see figure 13).[76] Lonely Whale found the campaign had wide appeal, but it particularly resonated with "SciFi Geeks," climate advocates, New York and Los Angeles popular culture fans, and those opposed to Trump, using the hashtag #resist.[77]

In the #StopSucking PSA, the celebrities begin by iterating the enthymematic hook: "I suck." They confess to sucking in general, sucking over time, and sucking with others, as well as their belief that most of us suck. At first, the hook is ambiguously tongue-in-cheek. (Audiences might wonder: Who sucks what? Is this sexual? Or are they admitting to having done something taboo?) Eventually it is revealed that they are sucking through plastic straws, which they note harms the ocean. The hashtag and PSA come with a viral challenge for spectators to create their own videos pledging to stop sucking, to share on social media with the hashtag, and to challenge someone else publicly to do the same.[78]

Ives explains: "Our #StopSucking social media challenge aims to make the ocean plastic pollution problem relatable (we all have used a straw), participatory (through a direct social media challenge), culturally relevant (a straw was featured in the most liked Instagram post of 2016), and emotionally resonant (who wants to suck?!)."[79] One creative director shared the following on the #StopSucking concept: "No one needs to see another PSA with some celebrity asking you to do something for a cause they care about. We've become completely desensitized to that technique. But a celebrity being self-deprecating—poking fun at themselves and admitting maybe they're not so perfect, that they in fact SUCK just like the rest of us—well, that just might get people to sit up and take notice. And thus, #STOPSUCKING was born."[80] The self-deprecating #StopSucking campaign turns away from the sincerity trope of cause marketing toward a more comedic one that includes successful people willing to admit imperfection and able to laugh at their own expense. This more humble and humorous approach to environmental messaging stands out from broader trends of celebrity environmental discourse, which previously tended to be framed as sincere and serious.[81]

FIGURE 13. Video stills of Kendrick Sampson (actor, top left), Xiuhtezcatl Martinez (hip-hop artist and climate justice advocate, top right), Amy Smart (actor, bottom left), and Van Jones (political pundit, bottom right), all admitting they suck, from Lonely Whale, "#StopSucking," August 8, 2017. *Source:* www.youtube.com/watch?v=Q91–23B8yCg.

On the choice to not invoke the sadness of 52 or the horrors of plastic pollution, Kane explains that the

> narrative of Doom and Gloom with climate change [was] not actually moving the needle; so, we . . . were the new kids on the block at the time—we had just started Lonely Whale—and, so, we figured: let's try something different. Let's see if this resonates when we . . . put a lens of humor and fun and levity on this. Not to take away from the reality again of just how angry the ocean is and how, you know, massive this problem of plastic pollution is. But just to say: "hey, like we're all in this. This is heavy. But we can do this if we work together." And so, I think when you offer a hand into the conversation—whether it's through humor or through joy or you know even through personal self-reflection—I think it's so much more welcoming in general into, again, when you think about building a movement as opposed to kind of just . . . shaming or scaring people into joining you.[82]

Kane's insights address an ongoing controversy in environmental communication over which frame is more appealing—and more importantly for advocates, which motivates social change, fear or joy?

Fear appeals—also called the "doom-and-gloom" frame—were the initial choice of most climate messages, with the most frequent condensation symbol of the crisis

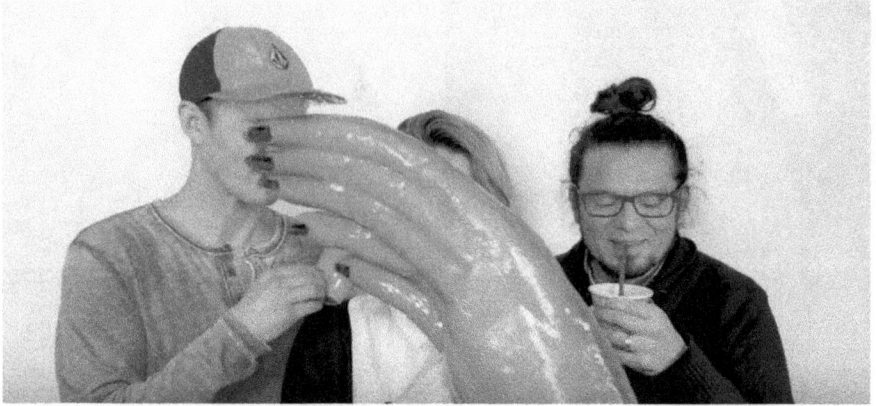

FIGURE 14. Video still from Lonely Whale's Chilean campaign, which featured a seal flipper, September 12, 2018. Lonely Whale has shared its tool kit globally, encouraging localized variations. *Source:* Ministerio del Medio Ambiente, "Campaña ciudadana Chao Bombillas."

being a polar bear seemingly stranded on melting ice. Yet increasingly, climate communication researchers and practitioners have argued that such an emphasis can emotionally shut people down, make us feel guilty, and motivate us to do little. Instead, while recognizing the heaviness of the issue, increasing numbers of climate communicators resonate with Lonely Whale's joyful appeal, emphasizing that the climate crisis is solvable and collective action can be consequential.[83]

The comedic, joyful atmosphere might feel like Lonely Whale is trivializing the seriousness of the risks of plastic production, consumption, and waste to ocean health and all life on Earth. Yet as Daniel C. Brouwer wrote of camp humor in HIV/AIDS zines, it is important to remember that what might appear to be small gestures of humor are not just "discrete acts" but rather "constellated and interanimating acts" that might "circulate oppositional understandings . . . , promote the occupation of resistant subject positions, and cultivate resistant frames . . . that can be preparatory for and concurrent with the improvement of economic conditions."[84] In this case, #StopSucking invites an understanding of the ocean as an animate, affective agent and marine life as species with whom we should (re)consider our "sucky" relations. The message was not a wagging finger of shame, but one of self-deprecation, suggesting none of us are perfect and we all can improve.

It then became a model of how Lonely Whale's messages could be adapted to a specific culture. When more politicians and advocates reached out to ask how they could create similar campaigns, Lonely Whale developed an online tool-kit to download with guidelines on the campaign's message and working with venues, customizable posters and logos, and advice on how to track impact.

Lonely Whale also worked with translators of the materials in twenty-five languages.[85] Each campaign is encouraged to adapt contextually. In Chile, for example, a version of Lonely Whale's sucker punch PSA "transcreated" with the Chilean government included the message "The Ocean Is Angry" ("El Océano Está Furioso"), but switched from an octopus slap to the more culturally relevant seal flipper (*y nos quiere echar la foca*) (see figure 14).[86] Sharing the open-source toolkit for free and encouraging communities globally to go viral with their own Strawless Ocean campaign suggests that Lonely Whale was striving for a greater impact for the public good globally. Further, it since moved beyond straws into other plastics.[87]

#StrawlessInSeattle: SPORTING FUN
IN "A CITYWIDE TAKEOVER"

Beginning on September 1, 2017, Lonely Whale then partnered with the City of Seattle and the company Aardvark Paper Straws to launch an "industry-led" month-long campaign. Everyday people were encouraged to use Aardvark straws, which are degradable and decompose, as well as other zero waste alternatives, such as reusable beverage containers, bags, and utensils. The Strawless in Seattle campaign included partnerships with locally recognized brands and icons, such as the Seattle Seahawks (the city's professional American football team), Mariners (the city's professional baseball team), Space Needle (the city's most iconic tourist destination, which includes a restaurant), and Port of Seattle (an organization charged with fostering economic, environmental, and socially responsible business growth). According to the *Seattle Times*, about two hundred retailers agreed to participate by replacing their plastic straws with compostable ones to phase them out.[88] In addition to social media content creation and circulation, Lonely Whale designed and coordinated events offline, such as pop-ups selling alternatives to single-use plastic. They also encouraged Seattle-identified celebrities, like Ellen Pompeo (star of *Grey's Anatomy*, a television show based in Seattle), to use their social media platforms to commit to the pledge and invite others to join.[89]

On Twitter, Grenier announced the campaign kickoff at a Mariners baseball game (see figure 15).[90] He had arranged to throw the ceremonial first pitch, a ritual in baseball of allowing a guest of honor to throw a ball to signal the beginning of a game. Literally reaching out to local sports teams, Lonely Whale's attitude of "sporting fun" is not explicitly named but connects with a playful and upbeat energy embodied by Grenier and the celebrity athletes the campaign features (once again, across a range of genders and ethnicities).[91]

Lonely Whale then created a #StrawlessInSeattle PSA video that showed celebrities in everyday scenes (dining at a restaurant, roller blading on the sidewalk, sitting in an airport, laughing at a bar, and juggling a soccer ball) leaning in to use a

Adrian Grenier ✓
@adriangrenier   ···

Kicking off #StrawlessInSeattle with the
@Mariners and @LonelyWhale.

🐋 ⚾ ✌️ #firstpitch

👤 Lonely Whale and 2 others

7:49 PM · Sep 1, 2017 from Safeco Field · Twitter for iPhone

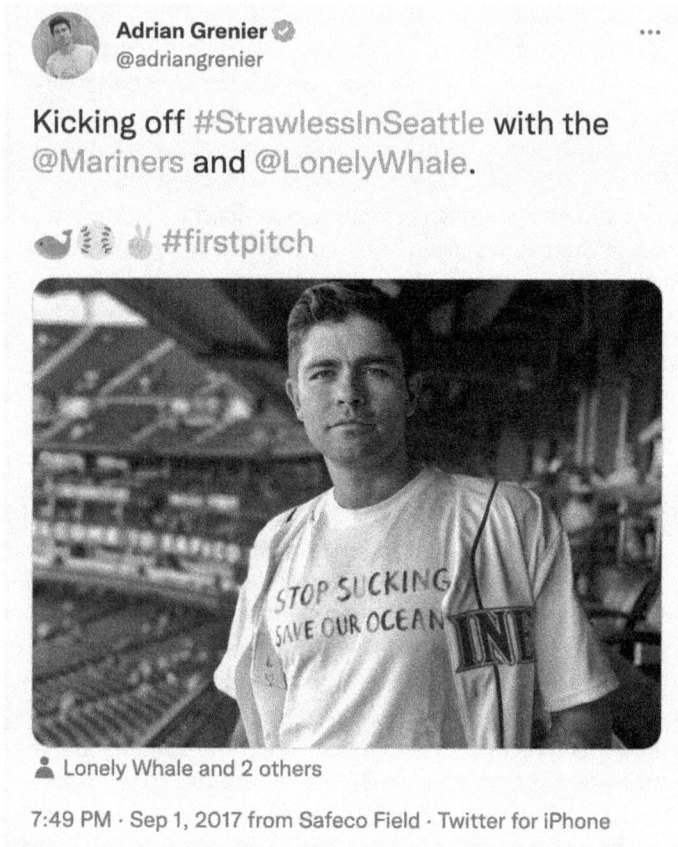

FIGURE 15. Actor and cofounder of Lonely Whale Adrian Grenier
(@adriangrenier) launching its Seattle-based campaign on social
media at a baseball game.

plastic straw and discovering their straws had been stolen—to their surprise. They
then shrug it off with affable attitudes, carrying on what they were doing by drinking without one. The soundtrack is a playful gameshow melody, and the video
ends with the text: "Place your bet and join them in going Strawless in September.
And maybe you'll find it's more fun to #StopSucking! Find out which celebrity
athlete is leading the charge by visiting: StrawlessOcean.org/Seattle." The invitation, then, is to imagine if plastic straws weren't a part of the taken-for-granted
scripts of everyday life and to guess which well-known celebrity was allegedly
"stealing" the straws. Rehearsing behavior change in these scenes through playful,
ephemeral acts troubles the habitus of everyday throwaway culture.

Lonely Whale describes the #StrawlessInSeattle PSA as full of "fun and joy," "present[ing] a serious topic with humor." In response to negative comments about the video online, which Lonely Whale claimed "are full of ignorance and intolerance" (and had been turned off), it argues: "People who don't seem to care about the environment at all, criticize those who care about it."[92] While this move did silence some backlash, there were other platforms it did not censor and, as discussed in the next chapter, Lonely Whale did listen to criticism aligned with creating a better world.

Eventually, the campaign reveals Russell Wilson, then the quarterback for the Seattle Seahawks, as "the plastic straw thief." Drawing on sports metaphors, Wilson states: "We are the first line of defense, and that is why I decided to accept Lonely Whale's challenge and lead our team to fight for our ocean." He then challenges other cities to join Seattle in becoming plastic straw free.[93] Using a Seattle-based celebrity spokesperson as a mystery created a digital story to develop throughout the campaign and connect to local culture. A popular sports figure can make refusing plastic straws feel more mainstream and less radical, reaching audiences that might overlap but exceeding those who primarily identify as environmentalists. Like Grenier, Wilson also embodies a handsome, more playful masculinity of a good sport, attracting a wide appeal for the progressive city. The #StrawlessInSeattle campaign linked a similar attitude of sporting fun to a specific place. Lonely Whale partnered with beloved local celebrities, restaurants, and city destinations to give the campaign more meaning for those living in Seattle.

Overall, Lonely Whale's #StrawlessOcean initiative proved to be popular, judged by social media impact across platforms and reduced consumption of plastic straws. Hashtagged social media increasingly circulated images of signs on which people were asked to refuse plastic straws in places from restaurants to aquariums. Others shared selfies or pictures of their kids pledging to use straw alternatives to plastic. Even US celebrity, producer, and philanthropist Oprah Winfrey named reusable straws in her popular annual list of "Favorite Things" in 2018, declaring: "Friends don't let friends use plastic straws!"[94]

As noted, the following year moved single-use plastic bans from emergent and marginalized to mainstream. *Single-use* became the word of the year. In 2018 Vanuatu became the first nation to ban single-use plastic straws, as part of larger efforts to mitigate plastic pollution using the hashtags #NoPlasticVanuatu, #NoPlastikBag-Plis, #TataPlastik, #NoPlasticStraw, and #StrawBan.[95] Taiwan also announced that it would ban plastic drinking straws the following year with #KTAXAxMSS, #Plastic-Straws, #PlasticFree, #SugarCaneStraw, and #甘蔗吸管.[96] Discussion of plastic bans in Bali, Britain, Jamaica, and the European Union also included straws.[97]

Nevertheless, there was backlash. The reaction of conservatives in the United States was immediate and did not die down before the election campaigns of US presidential candidates two years later.

#MAKESTRAWSGREATAGAIN: US CONSERVATIVE BACKLASH

In response to the growing popularity of single-use plastic straw bans, conservative news outlet *Fox News* published: "Taking away freedom for nothing in return is now a specialty of the environmental movement."[98] "Taking away freedom" here is imagined as an individual consumer's right to plastic straws, and the dig at the environmental movement reflected a broader pattern of backlash against environmental policy imagined as unnecessary overreach.

This conservative position articulates straws to a broader structure of feeling of American exceptionalism, which maps crisis differently than previous accounts. Consider, for example, then *Fox News* pundit Tucker Carlson's statement in 2019: "Well, millions of tons of plastics enter the world's oceans every year—and not from America. Instead, the vast majority of this garbage, that's toxic garbage, comes from Asia, in particular, China. . . . Of course, they're not planning to boycott China. . . . To America's environmentalists, America is the villain. China is not. Ever. So, the obsession over plastic straws, as you can see is a microcosm of how our rulers view the world at large."[99] Initially, his commentary offers common ground with other voices in this book. Carlson agrees that plastic pollution matters—in the ocean, as garbage, and as toxic. By calling plastic straws a "microcosm" of broader perspectives, Carlson also appears to agree with the premise of this book: plastics, while an environmental problem, had become an articulator of crisis.

Carlson and I part ways, however, on his belief that the United States too often is positioned by American environmentalists as a scapegoat, quickly blamed without just cause.[100] At the time, Ocean Conservancy advocated a similar perspective: blame China, Indonesia, the Philippines, Thailand, and Vietnam.[101] It is worth clarifying that while China has become the largest producer and consumer plastics of late, the US role is more than a straw man argument. Together, the United States and China more than triple the rest of the world in single-use plastic generation, and most notably, *US single-use plastic waste generation per capita is estimated to be more than double China's*.[102] Further, as the 2021 Minderoo Foundation Plastic Waste Makers Index reports, two of the three leading polymer producers are US-based companies and the major asset managers (global investors and banks) of single-use plastics, led by US-based companies.[103] Research also finds the United States is the largest producer of single-use plastic waste in the ocean.[104] Relatedly, the United States has long been the leading producer of global climate crisis emissions.[105]

Further, Carlson's indignant objection to the villanization of the United States constitutes a significant—and problematic—diagnosis of the times. This structure of feeling has been palpable to many of late, intertwining victimhood with an entitled sense of toxic masculinity and authoritarian nationalism. Karen Lee

Ashcraft identifies this mode of "aggrieved masculinity" as a lashing out due to feeling blamed unfairly as a particular performance of white, cisgender Christian masculinity.[106] Carlson's scapegoating of China further links this toxic masculinity with a broader anxiety about the fragility of American exceptionalism if China overtakes the United States as the leading world power.

Carlson's rhetoric resonated with the then US president's discourse. "Trump routinely praises his people," as Jennifer Mercieca argues, "as the smartest, best, most patriotic, hardest-working Americans. They are great and good and everyone else is not."[107] Paul Elliott Johnson argues this populist posture pivots on Trump's toxic gendered ethos, encouraging his audience "to imagine themselves as victims of a political tragedy centered around the displacement of 'real America' from the political center by a feminized political establishment," which he associates with "traits like empathy, an ethic of care" and more.[108]

Trump's xenophobic rhetoric of identification gained affective force through social media vernaculars at the time.[109] While American exceptionalism has been the norm for many political figures in the United States, as Jason A. Edwards argues, Trump's discourse included key shifts as his political platform grew. First, Edwards identifies moving from a belief in the United States as the leader of the so-called Free World from the post–World War II era to severing many global connections (which arguably contributed to China stopping its importing of US plastic waste). Second, Trump's discourse shifted conservatives to primarily promoting that, instead of ideals of freedom or democracy, "material wealth was the source of U.S. greatness."[110] Third, I would add that, during this time, Trump's use of Twitter to constitute and engage US publics was far more frequent and less restrained than any president before, signaling a shift from diplomacy as a key characteristic of the highest national office to injudicious bravado.[111]

In addition to scapegoating China to prop up American exceptionalism, Carlson's *Fox News* statement constituted another trope promoted by then US President Trump and the Republican Party: environmental regulation as overreach. For his audience, Carlson's framing of environmentalists as "rulers" had appeal for those who voted for Trump to "drain the swamp." Their discourse rationalized an extreme record during Trump's presidency of deregulating more than one hundred environmental protections within four years while increasing regulation of Muslims, transgender people, migrants, reproductive health, and more—often through bans. Trump's position, then, wasn't about whether we should have bans, but about which values to prioritize. Portraying environmentalists as elite rulers when none were in charge resonated with the aggrieved masculinity performance of victimhood, which deflected attention from Carlson's own media platform's power to sway public opinion and Trump's power as the US president.[112]

This convergence of conservative talking points remarkably enabled single-use plastic straws—of all pressing topics of the day—to gain attention in the 2020

US presidential election. With the rising popularity of single-use plastic straw ban hashtags on social media, it perhaps was predictable that Trump's popular campaign slogan, "Make America Great Again," became rearticulated to #MakeStrawsGreatAgain.[113] With a picture of a paper straw smashed into a plastic cup hole, Trump's then campaign manager Brad Parscale tweeted on July 18, 2019: "I'm so over paper straws. #LiberalProgress. This is exactly what they would do to the economy as well. Squeeze it until it doesn't work."[114] Soon after, Parscale tweeted "Making Straws Great Again" and a link to the campaign store.

Masterfully combining its penchant for profiteering and culture wars, Trump's campaign began selling "Official Trump Straws" in packs of ten red plastic straws for US$15, with his name engraved on them in white capital letters. The website shared the message: "Liberal paper straws don't work. STAND WITH PRESIDENT TRUMP and buy your pack of recyclable straws today." The straws were described in ways that reinforced the message: the environmental claims ("recyclable," "BPA free," and "Reusable & Recyclable") linked to national pride ("Made in the USA") on an excessive scale (nine inches long when most straws are seven to eight inches) promoted a candidate who was not claiming to be anti-environmental, but anti-regulation, pronationalism, and well-endowed. Trump straws quickly sold out, according to his campaign.[115]

Interestingly, like Carlson, Trump's position was not to defend plastics but to problematize the plastic straw as the focus of the crisis. The often-quoted talking point by Trump on single-use plastic straws during the time emphasized scale: "I do think we have bigger problems than plastic straws. You have a little straw, but what about the plates, the wrappers and everything else that are much bigger and they're made of the same material?"[116]

Posting across social media platforms (Facebook, Instagram, and Twitter), often with pictures of their purchases, those who bought Trump straws in 2019 repeated the campaign's message that paper straws—and therefore, liberal government regulations—do not work. Conservative hashtag activists riffed off the message linking a range of tags and topics. Consider this emblematic post: "I might as well stock up on plastic straws and make a good use of them before the next Democrat President bans them along with cars, oil and natural gas, guns, and red meat. Don't get me wrong I do care about the environment, but paper straws suck. #trump #TrumpStraws #Trump2020 #KAG #KeepAmericaGreat #PlasticStraws #PaperStrawsSuck #MakeAmericaGreatAgain #MAGA #QT #QuikTrip."[117] Another wrote: "Paper straws dont work. Get one that will stay intact, just like our borders. #TrumpStraws #MakeStrawsGreatAgain." Likewise, another posted: "They're here! #MAGA #MakeStrawsGreatAgain @realDonaldTrump @POTUS paper straws don't work like leftist, socialist ideals. Protect America and our #freedom #GoodVsEvil #FreedomVsSlavery." For their audience, the Trump straws were successful insofar as they sold out a relatable, affordable commodity that could articulate Trump's anti-regulation sentiments with a range of hashtagged positions.

In response to the "Official Trump Straw," a reusable metal straw company, Steel Straw, started selling blue metal straws with engraved text stating "Trump Sucks" or with the logo for Bernie (Sanders, a progressive candidate who ran against Trump in the election), costing US$7.27, with the promise of donating the maximum amount allowed to Bernie's campaign.[118] Elizabeth Warren (another progressive presidential candidate in the electoral race) made headlines as the first candidate with an ocean plan (A Blue New Deal) and for her response to the plastic straw question at the inaugural climate town hall candidate discussion during a US presidential election. There, she argued that straws were being used as a deflection:

> Look, there are a lot of ways that we try to change our energy consumption, and our pollution, and God bless all of those ways. . . . And I get that people are trying to find the part they can work on and what they do. And I'm in favor of that. . . . But understand, this is exactly what fossil fuel industries hopes we're all talking about. . . . "This is *your* problem." They want to be able to stir up a lot of controversy around your light bulbs, around your straws, around your cheeseburgers when 70% of the pollution— of the carbon that we're throwing into the air—comes from three industries.[119]

In contrast, the winning candidate of the 2020 election, former US vice president Joe Biden, supported a ban in the food service industry, stating at a town hall: "I don't think we should be using plastic straws anymore in restaurants."[120]

CONCLUSION

In the summer of 2021 I visited a local art exhibit of anti–plastic pollution advocates and researchers, *Life in the Plasticene*, which included art by Chris Jordan and cofounder of the 5 Gyres Institute, researcher and artist Dr. Marcus Eriksen.[121] Eriksen's section of the installation includes a large white cardboard silhouette of a camel framing a polybezoar from a camel's stomach, like a cat's furball but composed of undigestible plastic. One of the most moving pieces was a large photograph with white gauze over it. Next to it was a poster with a typed excerpt of an interview with Eriksen titled "feeling deeply," explaining that he had been invited to Dubai to witness camels, who gain a false sense of feeling full when eating plastics and then have died by the hundreds from starvation, as plastics are not the nutrition needed to sustain their lives.[122] Given the sacred atmosphere evoked, I told my child I should look first. As I lifted the gauze, it created an intimate space between me and the photograph. The photograph revealed itself to be a dead camel near a large amount of plastic waste in the desert. With cloth draping over and around us, I felt invited to sense the sacredness of this loss of life in an exhibit that we might have passed by otherwise without pausing. Witnessing required effort, even if minimal compared to traveling the world. And I wondered: What does it mean that a camel, which has evolved such sophisticated ways of staying hydrated and nourished over generations in the desert, has become tricked in a world gone plastic?

Lonely Whale's mission is motivated by a desire for reconnection with oceans and marine life. Its founding symbol of the lonely whale or 52 is an animal some long believed had a unique frequency and therefore was unable to be heard or to share meaningful dialogue with others. As shown through popular culture examples, feeling isolated, misunderstood, and unheard has resonated with many globally, even if not universally. Inviting a communal, affective connection with nonhuman kin mobilized a network of care that went viral, blurring social boundaries between elites and everyday publics.

To trouble throwaway culture's impact on marine life in the United States, Lonely Whale's Strawless Ocean initiative marketed a paper straw and nudged individual behavior modifications through articulating single-use plastic straws as a gateway into the bigger plastics crisis. This chapter focused on two campaigns: #StopSucking, including the Sucker Punch pop-up event and two PSAs to garner publicity for ocean conservation through self-deprecating celebrity humor and viral social media sharing, and #StrawlessInSeattle, including partnerships with the local government, restaurants, sports, and wildlife advocates, to create what Lonely Whale calls a "hyper-local" approach in the move away from single-use plastics through popular culture. The publicity generated about single-use plastic pollution had a remarkable cultural impact on public opinions and practices—one that remains heated years later, as this book goes to press.

Popular culture has long been an arena of hegemonic struggle. Following Hall, I have called for researchers "to continue to map how the spectacular is born not tabula rasa, but out of the connected complexities of everyday life, structural causes, and the extra-ordinary. To explore how the spectacle of popular culture also holds promise as a sign of the times and a terrain of struggle over values and practices on this planet that is our home."[123] In the case of Lonely Whale, despite the fun tunes, well-known spokespeople, and high-tech resources, its advocacy work involved negotiating a deeply ingrained sense of alienation from marine life, loneliness, and a collective recognition of humanity's imperfections, as well as oceanic feelings.

Rather than calling for the public to "Save the Ocean" or even "Save the Whales," the #StopSucking campaign appealed to those tired of celebrity culture, modeling a different style for acting human. That is, instead of promoting an arrogant anthropocentric perception of having all the answers and knowing what is best, there was a humility publicly performed both in claiming one sucks and in allowing a tentacle slap to hit one's body, to then share a slow-motion image of one's mushed, disheveled appearance to circulate online. Instead of promoting the hero or savior complex, this mode of engagement embraces self-recognition of harm, being able to laugh at oneself, and striving to be better. A brief social media gimmick may not exactly embody the intensive work Lugones calls for in "world-traveling," but it is at least—in a fast-paced, tech-infused culture—an embodied opportunity through which one might begin to transform one's sense of identification and model a different role for humanity in an age of planetary crises.[124]

Throughout this chapter, I have highlighted stories of carbon-heavy masculinity and now also the backlash's performance of aggrieved masculinity. In contrast, Grenier, Russell, and others in the #SuckerPunch, #StopSucking, and #StrawlessInSeattle campaigns perform a playful attitude of "sporting fun," one capable of traditional characteristics (athletic, attractive, driven) and what Beth Osnes, Maxwell T. Boykoff, and Patrick Chandler call "good-natured comedy," which "connotes both a mode of comedy that is good for nature, and also good-natured, meaning kind in intent—not seeking to shame or expose in a cruel or demeaning matter."[125] Lonely Whale's good-natured advocates model a figure that embodies resistance in contrast to carbon-heavy masculinity through humility and humor.

Choosing to focus on straws was impure politics for high stakes, provoking affirmative reactions and backlash among many online and offline. Nudges have been criticized for focusing on individuals instead of structural change.[126] The successful conservative rollback of national environmental regulations at the time is important context for plastic straw bans to become imagined as a gateway victory. For conservatives, 52 or "the lonely whale" did not appear to offer a compelling parable of their self-perception of their identities. Instead, the plastic straw became articulated to American exceptionalism as a rejection of listening to or caring about international relationships, an opportunity for a profiteer to uphold the value of material wealth above all other, and a sign of environmental overregulation. Popular discourse about straws became an opportunity once again for conservatives to scapegoat China and demean environmental regulation, rather than to accept US accountability for the plastic crisis. Although Trump supporters weren't Lonely Whale's primary audience, Lonely Whale took their backlash as a sign that at least plastic pollution was receiving attention, which had been lacking before.[127] While US culture increasingly has normalized single-use plastics over the past century, Lonely Whale's hashtag activism critically interrupted those taken-for-granted scripts.

The backlash from conservatives did validate plastic straws as a relatable entry point and popped social media filter bubbles with humorous hashtags and celebrity-endorsed content. Yet the campaigns also illustrated that exposure was not enough to move everyone. Instead, for some the idea of a plastic straw ban reinforced existing conservative beliefs about environmental regulation. This entrenchment in a political position despite exposure suggests digital media campaigns need to consider *how* different positions are shared, not just that they are shared.[128] Taking this backlash seriously is not an endorsement of both-siding the issue (i.e., claiming that all positions are equally valid), but might move us to reconsider what common ground might exist and to make grounded judgments about divergent perspectives.[129]

As has been reported, in this context, "Single-use plastic straws have become a lynchpin in the culture war between conservatives and liberals."[130] But this heated controversy has more than two sides. While speaking for the ocean and nonhuman life continues to be a fundamental role for environmental advocacy, speaking for humans remains deeply fraught, which brings us to the next chapter, on disability backlash.

5
———

#SuckItAbleism Intervenes

Eco-normative Shaming, Voicing Justice, and Planetary Fatalism

When is a straw more than a straw? It depends on who you ask.
—ALICE WONG, FOUNDER AND DIRECTOR,
DISABILITY VISIBILITY PROJECT

Despite conservative backlash, Lonely Whale helped transform plastic straws from something largely taken for granted in US culture to something worth questioning. Lonely Whale's message aimed for humility: beloved celebrities claimed they "sucked" for having used single-use plastic straws, playing off the sexiness of the double entendre of sucking and nudging instead of shaming. When that message was generalized to everyday people, however, no exception initially was made for medical uses or what disability justice calls more broadly "access." As noted, bendy or flexible plastic straws were inspired by a child drinking a milkshake in her uncle's San Francisco store; nevertheless, the mass marketing initially targeted hospitals, and plastic straws have become incredibly important for improving the lives of many people with a range of chronic disabilities and temporary health needs.[1] In response to the campaign's erasure of disability and everyday people shaming the lived experiences of disabled people, another backlash occurred, appropriating the popular #StopSucking trend.[2]

Prompted by the growing momentum of single-use plastic bans in the United States, the well-established Bay Area–based disability advocate Alice Wong (see the epigraph) started the #SuckItAbleism hashtag, riffing off Lonely Whale's hashtag, to critique the focus on plastic straws.[3] Resonating with a broader feeling at the time, Wong's hashtag quickly connected many disabled people and accessibility advocates online within her networks and beyond.[4] Some of the counterpublic discourse was internally focused: disabled people exchanging their feelings and experiences with other disabled people about reactions to plastic straw requests.[5] As one writer claimed: "Often without noticing that they're doing it, able-bodied and -minded people often lampoon or shame people for using products that make the lives of the disabled easier, dismissing them as unhealthy, unnecessary, or a waste."[6] Some of the hashtag activism also became linked to an external critique of ableism or, more specifically, eco-ableism, in US plastic straw campaigns and policies.

Eco-ableism refers to the marginalization of disabled people through environmental design; the exclusion of disabled people in environmental decision-making; and the discrimination against disabled people through environmental discourse, beliefs, and attitudes.[7] Relatedly, Di Chiro cautions that such discourses can produce what she calls "eco-normativity," an ableist, evaluative standard of what constitutes a body—a concept she develops in examining science communication about the impact of toxic pollution on sex and gender norms.[8] While environmentalists continue to try to hold corporations and governments accountable for toxic pollution, Wong, Di Chiro, and others have emphasized the importance of not further stigmatizing individuals already marginalized in US culture.

This chapter contextualizes the #SuckItAbleism trend through considering the significance of voice to both disability and environmental justice movements. Next, it turns to a straw man argument related to this hashtag controversy: that of planetary fatalism, which falsely assumes that all environmental advocacy is futile. Then, the chapter focuses on three critiques from disabled people and accessibility advocates. Following that, I share the impact of #SuckItAbleism activism, including how disability activists successfully shifted Lonely Whale's messaging and related policies. In addition to sharing more from my podcast interview with Emy Kane, this chapter includes insights from Mia Ives-Rublee, a disabled advocate for accessibility who worked with Lonely Whale.

As suggested in the introduction, impure politics is vital to a more inclusive approach to resisting plastic pollution. Since mistakes inevitably are made, the praxis of transformative justice is important to engage. Hashtag activism is not just an opportunity to become more entrenched and polarized in public controversy; it can create possibilities for more intersectional advocacy and for broader coalitions to emerge.

ON SPEAKING UP AND BEING HEARD

Voice long has been significant to disability rights advocates globally. James I. Charlton, for example, recalls the first time he heard a key slogan of disability rights, "Nothing About Us Without Us": "I first heard the expression 'Nothing About Us Without Us' in South Africa in 1993. Michael Masutha and William Rowland, two leaders of Disabled People of South Africa, separately invoked the slogan, which they had heard used by someone from Eastern Europe at an international disability rights conference. The slogan's power derives from its location

of the source of many types of (disability) oppression and its simultaneous opposition to such oppression in the context of control and voice."[9] The argument of
"Nothing About Us Without Us" reflects a global movement linking one's lack of
voice in the decision-making processes of a culture or community to one's disen-
franchisement or empowerment. Thinking through multiple disabilities, "voice"
might be imagined through a wide range of expressions, including speaking, typ-
ing, audio-to-text, signing, and more. The broader procedural justice ethic is to
include disabled people in decisions that impact disabilities, affirming experien-
tial expertise and recognizing the dignity of each person.

The environmental justice movement also has long underscored the importance of voice to environmental decision-making. In 1991 the Principles of Environmental Justice were articulated and adopted at the First National People of
Color Environmental Leadership Summit in Washington, D.C. This gathering included participants from all fifty US states, plus Puerto Rico, Brazil, Canada, Chile, Ghana, the Marshall Islands, Mexico, and Nigeria. The preamble and
seventeen principles identify fundamental environmental priorities and beliefs,
including rights of Mother Earth and the need for all to have the right to protection from environmental harms, including toxins, hazards, and radioactive waste,
as well as discrimination, destruction, and exploitation.

A vital part of the emergence of the movement for environmental justice has
been procedural justice, which suggests decision-making processes for collective
decisions through respectful and transparent (right to know) practices, including
the opportunity for those most impacted to be given the opportunity to share lived
experiences as vital expertise. Principle 7 explicitly states: "Environmental Justice demands the right to participate as equal partners at every level of decision-
making, including needs assessment, planning, implementation, enforcement and
evaluation."[10] As one environmental justice activist leader, Dana Alston, emphasized in 1990: "We speak for ourselves."[11]

The Jemez Principles for Democratic Organizing, written in 1996, again
emphasize voice in "#3 Let People Speak for Themselves," which states: "We must
be sure that relevant voices of people directly affected are heard."[12] In 2002, at
the Second National People of Color Environmental Leadership Summit, the
Principles of Working Together also affirmed voice as a core value: "2.C. The
Principles of Working Together reaffirm that as people of color we speak for ourselves. We have not chosen our struggle, we work together to overcome our common barriers, and resist our common foes."[13] Again and again, the movement
was mobilized when those most impacted by environmental injustices found their
voices were excluded from environmental decision-making.

Voice, in this sense, is not just having an opportunity to speak, but it requires
the right to be heard when one's story might be able to shape collective decision-
making that impacts one's life. Eric King Watts theorizes this communal

understanding of voice by underscoring how digitally connected technology can hamper voice: "The challenge . . . over the horizon is to find new ways to 'keep it real' in a fast-approaching virtual reality. As we all increasingly feel displaced from one another, occurrences of 'voice' represent events in which we can characterize our commitments and sentiments towards our social spaces. 'Voice' grounds us by reminding us of our situatedness. We are also reminded of our needs. Such an interest for criticism serves to return us to an examination of the basic requirements *for living with ourselves and with others.*"[14] Watts identifies the fast pace and proliferation of communicative channels as increasing humans' capacity to perhaps become more connected while simultaneously feeling more isolated. "Having a voice" in public culture, then, is more than having the technology and the capacity to share opinions online. In these exchanges, what is required *for living with ourselves and others* is not a foregone conclusion. Voice ideally involves opportunities to have a role in decisions that impact one's life, a praxis that requires ongoing relational care. Although results are not guaranteed, as Stacey K. Sowards argues: "Empowerment comes through using one's own voice to speak, a kind of embodied enactment."[15]

In the United States, the Americans with Disabilities Act (ADA) was not passed until 1990. Legalizing the civil rights of people with disabilities resulted from over three decades of organizing. Disability civil rights leader Judith Heumann emphasizes interdependence in her memoir, stating repeatedly: "Everything I've done in my life, I never would have been able to do alone—whether it was my mother and father or my brothers, my friends from school, or my fellow activists helping, listening, laughing, and leading."[16] In a film featuring Heumann and other activists, emphasis was placed also on alliances, including the support of the Black Panthers, United Farm Workers, International Association of Machinists union, and Butterfly Brigade (an LGBT organization).[17] Yes, there now are legal architectural requirements in the United States for new buildings (such as wheelchair ramps and elevators), but with that also had to come a cultural recognition that recognized disabled people as belonging in public schools, grocery stores, workplaces, and entertainment spaces. Disabled people have advocated for civil rights in ways that can claim success since the mid-twentieth century, though laws remain unevenly enforced and understood.

Heumann describes "disability culture" as one that prioritizes "the humanity in all people, without dismissing anyone for looking, thinking, believing, or acting differently"; disability culture, she emphasizes, includes a communal sense of voice: "Slow down enough to listen and truly see each other. Ask questions. Connect. Find a way to have fun. Learn."[18]

The disability justice movement also affirms the importance of voice in communal decision-making.[19] The "10 Principles of Disability Justice" by Patty Berne, edited by Aurora Levins Morales and David Langstaff on behalf of the disability justice organization Sins Invalid, also affirm the significance of voice. They include "Leadership of Those Most Impacted," following the insight of Morales, who argues: "We are led by those who most know these systems."[20] By "know these systems," Sins Invalid argues for intersectionality, understood through the work of Audre Lorde (frequently quoted for her speech affirming that "we do not live single-issue lives"); it also emphasizes the ways disabled people and access caretakers have expertise in the care work needed for resilience.[21] Aimi Hamraie calls this epistemology "access-knowledge," which "emerged from interdisciplinary concerns with what users need, how their bodies function, how they interact with space, and what kinds of people are likely to be in the world."[22]

Having a voice matters to an ethic of care. Care work is defined in the Feminist New Green Deal as "the market and non-market work that all of us engage in to sustain life."[23] The voices of disabled people are particularly well poised to provide expertise on care work, as a result of intimate knowledge of how to navigate what Mel Chen calls "the fiction of independence and of uninterruptability."[24] Consider, for example, Leah Lakshmi Piepzna-Samarasinha's account of how online discourse emerging after California wildfires found disabled people sharing expertise in resilience: "Who were the people who already knew about masks, detox herbs, air purifiers, and somatic tricks for anxiety? Yeah, you guessed it. Over and over, it was sick and disabled folks."[25] Likewise, "disabled people," as Wong writes, "are creatures of adaptation that design and build worlds that work for them. The skills that we have reimagining/hacking/surviving hostile ableist environments would serve us well in any dystopian future."[26] The sharing of information, resources, and energy needed in times of disasters and as part of our everyday lives underscores the value of including disabled people in building networks of care not just for the sake of disabled people, but for everyone.

Beyond moments of disaster or crisis, accessible design as a cornerstone of imagining infrastructure also involves an aspirational value of universal inclusion. Consider, for example, the quintessential example of the universal design principle "the curb-cut effect" (see figure 16). By creating curb cuts or sidewalks that are designed to lower gradually like mini ramps so that people in wheelchairs or using crutches or walkers can cross a street more easily, the sidewalk becomes more accessible to a wider range of activities, from babies in strollers to people riding bikes, from tourists with suitcases to workers moving packages on a dolly.[27] Another example of universal design is how closed captioning in classroom videos can assist not only people with hearing impairments but also students for whom English is a second language or a student who is a parent with a kid running around in the background during a remote class.[28] Of course there are some technologies, like straws, that are helpful to some people and unnecessary for others, which is why some prefer the emergent language of "pluriversal design."[29] The point is that, like designing infrastructure, campaign designs also

FIGURE 16. The "curb-cut effect" or "curb effect" is an exemplar of universal design because it increases mobility and access for the lived experiences of many. *Source:* UK-based artist Jono Hey, https://sketchplanations.com/the-curb-cut-effect. Reprinted with permission.

involve embodied assumptions when considering audience, messages, media, and goals. Which voices are included or excluded matters.

If bendable plastic straws are necessary for some to live, then, why ban them at all? What difference do single-use plastic bans make?

THE PITFALLS OF PLANETARY FATALISM

Fatalism is the belief that what we do does not matter. It provides cynical, straw man readings to claim that the future is determined, and therefore we everyday people might as well do whatever we want. Planetary fatalism falls back into false choices (do all or nothing) of straw man arguments and can set up purity tests for environmental advocacy. Although plastic straws were chosen tactically by Lonely Whale as a gateway plastic, and while consumer advocacy has its limitations, there is a slippery slope that can occur when some dismiss the value of an impure environmental choice: if (fill in the blank) is not going to solve the plastics (or climate or whatever) crisis, why should one individual bother to do anything at all?

As Samantha Hautea and colleagues describe in their study of affective publics on TikTok, resignation is a common trope on social media today, exemplified in the following: "Right now, the Amazonian rainforest is on fire, and scientists agree that unless every single country agrees to fight climate change, we're all heading towards a global extinction. And there's nothing you can do about it, 'cause you're only one person. So go ahead, use that plastic straw: we're all as good as dead anyway!"[30] The video appears sarcastic in tone, but it also reflects a larger discourse about how plastic straws became an articulator of the crisis, in this case of the lack of agency some individuals feel in enacting systemic environmental change.

The appeal of planetary fatalism is that we can rationalize any behavior we desire. Economic and political elites of the Global North are particularly adept at rationalizing consumption, sometimes even wrapped in the discourse of "self-care." Further, blaming individuals without a systemic critique falls back into previously critiqued neoliberal politics of corporations that use individual consumption to distract from their own accountability.

The drawback of planetary fatalism is that it provides no recognition that cultural shifts are a constant or that individual agency has mattered throughout human history to transforming social structures. Planetary fatalism proves disastrous because doing nothing other than maintain the status quo guarantees an uninhabitable planet. Climate scientist Michael Mann calls these "too late" narratives, which he argues "unwittingly do the bidding of fossil fuel interests by giving up."[31]

Likewise, resistance to plastic bans through planetary fatalism does the bidding of the plastics-industrial complex. Consider this talking point of the ACC's undermining of a plastic ban in 2010: "Instead of wasting time and telling us how to bag our groceries, lawmakers should be working on our real problems, including a huge budget deficit, home foreclosures, and millions of workers without jobs."[32] This perspective resonates with Trump's "we have bigger problems than straws" discourse, noted in the last chapter. The pattern of the planetary fatalism fallacy trivializes the significance of single-use plastics bans—as if single-use plastics have no impact or have no relation to so-called larger crises, such as climate, budget deficits (hello, fossil fuel subsidies), or jobs. Sometimes those deflecting accountability for plastic straw consumption then also fail to take more systemic steps because they can rationalize dismissing plastic straws or, more broadly, giving up on all advocacy since no one action will solve the "real problems."

Plastic bans articulate an impure politics that denies these fatalistic discourses by attempting to improve planetary health through structural changes. As noted, plastics do create harms across their lifecycle, including toxic extractivism, production, transportation, consumption, and disposal. Most backlash against plastic straw bans only focuses on disposal, ignoring frontline communities. Even then, it tends to downplay the significance of plastic straws to the plastics crisis.

That is, statistics that plastic straws are a very small part of the plastic waste crisis perform a trick of scale. First, straws do not weigh much, so, if we are quantifying straws by weight, they will appear to matter less. Consider, however, even the commonly cited UN statistic that plastic straws constitute 0.025 percent of ocean plastic, which is drawn from a report that identifies ocean plastic at 13 million tons total.[33] Doing the math (yes, I had to pull out a calculator): 0.025 percent of 13 million is approximately 3,250 tons (7,165,024 pounds or more than thirteen Statues of Liberty)—every year. While not the largest source of plastic waste, it is a monumental amount to create annually (especially recalling that those statistics only account for those that end up in the ocean, and some sources estimate plastic straws have an infinite life postconsumption). This point about single-use plastics and magnitude is embodied in variations of the joke: "'It's only one . . .' said 8 million people. #SkipTheStraw #TurtleTuesday #StopSucking."[34] Willingly or not, claims that individual plastic straw consumption doesn't matter at all can set up a straw man argument of plastic straw–ban advocates to rationalize that throwaway culture in the Global North is not a problem, because it's not the biggest or only problem.[35] Sure, there are other plastics to focus on as well (small ones, like balloons, and larger items, like fishing nets), but I have found no plastic ban advocate who has claimed straws are all that matter.[36]

To be clear: one person skipping a plastic straw when ordering a drink is not all we need to accomplish to address the plastic crisis (or the climate emergency or pollution colonialism or . . .). As I wrote in chapter 2, to address the climate emergency, ending fossil fuel subsidies, as well as banning private jets, superyachts, and space tourism, would have far more significant impacts overnight.[37] But listening to people in the Global South and witnessing impacts on marine life, structural decisions to ban plastics, including single-use straws, matter.

Mary Annaïse Heglar writes: "We don't have to be pollyannish, or fatalistic. We can just be human. We can be messy, imperfect, contradictory, broken."[38] The question, then, is this: How can we find a way to hold both a desire to care for/as disabled people and Earth's climate, as well as marine life and people in the Global South who bear a disproportionate burden of the costs of plastics?

ON #SuckItAbleism AND #StrawShaming

Although a great deal of care work happens in person, online disability networks of care are increasingly important to witnessing stories of trauma and to sharing resources for those previously unconnected to connect and build community. Like other social movements globally, hashtags have become a key tool for contemporary disability activists to find each other, amplify common experiences, and articulate compelling critiques.[39] In her analysis of #CripTheVote, Heather Walker has argued that online communication is "less risky because it allows for a decrease in form-based prejudice and discrimination" and "also allows for greater flexibility in identity and self-image construction."[40] Likewise, Elizabeth Ellcessor claims that the "formation of a group identity, often via multi-voiced blogs and social media spaces in which varied perspectives on disability are evident, facilitates new forms . . . that operate outside the limitations of what is culturally legible to mainstream audiences."[41] Hashtags, though not outside dominant media practices, provide a technology to enable strangers to connect, to commiserate, and to rehearse collective, polyvocal critiques.

Two years prior to #SuckItAbleism, Wong was one of three organizers of #CripTheVote, another relatively well-known disability hashtag activist network in the United States. The hashtag raised the visibility of disabled people as voters who matter to elections. "#CripTheVote represents a 'watershed moment' for contemporary disability studies," according to Heather R. Walker. "Within the tweets authored by these activists is a demonstrated knowledge of intersectionality, systems-based discriminatory practices at both state and institutional levels, and of ableist ideologies."[42]

Consider also the annual #MillionsMissing protests that began in 2016, which aim to "raise awareness and fight for recognition, education and research for people living with Myalgic Encephalomyelitis (ME or ME/CFS)."[43] They use online advocacy to publicize their voices and create protests of empty shoes in public spaces to mark how they are too often absent from public decision-making.[44] Writing about these actions, Jennell Johnson points out that digital media networking may be more accessible to some people with some disabilities and less accessible to others, depending on the disability: "For public events and campaigns that begin on social media, then, it may be helpful to offer alternate ways of accessing information (such as e-mail) and circulating opinions (such as petitions and letter writing), or even providing the opportunity for anonymous participation (which lowers the stakes), which all allow public participation with differing levels of appearance."[45] In the straw controversies, it is perhaps unsurprising that online disability activism does move between offline and online actions.

Though Wong coined #SuckItAbleism, it quickly went viral, reflecting horizontal organizing among disabled people, which exceeds any one speaker or organization. Although the trend might at first appear to set up a binary of pure politics, many disabled people offered nuanced positions that reject ableism but embrace concerns of ecological crisis. The following analysis is based on social media (mostly Twitter and Instagram), interviews, articles, and blogs primarily written by disabled people from 2018 to 2021; my direct quotes favor published statements on blogs and news articles because there was a sufficient archive to illustrate the counterpublic's main points without quoting unverified social media accounts. Even though legally, unlocked social media—especially using a hashtag(s)—is searchable by design and circulates publicly among strangers, I took precautions because being quoted in a book can lead to increased harassment. Nevertheless, the vernacular and grassroots perspectives amplified here are readily found online through the hashtags noted. This chapter's archive predominantly includes arguments from people in North America since those were most vocal in this backlash. The arguments are organized along the two main hashtag trends of this backlash: #SuckItAbleism and #StrawShaming.

ON VOICE AND RESISTING TWO-SIDED THINKING: #SuckItAbleism

Consider Wong's narrative of how she came to spend much of the summer of 2018 engaging the "Straw Wars":

> I think as I started reading more and more about the ban in Seattle, the ban in Vancouver, you know, I really listened to what disabled people are saying. And on the ground, people were saying how a lot of the decision makers at the city level were just not really paying attention to what concerns people with disabilities are having. You know, each kind of ban is different; some have an exception for people with disabilities, and even that is problematic in itself because of how these exceptions are written and enforced. Again, you know, bans are, I think they really need to be carefully constructed. And based on my observations of a lot of these bans and this trend toward phasing out plastic straws, a lot of it has been just in reaction to PR campaigns by environmental groups who are just like seizing this moment, using celebrities, and raising awareness without really thinking about the implications on actual marginalized people.[46]

Wong heard of the rising number of plastic straw bans in North America and listened to disabled people, including herself. She emphasizes concerns of voice: that disabled people were not being heard by changemakers and policymakers as "actual marginalized people."[47]

In response, Wong began to publicize accessibility concerns using a hashtag she created: #SuckItAbleism. The play on Lonely Whale's slogan appropriated #StopSucking by rearticulating "sucking" to "ableism" and by disarticulating "sucking" from plastic straws. She argues the intervention was to trouble the two-sided thinking that had emerged: "The entire conversation about plastic straws is about power: who knows best, who decides how change is made, who is centered in all of these activities. One example of this power is the moral reframing of plastic. The simplistic binary of plastic = bad/compostable = good obscures the intimate complexity of plastic in every person's life, including mine."[48] As Wong observes, "the moral reframing of plastic" was what was at stake in these policies and controversies. By excluding disabled people who needed plastic straws from decision-making about policies that would impact them most, the epistemic and procedural justice dimensions of voice—"who knows best, who decides how change is made"—arose as salient. Wong then encouraged a critique of the emergent frame, as well as to trouble the assumption that all backlash was anti-environmental.

Wong's hashtag became popular because she had an existing online base, and she was naming a sentiment at the time that clearly resonated among many. Consider an excerpt of this guest blog for Greenpeace USA (a Canadian-founded environmental NGO) by Rev. Theresa I. Soto, whose bio reads: "a Unitarian Universalist minister and liberation worker. Theresa identifies as Latinx, non-binary queer, and disabled, all of these all the time. They believe in the abilities of movements to increase their inclusivity as we move toward solutions":

> **All of us need all of us to make it.**
>
> One of the flaws of the straw problem being described as a debate is that it makes two camps: straw users and non-straw users. This structure and language push disabled people into a category of plastic lovers, inattentive to the burden they put on the environment, as many other good people struggle to save the planet.
>
> People with disabilities are not enemies of the environment. One of the questions environmental movements are faced with at this time is this:
>
> **When movements (unintentionally) embrace solutions that leave people behind, what should a justice-based response be?** . . .
>
> Straws only comprise .03% of the plastic in the ocean. That doesn't mean that movements shouldn't turn their attention to straws. Rather, it means that if we use methods that rely on leadership by those the most impacted, we will be leading across movements in ways that leave room for nuanced solutions.[49]

The call for having a voice and centering care as an ethic reminds us that people with disabilities and accessibility advocates are vital teachers for surviving crises through interdependence, necessary for successful environmental advocacy and a more viable future for all. Having a voice enables people to influence decision-making and build trust across environmental perspectives.

Some made similar arguments by sharing their intimate relations with plastics. For example, Rae DeFrane, a writer who describes herself as "passionate about advocating for the health of our oceans, sustainable living and conscious-consumerism, and strives to be an activist in her eco-related works and sustainable crystal and upcycled jewelry designs," elaborates:

> I care about the environment: I make every effort I can to align my actions with the conservation of the planet's resources. But I need plastic to live—moreover, I need a lot of plastic to live. As someone who has had Type 1 Diabetes since age three, the

FIGURE 17. Cartoon created by Hong Kong–based animal rights artist Joan Chan (@justcomics_official) to raise plastic pollution awareness beyond straws. Chan was inspired by the Ocean Conservancy ("Trash Free Seas"). *Source:* Posted on Instagram with #StopSucking. Reprinted with permission from the artist.

> amount of glass vials, syringes, lancets, plastic infusion sets, tubing, cannulas, and sterile packaging . . . used in my lifetime is incalculable. Without it, coupled with the refrigeration technology needed to cool and store the insulin that my body cannot make, I would be dead within 48 hours. The amount of medical waste needed to keep me alive is an albatross around my neck.

Here, DeFrane articulates the relevance of plastics from single-use to refrigeration, as well as feeling a burden about creating waste. DeFrane then declares: "It's time to shake our heads and realize that we cannot move forward with a monolithic view of single-use plastics in the future and cannot ethically justify negating human rights and lives for one single way forward."[50] This call resonates with pluriversal design, suggesting multiple paths to a more just and sustainable world.

Still others pointed out that abandoned and lost fishing nets, often called "ghost gear," are widely recognized by ocean advocates and scientists as the most widespread and harmful plastic debris for marine life (see figure 17). Part of this backlash, then, was that people who identified as environmentalists and animal welfare advocates were wondering: Why focus on the gateway plastic rather than focusing on the bigger sources of marine pollution?

ON #StrawShaming AND ALTERNATIVE STRAWS: "WHY CAN'T YOU JUST USE ____?"

As bans grew in popularity at companies, in cities, and across states in the United States after 2017, handing out plastic straws was no longer part of the taken-for-granted habits. Disabled people increasingly found themselves in public spaces offline and online asking for plastic straws and being denied or asked why they could not use an alternative straw made of paper, stainless steel, and so forth. Lots of stories were shared on social media with #StrawShaming; most narrated going to restaurants or other public spaces, asking for a plastic straw, and being met with some resistance from servers asking why a plastic straw was needed and claiming "we care for the planet" or "care for the turtles."[51]

In person, disabled people shared common experiences of being asked to defend requests for plastic straws because of the growing public awareness of single-use plastics straw campaigns; online, those affirming #StrawShaming were not as popular, as the hashtag overwhelmingly was used to mock the practice. There were some exceptions, such as a few people who posted selfies and said they were #StrawShaming themselves; otherwise, most #StrawShaming posts by people doing the shaming appeared unpopular on social media.

Presumably, the goal was to educate customers that plastic straws have an impact. Like the planetary fatalism discourse, embracing straw shaming on social media—when it occurred—appeared to come with a sense of humor from the perspective of those doing it. Nevertheless, as Ochieng cautions when considering shaming in restorative processes: "Shaming perpetrators is a particularly problematic form of punishment and may undercut deeper articulations of justice. This is because shaming may moralize . . . rather than offering explanations for these occurrences."[52] Stigmatizing, then, can deflect attention from why someone might want or require a plastic straw.

Of course, shaming can work pragmatically to induce desired responses. Arguably, for environmental advocates shame has worked less well on individuals and better on corporations or public figures (such as politicians or celebrities). In *Is Shame Necessary?,* Jennifer Jaquet identifies multiple times corporations have been shamed publicly, leading to a consequence in policy or practice. One example was in 2012, when the Susan G. Komen for the Cure Foundation announced withdrawing funds from Planned Parenthood for breast cancer screening and education; the public expressed outrage online at approximately five tweets every two seconds, and Komen reversed that decision within twenty-four hours after the news broke.[53]

Perhaps the quintessential corporate shamer of the environmental movement is the NGO Greenpeace, which, for example, has targeted large retailers of Chilean sea bass rather than individual consumers, including shaming Trader Joe's to improve its seafood sustainability using scorecards. It also has established shaming as public spectacle or image event, including flying a blimp over Costco's headquarters and then posting those photographs online to be circulated and used to shame digitally.[54] "It's not that shaming is preferable," Jaquet argues, "it's just that, in some cases, shaming is all we have."[55] That rationale for shaming corporations and public figures, I argue, is an exemplar of impure politics. Shaming is not a first resort, just one appeal that might (or might not) successfully mobilize counterpublics to leverage numbers and social norms against corporations and public figures.

"Acceptable shaming," Jaquet argues, "tends to focus on the powerful over the marginalized."[56] This assumption might be why most experts caution environmental advocates against using shame as an appeal to motivate individuals to change behavior, as it can backfire. Shaming may be ineffective, creating a sense of catharsis for the shamer without changing the behaviors of those who have been shamed. Shaming also may negatively direct the shamed individual's focus inward and, instead of leading to the acceptance of responsibility, further isolate people. Moreover, it probably is unsurprising that experts find women are quicker to feel shame than men and adolescents more than adults, leading to "low self-esteem and depression."[57] Brené Brown suggests we need to move away from "shame as a tool for change" because it not only can create a sense of feeling "unworthy of love and belonging" but also may generate a "fear of disconnection" that "can make us dangerous."[58]

Returning to #StrawShaming, when disabled people began to be told publicly that the plastic straws they requested in public spaces were unnecessary, it erased their expertise in knowing their own lived experiences best. These scenarios have exceeded any one individual's feelings and became narrated as a collective effect of shaming shared by ableist social formations. As noted, disabled people have a long history of being treated as less than human and as unable to make decisions for themselves, so any public questioning about accommodations can trigger that (ongoing) legacy of trauma of being silenced or second-guessed about one's own body.[59] Perhaps most obviously, in these contexts, some acts of shaming place responsibility on individual maintenance and discipline (in this case, "why not bring your own straw?") instead of on systemic change.[60]

People who shamed—and continue to do so, as this book goes to press—by suggesting (sometimes adamantly) that there were/are alternatives to plastic straws seem to assume an information deficit model of communication, which erroneously assumes the issue is a lack of knowledge rather than a difference of opinion, values, or experiences. The need to teach ableds why alternatives to plastic straws did/do not work for all has appeared necessary again and again—in restaurants, online, and with legislators advocating for bans without required access accommodations.[61] This repeated script of framing plastic straws as unnecessary and alternatives as viable eventually led to many disabled people and accessibility advocates sharing their reactions online, from sarcastic and frustrated tweets to humorous videos.[62]

One trauma and disability educator, Kaalyn, describes her disabilities as erythromelalgia (also known as "Man on Fire"), EDS (Ehlers-Danlos syndrome), POTS (postural orthostatic tachycardia syndrome), and MCAS (mast cell activation syndrome), and as a wheelchair user.[63] To correct ableist assumptions that assumed disabled people were using single-use plastic straws due to an information deficit, Kaalyn created a chart outlining the reasons bendable plastic straws were the safest option for some people. This intervention became circulated widely online to help summarize disabled expertise on multiple disabilities and multiple straw types. Kaalyn's chart provided a pivotal tool for changing public attitudes from assuming disabled people had a lack of knowledge about alternative straws to suggesting perhaps abled people had a lack of knowledge about disabilities (see figure 18).

During this time, online discourse appeared to move from testimony of feeling shamed to affirmation of the disabled community. The slogans "Straws are Access" and "Ableism is Trash" emerged, linking sucking to access and ableism to trash. The disability hashtag activism also became commodified relatively quickly, with online shops selling stickers, pins, masks, and mugs, presumably to financially support the advocacy work and to unburden disabled people from having to explain requests every time.[64]

Again, the goal of the major environmental organizations that advocated for plastic straw bans did not appear to be shame—consider Lonely Whale's whole approach to *avoid* "shaming or scaring" by promoting humorous and joyful sporting fun. In addition, two major environmental organizations in North America both published stories on the importance of being inclusive of people with disabilities, including the Greenpeace USA blog post by Rev. Soto quoted earlier. Likewise, San Francisco "supervisor Katy Tang, who authored the [city's plastic ban] legislation, said the purpose was not to shame straw users, but to get both consumers and sellers to reflect on the waste they create."[65]

Nevertheless, like all movements, there are people who supported plastic straw bans who decided shaming would serve the cause. Many claims of shaming online tended to be general, which is not to dismiss them but to emphasize the earlier point that when one culture transforms the way an object is signified

"Just use ___ straws!"

	Allergy Risk	Choking Hazard	Injury Risk	Not Positionable	Not Hot Liquid Safe	Dissolve w/ Long Use	High Cost
Metal			X	X	X		X
Paper		X		X	X	X	
Glass			X	X			X
Silicone				X			X
Acrylic	X		X	X	X		X
Pasta/Rice	X	X	X	X	X	X	
Bamboo	X		X	X			X
Single Use							

❖ Many disabled individuals require straws for food, meds and to be social with friends. ❖
We can ALL reduce plastic use, but banning items many depend on harms a very vulnerable population.
Pressure companies to make safe alternatives available to all and reduce waste in larger ways.
Hurt turtles are devastating. So are children and adults aspirating liquid into their lungs.

Love, Hell on Wheels

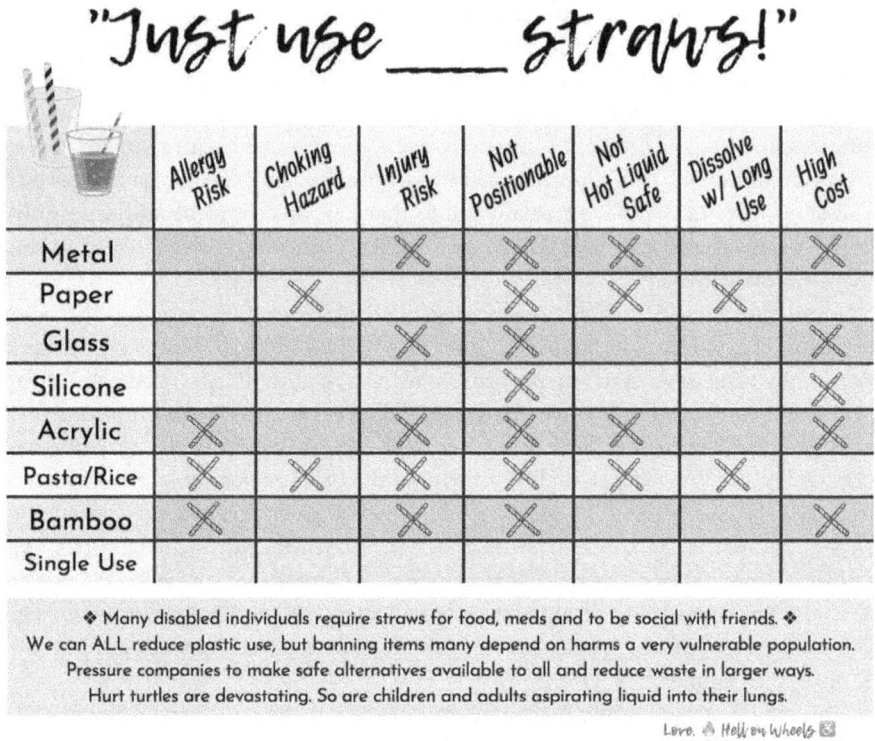

FIGURE 18. Chart comparing problems with straws. In response to eco-ableism, disability justice advocates became exasperated by being second-guessed and having to explain why alternatives to single-use plastic straws can cause health risks. This chart was circulated widely as a turning point in the debates. *Source:* US-based artist Hell on Wheels (@rollwthepunches). Reprinted with permission.

or prioritized, that element can make a whole cultural system feel like it's shifting. Even if advertising that "straws suck" is playful, if one needs a plastic straw, then after seeing such an ad one can reasonably feel like a fallacy is set up that anyone who needs a plastic straw sucks. Whether or not stigmatizing disabled people using a straw was the goal, the unintended consequence landed on many as eco-ableism.[66] The critique was not ignored by all.

ON IMPACT: TRANSFORMATIVE JUSTICE AND THE IMPURE POLITICS OF BANS, REVISITED

Dividing opposition, as noted, is a classic strategy of economic and political elites of the Global North; this does not mean progressive allies cannot raise criticisms, but the difficult labor of collective action often requires digging deeper to consider desired coalitional goals. As adrienne maree brown writes in the brief but powerful and timely book *We Will Not Cancel Us*, sometimes calling people or institutions out online can be a powerful tactic when there is a deep power imbalance. "But," she argues, "call outs need to be used specifically for harm and abuse, and within movement spaces they should be deployed as a last option." She adds: "I don't think it is transformative to publicly call people out for instant consequences with no attempt at a conversation, mediation, boundary setting, or community accountability." Brown asks: "I can't help but wonder who benefits from movements that engage in public infighting, blame, shame, and knee-jerk call outs? I can't help but see the state grinning. . . . Meanwhile, the conflicts are unresolved, and/or harm continues." The goal of transformative justice, she emphasizes, is to name and end harm.[67]

In the #SuckItAbleism trend and related hashtag activism, disabled people clearly named the harms caused by exclusion and erasure from straw ban advocacy by everyday shaming. While transformational justice may be uneven, Lonely Whale's messages were changed as a result. This transformation seems important to recognize as a possibility of the wider arena of hashtag activism and plastics.

In September 2018, a year after Lonely Whale launched its plastic straw campaigns, then Lonely Whale digital strategist Emy Kane was spotlighted on stage with Mia Ives-Rublee, now director of the Disability Justice Initiative, then a North Carolina–based public speaker and policy expert who has consulted with organizations from the Ford Foundation to AirBnB to the Women's March.[68] The two were invited to be showcased as a model of how movements that clash might find common ground (see figure 19).

At this public event at which both parties agreed to speak, no mention was made of Lonely Whale's exclusion of disabled people in the initial launch design or publicity. Ives-Rublee does emphasize the importance of a range of policymakers and organizations needing to value the voices of disabled people: "Let's slow down, let's talk to the disability community, let's include the disability community so that we can have these conversations before we start creating bans."

Ives-Rublee then gave the example of asking for a plastic straw at a national park and being denied. As I noted, after the initial plastic bag ban, Kenya passed a second ban focused on protected areas, which seemed to some like an easy first step in expanding plastic reduction because presumably people visiting public parks have some care for them as environmental destinations, and those public lands are set aside by countries as public goods. Although the Trump administration forbade the restriction of the sale of disposable plastic water bottles in US national parks, Public Employees for Environmental Responsibility (PEER), GreenLatinos, and Beyond Plastics all are currently seeking to overturn his decision.[69] Ives-Rublee's story, however, raises a different inflection on the public

FIGURE 19. Emy Kane (left) listening to Mia Ives-Rublee (right) at the Movement Makers Summit, The Times Center, New York City, September 24, 2018. *Source:* Noam Galai/Getty Images North America. Reprinted with permission.

good, focused not on the place but *on the people imagined as valued members of the public.* This suggests a need to balance the protection of public parks with that of people, who constitute public life in the first place.

In response to Ives-Rublee and the larger disability critique, Kane reiterated Lonely Whale's new talking points on accessibility in relation to the #Strawless-Ocean initiative:

- Recommending the language of voluntarily reducing plastic straws instead of a "ban."
- Noting "accessible alternative must always be available on-demand."
- Making sure plastic straw are available on request without questions.
- Recognizing "flexible plastic straws or compostable plastic straws" might be the only useable option currently for people with disabilities.
- Including a statement about disability explicitly ("*For example*: 'As part of our voluntary campaign for a #StrawlessOcean, we're asking those that can forego a plastic straw to do so. We recognize and champion the needs of all, including our friends in the disability community'").[70]

These talking points recognize that the language of a "ban" does sound totalizing and that some "friends" need plastic straws. More than just asking organizations and businesses to have them on hand, Lonely Whale emphasizes that all people should be able to ask for a single-use plastic straw "without questions." This point addresses the acts of shaming noted earlier, wherein people with disabilities felt the act of *being asked* why they needed a straw was normative and unnecessary. These shifts in Lonely Whale's discourse happened after the public critique and subsequent dialogue that I now want to share.

Kane also emphasized that the language of a "ban" sounded like an unproductive, all-or-nothing purity test: "We don't like to use the word 'banning' plastic straws. I think when you start to get into an outright ban, you start to lose the nuance behind . . . the movement, right? So, that's when you start to get into issues of accessibility. So, a lot of work that we do is voluntary reduction of single-use plastic straws . . . while still offering one upon request." Lonely Whale does claim the Seattle straw ban in its materials in multiple ways as an indicator of impact and success, but it has explicitly shown how it has reflected on the critique of eco-ableism as it could apply to its campaigns, tried to repair its error by paying for disability consultants to help teach it as well as attending disability events to learn, and changed its behaviors as a result.[71]

Kane's argument for a more nuanced understanding of strawless initiatives seems critical to showing the organization's self-reflection and modified messaging. Subsequent to the public critique of disability advocates, as well as private conversations and consultations, Lonely Whale's website began displaying the following statement on multiple pages: "Some of our allies in the disability community require a straw to drink. Please always ensure appropriate alternatives are available."

For accountability, disability justice scholar and advocate Mia Mingus calls for self-reflection, changed behavior, repair, and apology.[72] Lonely Whale is not just claiming disability advocates as allies but is also working on altering behavior by advocating for making plastic straws available and not questioning people when plastic straws are requested, which feels vital to recognizing the dignity of all and deterring the shaming rhetoric noted earlier. Further, Lonely Whale since has worked with policymakers to encourage disability accommodations in plastic ban policies now enacted in places such as New York City and Los Angeles, as well as Canada.[73]

Starbucks also clarified its position on plastic straws in response to disability hashtag activism and the broader critique. Its statement is more minimalist than Lonely Whale's, but still clarifies: "Starbucks offers, and will continue to offer, straws to customers who need or request them in our stores. Starbucks recent announcement about straws will not impact the ability of those who need straws to access them."[74] It also reaffirms its commitment to universal design, highlighting its partnerships with disabled people and access advocates globally.[75]

A "ban" does mean to legally prohibit, but an increasing number of bans are making the impure politics more explicit in their messages and are writing them into policies that previously did not have an accommodation clause. For example, the Seattle city government notes in small print on a public explanation of the ordinance: "Accommodations for people with disabilities provided upon request." That same month (September 2018), it was announced that the nearby US state of California was planning on becoming the first state to ban plastic straws, "unless requested by the customer."[76] This more nuanced message about the intent of California's plastic straw ban was not lost on many news headlines. Consider, for example: "California just became the first US state to ban plastic straws in restaurants—unless customers ask"; "Want a plastic straw in California? You'll have to specifically ask for one"; and "California Bans Restaurants from Automatically Giving Out Plastic Straws."[77] This shift in the frame of a plastic straw ban aims to deter the second-guessing of disabled people when plastic straws are requested, which feels vital to transformative justice as an ongoing practice.

Likewise, Milo Cress, the person whose statistic of 500 million plastic straws a day (which he calculated when he was nine) is repeated often by Lonely Whale and other anti-plastic advocates, long has supported "offer-first" policies. Cress explains: "The offer-first policy is where a server comes up to your table, takes your order for a drink and asks whether you'd like a straw with your drink. And just by doing that, a restaurant can actually cut down on their straw use by 50-80%. . . . I think often it's assumed that we don't need to use straws at all. Which may be true of some of us, but it's not true of all of us."[78] This shift from a *plastic ban* to an *offer-first policy* is appealing to some insofar as this statement critically interrupts taken for granted narratives and practices of throwaway culture, such as the habit of giving and accepting single-use plastic straws for those who do not need them.[79] Offer-first also appears more inclusive of disabled people, who should be part of any conceptualization of who "us" is. Further, it reduces single-use plastic waste without falling into the planetary fatalism trap that consumer choices don't matter.

Of course, some disabled people and accessibility advocates may never forgive the campaigns for not inviting them to the initial decision-making tables and/or for stigmatizing single-use plastics without including disabled representations or caveats. And some may conflate organizations that did listen and take in the critique with the individuals who shamed them. Healing takes time, and I do not mean to suggest that all is forgiven or that the critique is over. The experts on whether these statements of accountability have repaired and healed are those who were harmed the most, not I. Cultivating access as a value will require ongoing, challenging conversations between disabled and abled people.[80]

CONCLUSION

James Wakibia (see chapter 3) interviewed Mimi Legesse, cofounder of Teki Paper Bags, which is based in Addis Ababa, Ethiopia.[81] She founded the company, which makes partially recycled paper bags that can be used more than five times, to support women who are Deaf, like her, to be empowered to have careers and to reduce plastic pollution. With the motto, "Signs Speak Louder than Words" and using the hashtag #WeAreTeki (with the handle @TekiPaper-Bags on Twitter and TikTok), the company claims to have replaced 1,129,000 plastic bags, created twenty-five jobs, and employed eighteen Deaf individuals. It has begun to partner with larger companies and is lobbying the government to invest in creating careers for people with disabilities as part of the country's transition away from plastic bags. Teki's testimony illustrates how thinking through the lived experiences of disability and knowledge from disabled advocates can enable more inclusive, just, and ecologically sound futures: "The way we work gives deaf mothers enough time with their children and allows others to further their education. At Teki we cover transportation expenses fully to create a fair salary system for all. These values turn our family members into the best ambassadors in our fight against plastic bag."[82] Teki realizes a workplace that recognizes fuller lives, including meaningful labor, time with family, a space to practice Ethiopian sign language, and transportation as part of one's employment benefits. Legesse's mission illustrates the more inclusive and holistic thinking required to create a more sustainable future.

Plastics are complicated. Choosing plastic straws as a gateway plastic was a tactical choice for Lonely Whale, given backlash against plastic bag bans previously in Seattle. Yet in response to its having excluded disabled people in the decision-making or from visible representation in the campaigns, disabled hashtag activists claimed the message reinforced eco-ableism and eco-normativity in public spheres. "Voice," broadly defined, long has mattered to justice movements because exclusion often leads to marginalization and oppression of anyone not invited. Further, as noted, disabled expertise could nuance US culture's understandings of care, including interdependence.

Planetary fatalism perpetuates the straw man argument that nothing any one of us does matters. In this context, reflecting on individual plastic consumption is futile because environmental advocacy has yet to turn the tide on the plastics-industrial complex's hegemonic control. Yet plastics add up, and challenging the industry requires a cultural transformation of our relationship with plastics—which requires structural policies and a change of individual attitudes and behaviors. Even if we turn attention toward corporations, those decisions still will have ramifications for access; social structures and individuals are mutually constituted. It is not that neither matters but that both matter. Impure politics offers a way to reimagine collective action as imperfect but necessary.

While Lonely Whale, Starbucks, and other plastic ban advocates did not explicitly criticize people who needed straws for access, the exclusion of accessibility considerations still inspired shaming. Further, individual acts of shaming were felt affectively, through the collective history of being disabled in an ableist society. That felt experience generated backlash. From blog posts and charts to #SuckItAbleism and #StrawShaming, disability hashtag activists successfully networked to express care for each other and the broader environments of which humans are a part. In the future, to avoid alienating potential allies, it seems more plausible that disabled people might be hired behind and in front of the cameras of environmental advocacy.

Considering which plastics are necessary for some versus which plastics we can live without remains central to negotiations over plastics, from global policy controversies to everyday interactions. The shift in public opinion has created incentives in the market for straw alternatives; while some businesses are using offer-first policies, others have begun to provide a range of new straws, from better paper straws to straws made of biodegradable, plant-based materials, such as cornstarch and sugar. Social media about straws continues to elicit a spike in attention over disagreeing perspectives. In its best moments, the anti–plastic pollution movement's goal might best be stated as *every person is necessary, every plastic is not.*

As hopefully is apparent by now, the significance of single-use plastic straw bans depend on one's lived experiences in everyday life. As executive director of the Just Transition Alliance, José Toscano Bravo, shares: "I attended a meeting in Houston, Texas, about three years ago where they were talking about, you know, removing straws from Starbucks. Okay, I'm good with that, but there, in the neighborhood, the *barrio* where I come from, there are no Starbucks."[83] Disability studies and movements themselves have been critiqued for privileging Global North perspectives, particularly of the United States and Europe.[84] Conversely, Shaun Grech and Karen Soldatic argue that disability too often is invoked solely as a metaphor in postcolonial scholarship: "While postcolonial theory and associated fields (e.g. critical theory and cultural studies) have engaged with race, gender and ethnicity in the exploration of themes of identity, representation, space, historicity and the neocolonial, they have almost wholly bypassed disability— paradoxically limited to the historical subjectification of the able-bodied, or rather disembodying colonialism and the postcolonial terrain."[85] Although this chapter primarily focused on a US-based critique of eco-ableism, the intervention—or perhaps what Grech and Soldatic call "points of emancipative disruption"—has relevance globally for single-use plastics policy and politics, including whether disabled people living in the Global North should be considered part of the Global South.[86] Given the importance of voices of Southeast Asia to critiques of plastics and that the anti-plastic pollution movement is not solely focused on consumption, the next chapter turns attention toward production and Vietnam.

6

Creating #ToiChonCa (#IChooseFish)

Trauma, Affective Art, and Big Tech Dominance

In Vietnamese, đất nước, the term for country, designates "land" and "water,"
but just saying "nước" or "water" already refers to a country (for example
"nước ta" means both "our water" or "our country").

—TRINH T. MINH-HA, AUTHOR AND FILMMAKER

In what was called a "marine life disaster," hundreds of tons of fish died in the seas along more than 125 miles of four central Vietnamese provinces in 2016.[1] For a culture consciously created from water and land, as Trinh T. Minh-ha suggests (in the epigraph), the impacts were traumatic.[2] Thousands of fish died, floating on the water's surface, washing ashore, and being caught in fishing gear. Fisherfolk lost income and jobs. People went hungry, and dozens died. In response to the rhetoric of the perpetrator, the Taiwanese-based Hung Nghiep Formosa Ha Tinh Steel Co., Ltd., a subsidiary of Formosa Plastics Group (Formosa Plastics), a hashtag started trending: #ToiChonCa ("#IChooseFish").[3] The Vietnamese government downplayed and delayed its responses. The human rights of protestors online and offline were violated through beatings, censorship, detention, and arrest. Eventually Formosa Plastics admitted culpability and reached a state settlement. Still, today many residents have not received compensation and their losses have not been acknowledged, despite the wide recognition of this event as one of the worst environmental disasters in the nation's history.[4]

Doan Trang reports: "Worse, and most importantly, the marine life disaster involved a humane and political crisis when the Vietnamese government fails to ensure relevant compensation for the victims; they even go further by brutally suppressing voices of dissent. Widespread human rights violations have made 2016 and 2017 dark years for democracy and freedom in Vietnam, characterized by arbitrary detentions, police violence targeted at civilians, and increasing imprisonment of peaceful activists."[5] Given Vietnam's one-party authoritarian government, the repression of "voices of dissent" through surveillance and detention, as well as the lack of fair compensation or voice of those most impacted by the decision, these events created controversy in Vietnam. As the Vietnamese environmental NGO Green Trees argued, "This event became a true crisis. It reveals how the current Vietnam political system has come to a deadlock regarding the management and handling of environmental disasters, as well as the protection of the Vietnamese people's rights and interests."[6] This chapter therefore revisits how human rights and marine life endangerment are intertwined through plastics.

The #ToiChonCa/#IChooseFish trend also stretches a book on plastics-related hashtag activism in some compelling ways. There are deep connections among industries of extractivism, including the "holy trinity of metal, petro-chemicals, and plastics."[7] In this case, the name of the company and much of its profit come from plastics. Beyond my bias to consider a hashtag trend that tagged me, the movement to #StopFormosa is a coalition that bridges Global South and Global South of the North communities through networking solidarity against a transnational corporation. Following environmental justice activists' lead by including this story enables taking a step back from single-use bans to make broader connections across scales of harm and resistance, as well as land and water. This chapter also implicates Big Tech by reflecting on the intertwined development of American imperialism and communication technologies.[8] Although used initially as platforms to network, publicize, and campaign, Big Tech also may become a barrier to publicity and human rights. In this case, Facebook (Instagram/Meta) and YouTube (Google/Alphabet) appeared to work with the government of Vietnam to quell dissent.

To explore these dynamics, I provide the context of Formosa Plastics, focusing on how its origin is upheld as a tale of rugged masculine rags to riches, as well as global frictions of transnational resistance. Then I engage the #ToiChonCa/#IChoose-Fish trend. Despite traumatic barriers, residents creatively built networks of care to publicize offline protests and online dissent through affective art. In closing, I turn attention toward Big Tech and its complicity with the state.

In this chapter I aspire to trouble transnational power inequities rather than reify them by following Kuan-Hsing Chen's "call for critical intellectuals in countries that were or are imperialist to undertake a deimperialization movement by reexamining their own imperialist histories and the harmful impacts those histories have had on the world."[9] While the focus remains on amplifying voices of people living in Vietnam and from the Vietnamese diaspora, I also ask what harm US imperialism and a lack of transnational corporate accountability have caused in relation to the 2016 marine life disaster and subsequent advocacy for environmental justice. How can a *deimperialist approach* enable the Global North to rethink *detoxification* as part of the story of *regeneration* that should be included in broader accounts of anti–plastic pollution advocacy?

FORMOSA'S FRICTIONS

Globalization is neither an even nor a universal flow moving throughout the world. Anna Tsing "reminds us that heterogeneous and unequal encounters can lead to new arrangements of culture and power" and suggests thinking of these relations as friction.[10] "Friction," she cautions," is not a synonym for resistance. Hegemony is made as well as unmade with friction."[11] While I analyzed posts in English and Vietnamese to attend to counterhegemonic frictions, I drew on translations from secondary sources for the latter.

Even a brief history of Formosa Plastics reveals once again how colonialism and US imperialism provide the conditions of possibility for the plastics-industrial complex to thrive. In the 1500s, Portuguese mariners named the place now known as Taiwan "Formosa," meaning beautiful island. Nine Indigenous cultures lived there.[12] Subsequently, the island was colonized by the Dutch (1624–1661), China's Qing dynasty (1683–1895), and Japan (1895–1951). According to Taiwan's government, over a million Chinese moved to the island during the 1940s and 1950s to establish Taiwan as part of the Republic of China.[13] This period of time involved many militarized struggles between Taiwan and/as China, which garnered US military attention and continues to matter for global geopolitics more broadly.[14]

In 1954 Wang Yung-ching (also known as "YC Wang") negotiated a US$680,000 loan from the US government as part of an economic aid mission to support efforts against communism. He combined that loan with a half million dollars from his timber business and a Japanese plastics technology license for polyvinyl chloride (PVC) to found Formosa Plastics.[15] Perhaps Wang was chosen in part because in 1944 a US bombing raid had leveled his family's rice mill, and therefore this loan could become part of a story of reconciliation.[16]

In an often-cited profile of Wang published in *Forbes*, Wang is portrayed as a Great Man: "Wang was born in . . . 1917—the Year of the Dragon, an auspicious zodiac sign for males that produces energetic, stubborn and capable people." Despite the unmentioned US subsidy, the profile portrays a self-made, rags-to-riches tale, from his meager childhood because his father was "a poor tea merchant," to how he worked long hours with only cold showers, to his now owning an "empire."[17] An anecdote of his rugged (working-class) masculinity and gendered normativity is recounted as a quintessential part of his ethos: "Wang sometimes shows his peasant origins with what Americans would regard as vulgarity. . . . At one of Wang's recent penthouse dinner parties the host eyed the American's corpulent wife and loudly asked the banker a personal question about the couple's sex life. But mainly his energy and charm captivate rather than repel." In between accolades over Wang's work philosophy, the journalist also quotes "one American petrochemicals man in Taiwan" criticizing Wang's lack of care about environmental regulations: "He doesn't play by the rules. If he gets caught polluting or evading taxes, he bargains with the [Taiwanese] government. He can get away with it here, but not in the U.S." Overall, however, the profile positions Wang's tale as one US businesses should heed because they "have grown complacent" with "moral decadence."[18] Eventually YC Wang became one of the richest people in Taiwan and one of the two hundred richest people in the world, until his death at ninety-one years old in 2020, when the *New York Times* claimed he was "known widely as the God of Management."[19]

Today Formosa Plastics is a typical transnational industry actor, insofar as corporate scandals have made regular headlines, and it is, in economic speak, "well-diversified" in its holdings, including over one hundred subsidiaries, from petrochemical plants to educational and medical organizations.[20] Its primary business is "the manufacture and sale of plastic raw materials, chemical fibers, and petrochemical products."[21] Formosa Plastics can be found in "packaging materials, bottles, articles of clothing, toy dolls, medical materials, sports equipment, sporting goods, auto parts, generation timber, motor parts, wires and cables, ships aerospace, paint coating, chemical building materials, waterproof insulation, furniture furnishings, tool parts, 3C parts coated, 3C shell timber, masterbatch pigments, additives, clean bleach, bottle box barrels, and dry water retention."[22] Formosa Plastics reminds us that the plastics-industrial complex far exceeds single-use plastics.

In 2021 the Center for Biological Diversity and Earthworks published a report based on six decades of Formosa's practices, embracing a judge's ruling that the company was a "serial offender" of "human health, local ecosystems, and the global climate." The report notes that the US\$569 million the company has been fined over the years is little in comparison to its annual profits of more than US\$103 billion.[23] And yet *Fortune* magazine ranked Formosa Plastics only as the 492th company in the Global 500.[24] Formosa Plastics, then, is both remarkable as a corporate polluter and unremarkable insofar as it is not one of the largest petrochemical companies.

Resistance against Formosa Plastics has been growing. Consider the previously noted virtual toxic tour of the #StopFormosa and #StopFormosaPlastics global network, bridging the Global South and the Gulf South (see figure 2 in the introduction). Each community on the Gulf Coast now has a victory to share. In 2019 in Point Comfort, Texas, a coalition led by Diane Wilson, a retired shrimper and environmentalist, won the largest citizen environmental suit settlement in history to date, US\$50 million, for violations of federal clean water and air laws based on plastic pellets and other pollutants from the nearby Formosa Plastics plant.[25] In 2021 Formosa Plastics also was fined \$2.8 million by the US EPA for air pollution.[26] That year, in the area of southern Louisiana known as "Cancer Alley," a coalition led by environmental justice leader Sharon Lavigne of RISE St. James successfully campaigned to have the US Army Corps of Engineers provide an environmental impact statement for a new Formosa Plastics plant, placing the US$9.4 billion dollar facility on hold.[27] The next year a judge ruled against the building of the new Formosa Plastics plant.[28]

Lesser known in the United States is that since 2016 advocates from Vietnam have networked in global solidarity with these Gulf Coast fenceline communities that have been most directly impacted by Formosa Plastics's pattern of ecological degradation, including disregard for human life.[29] The global #StopFormosa movement continues organizing tours and filing court cases, deepening its networks and expanding its reach.[30] This chapter turns attention toward Vietnam's unfolding story to demand accountability from Formosa Plastics, as well as the role of Big Tech.

ON BIG TECH'S DOMINANCE AND EMERGING HASHTAG ACTIVISM IN VIETNAM

When it first emerged, Big Tech was popularly imagined as media platforms for the people: anyone could go online, use a hashtag to try to start a trend that would mobilize more people, and critically interrupt hegemonic notions of the public sphere. Yet Big Tech increasingly has become criticized as a hegemonic force of its own. Today, three Taiwanese companies top Formosa Plastics in market capital, all related to information and communication technologies: TSMC (which makes semiconductors), Foxconn (an electronics contract manufacturer), and MediaTek (a digital media technology manufacturer).[31] Globally, "Big Tech" refers to the five largest media technology companies in the world, all started in the United States by men: Amazon (founder: Jeff Bezos), Apple (founders: Steve Jobs, Steve Wozniak, and Ronald Wayne), Google (now Alphabet, which includes YouTube and Android; founded by Larry Page and Sergey Brin), Facebook (which also owns Instagram, WhatsApp, and Messenger; founded by Mark Zuckerberg and four male classmates at Harvard), and Microsoft (founders: Bill Gates and Paul Allen). Bezos, Page, Brin, and Zuckerberg currently rank among the ten wealthiest people in the world. According to Oxfam International's executive director, Gabriela Bucher, "If these ten men were to lose 99.999 percent of their wealth tomorrow, they would still be richer than 99 percent of all the people on this planet."[32] Given that these economic elites of the Global North are the Big Tech leaders, it perhaps is unsurprising that the industry remains dominated by hegemonic gendered patterns.

Through carbon-heavy masculinity, Big Tech creates global frictions for the climate emergency. Despite promises to improve, Big Tech has had a large climate footprint as an industry that initially just bought carbon credits to allegedly offset its massive energy demands. Further, as InfluenceMap reports, "while Big Tech holds positive positions on climate policy, its support is not backed up by strategic advocacy. In addition, the companies' direct influence is overshadowed by the highly strategic and anti-climate advocacy of their industry associations in the US and abroad."[33]

Vietnam is a profitable market for Big Tech. Wayne Ma reports that even though neither Google nor Facebook has full-time employees in the country, "Vietnam is the[ir] most important market in Southeast Asia" as "the biggest country by revenue." Experts estimate that in 2018 Facebook profited $1 billion in Vietnam and Google $475 million, primarily from YouTube advertising.[34] Further, as is well established now, these platforms are "free" to users because the companies economically exploit what is posted as data to economic and political elites of the Global North that can afford to buy it.[35]

A damning 2019 Amnesty International Report, *Surveillance Giants: How the Business Model of Google and Facebook Threatens Human Rights*, concludes:

> But despite the real value of the services they provide, Google and Facebook's platforms come at a systemic cost. The companies' surveillance-based business model forces people to make a Faustian bargain, whereby they are only able to enjoy their human rights online by submitting to a system predicated on human rights abuse. Firstly, an assault on the right to privacy on an unprecedented scale, and then a series of knock-on effects that pose a serious risk to a range of other rights, from freedom of expression and opinion, to freedom of thought and the right to non-discrimination.[36]

This complicated Big Tech trade off remains unresolved regionally and globally.

Despite growing global awareness of drawbacks, the increase of social media in Vietnam has dramatically risen since 2000, among the highest in southeast Asia. One report summarizes: "In about 15 years from 2000, the number of Internet users skyrocketed from 200,000 to 40 million, representing more than 40 percent of its 93 million people. According to Sweden-based Civil Rights Defenders, the rise of the Internet and rapid uptake of smartphones—there are more than 130 million active mobile subscriptions in Vietnam—"dramatically" lowered the barriers for human rights advocates to disseminate information, which hitherto was both dangerous and costly."[37] The popularity of social media in Vietnam is often linked to a lack of voice or representation in mainstream media (newspaper, radio, and television), which the authoritarian government controls.[38] For example, Tuan Nguyen, a Vietnamese human rights advocate claims: "Social media now creates the only free public sphere for people to not only discuss all contemporary social and political issues, but also to encourage civil participation into politics."[39] Likewise, Ashley Lee argues: "Dense, overlapping layers of personal and organizational networks may be well suited to amplifying voice and growing movements in a repressive regime."[40] Unfortunately, not only has social media been a response to repression, but it also has become a contested platform for repression. Some argue, for example, that the 2007 bloggers' protest on the Yahoo! 360° social network against the governments of China and Vietnam was when "a confrontation between the authorities and pro-democracy bloggers started."[41]

Focusing on environmental hashtag activism in Vietnam, Quang Dung Nguyen emphasizes the significance of how human rights and ecological degradation are entangled: "Environmental activism for environmental commons in Vietnam has been extended beyond the scope of environmental justice for humans to include non-human environments, where ecological and social spheres are intertwined. In other words, such activism corresponds to the rights of humans and nature, in which ecological degradation goes hand-in-hand with social destruction."[42] Nguyen claims the rise of social media activism correlates with a growing environmental consciousness and movements in Vietnam. He provides three exemplars. First, he analyzes the 2009 anti–bauxite mining movement, which bridged offline activities (such as workshops and seminars often based at universities) with online resources (articles, petitions, and reports), as well as intellectuals ("economists, environmentalists, experts on mining technology, and scholars of cultural and social studies"), bloggers, reporters, politicians, and government scientists. "For the first time in Vietnam," Nguyen argued, "digital networks could mobilize collective sentiments and actions for environmental and ecological justice."[43] Second, using #SaveSonDoong, people who created a Facebook page (and eventually other social media, including with celebrities) catalyzed a successful environmental campaign in 2014 to protect the world's largest cave from cable car traffic. The third campaign Nguyen highlights is #ToiChonca/#IChooseFish.

The following account begins with social media using hashtags in both Vietnamese and English related to the marine life kill. On Instagram, I analyzed over 400 posts.[44] On Twitter, I analyzed over 160 posts and articles using the two hashtags (though some repeated, using both tags). I also read hashtagged posts on Facebook, which I note later. Posts were created by individuals, well-known bloggers or social media influencers, and organizations. I focus on 2016, but the story is ongoing, and the timeline exceeds that year of acute trauma. The archive I examined was image rich and generally contained fewer words in Vietnamese, likely to navigate censorship and/or to mobilize transnational solidarity.[45]

Reading an archive that has been censored has moved me to avoid revealing handles, unless they already have been recognized in mainstream venues. My archive is less accessible on Facebook today, since some of what I reference has been censored or removed and does not exist online as this book goes to press. I have kept records of the media cited but share items anonymously, as the silencing of these voices appears to be without their consent and I do not wish to contribute to increased surveillance or policing. To verify my archive, I read secondary research on media, NGO reports, and scholarly studies in English. For the *Communicating Care* podcast, I also interviewed Nancy Bui, vice president of the

NGO Justice for Formosa Victims (JFFV), to discuss events, activist tactics, and reception.[46]

Mei-Fang Fan, Chih-Ming Chiu, and Leslie Mabon argue that the marine life kill is helpfully analyzed through an environmental justice framework because it provides a compelling "example of claims to injustice between a higher-income east Asian and mid-income southeast Asian nation" and "a valuable case to evaluate the role of transnational alliances in supporting claims-making around environmental injustice for victims living in an authoritarian context."[47] The following pages recount the controversy as a national and transnational story of environmental justice chronologically, revealing the back and forth of the contested struggle between those seeking redress and the institutions trying to maintain social control.

CORPORATE COMMUNICATION AND DISSENT

Formosa Steel, a subsidiary of Formosa Plastics, was built over many years. In 2004, it was proposed to be built in Taiwan; however, the environmental impact assessment was negative, and the company worried about local opposition.[48] Within a decade the Vietnamese government turned over 2,000 hectares of land and over 1,200 hectares of water surface to the company, displacing local residents and businesses, which impacted "3,000 families . . . 15,000 graves and 58 churches."[49] Despite protests, the government also provided "the highest incentives on import taxes on machines, equipment and materials as well as on taxation and land."[50] Perhaps this warm welcome was due to a broader pattern that Tony Tran addresses, in which "the rules of neoliberal capitalism quickly superseded—but did not eradicate—most of the political ideologies that the Vietnamese state held regarding its former enemies."[51]

A groundbreaking ceremony occurred in December 2012 with the Vietnamese prime minister present.[52] In 2014 a newspaper headlined a story claiming that more than three thousand Chinese workers were laboring at the plant without permits.[53] In 2015 over a dozen people were killed and over another dozen injured at one of Formosa's construction sites.[54]

Some expressed concerns with Formosa Steel's pipeline off the coast of Vietnam before others. By April 2013 it became widely evident that something was wrong in the water. A repertoire of reactions unfolded. Posts began appearing on Facebook on April 2 about the appearance of dead fish across many miles, and some fisherfolk began sharing the ecological trauma they witnessed from a yellow discharge from the Formosa pipe.[55] A diver reported seeing a discharge pipe with yellow liquid on April 3. Thousands of dead marine animals began washing up on the shores, showing up dead in nets or cages and floating on top of the water: red snapper, sea bass, cobia, and grouper. Divers and local fishers began to feel sick.

A research group confirmed "toxic agents in the water" by April 11. More dead marine life kept appearing, and more people became sick or died. People continued to become ill, reporting stomachaches, nausea, and diarrhea. The Vietnamese government promised to study the waters, but the official position remained that farmed marine life "are growing normally" and "people can swim in these waters without feeling worried."[56]

On April 25 Chu Xuan Pham, the director of Formosa Steel's local external relations department, made a statement to a local youth newspaper:

> It is impossible to build a steel plant here without leaving any impact on fish and shrimp. Of course, we try to build a plant that meets the State's requirements. Yet it is normal to lose some things as you gain some things. . . . *Between these two things, we must choose one, whether I want to catch fish and shrimp or I want to build a modern steel industry?* When this area was cleared the local authorities already made a plan to support fishermen to switch jobs, why do they need to keep fishing in this area? *Do you want to keep fishing or do you want to keep the plant? Go ahead and make your decision.* If you want both, even the Prime Minister can't satisfy you.[57]

Still not accepting responsibility, or expressing regret, the statement suggested that the marine life kill was a reasonable business choice about what Vietnam should prioritize: steel, not fish.

Pham misjudged the structure of feeling in Vietnam. Nancy Bui shared that in the perception of many people in Vietnam: "The statement from the Formosa external relations manager was cruel and ignorant."[58] A communal affinity with marine life is not a surprise to most familiar with the social formations of Vietnamese culture, including its pervasive and iconic *nuoc mam*, fermented fish sauce. Millions in Vietnam depend on fishing for employment and food security. In 2016 the nation was "the fourth major producer of fishery and aquaculture in the world with a total production of 6.4 million tonnes."[59]

Perhaps predictably, Pham's statement immediately sparked an online trend with #ToiChonca. That day, a graphic circulated with a healthy fish over a wave in front of the Vietnamese flag with the words: "Tôi là người Việt Nam TÔI CHọN CÁ" ("I am Vietnam I CHOOSE FISH") (see figure 20). This post clearly articulates, or links, Vietnamese cultural pride (implicitly contrasting with the Taiwanese company), being human (which perhaps suggests vulnerability), and choosing fish (challenging Pham's foregone assumption).

The linking of "I am human" and "I choose fish" suggests an understanding of the mutual survival of each, especially within an authoritarian Eastern culture in which communication might be less direct. That is, if people are sick and dying, like the fish, then this hashtag can be interpreted as choosing fish over steel, people over profits, human rights over corporate rights, and life over death. This articulation is not a pure binary. Most of the people posting in Vietnam and all

FIGURE 20. Graphic circulated on social media in response to the presump-
tion that steel was more important than fish. Initially, the message was in
Vietnamese: Tôi là người (I am human) Việt Nam (Vietnam) TÔI CHỌN CÁ
(I CHOOSE FISH).

the concerns expressed over the fishing industry do involve killing fish to eat them
and making a livelihood from that set of relations. Again, the impure politics of
living, eating, and laboring become intertwined in response to being treated as
expendable.

Within twenty-four hours, using Facebook and Instagram as the predominant platforms, people began posting with #ToiChonCa (#IChooseFish). Documentation of trauma circulated, including those initial videos of pollution and
dying marine life. Affirmation of fish as part of everyday lived experiences also
was portrayed, even in these initial posts. Some visual images on social media,
for example, included typical foodie shots, like a person holding *taiyaki* (a fried
filled pastry in the shape of a fish) or cheese and meats cut out in fish shapes
with hashtags for foodies. One posted fish-shaped flip-flops. #ToiChonCa became

articulated to a range of vernacular expressions of how fish were/are a part of a whole way of life.

Hashtag activism had immediate impacts. Headline news picked up on the social media critique. Within twenty-four hours Chu was given notice that he was fired, and the company issued a statement of apology. Formosa Steel claimed Chu's comments did not represent the company, its relationship with the government, or its commitment to the environment. Chu himself stated: "I have frustrated the public and my actions have affected the corporation."[60] However, on April 26, "the Ha Tinh newspaper [also] published an article titled 'The sea is now clearer, the environment is no longer polluted' . . . [which] was criticized by even mainstream press."[61]

The hashtag activism continued. In English, #IChooseFish emerged as public dissent arguably aimed at increasing publicity and reaching a transnational audience.[62] On April 26, for example, an anonymous person named T. N. claiming to be from Ha Tinh created a petition on the Obama administration's "We the People" website, gaining more than one hundred thousand signatures.[63] Many people in Vietnam sought international solidarity almost immediately (see figure 21).

While some hashtags were used once or only a few times, such as #IChooseFishing, #IChooseFishTaco, and #IChooseFishes, #ToiChonCa/#IChooseFish remained the predominant trend. Linking the hashtag(s) with already trending memes expanded the publicity of dissent, as well as using an affirmative frame about fish (rather than a direct critique of human rights), arguably to avoid legal detention by the state.[64] Visual choices linked to the hashtags continued to range from legal and scientific documentation, including photographs, to creative artistic sketches and popular social media memes. Some continued to share images of fish as food, including sushi. Many began to feature fashion, including nail art of fish, the torso of a person in a polo shirt with a fish pattern, jewelry featuring fish (necklace, bracelets), and a celebrity wearing a dress with fish on the red carpet. Other hashtagged posts shared art, such as sketches of fish (abstract, surreal, and/or realistic), digital Google 3D fish in public spaces, murals, and sketches mourning the disaster. Some appeared to be commenting on censorship, wearing face masks with drawings of fish on them. Also circulating were images of offline protests, which began in May, including police beating intergenerational street protestors with signs, artists using body paint to publicize the disaster in public spaces, and kids smiling while holding posters of fish and boats.[65] Some visuals featured the ecological impacts on marine life: dead fish on the shore, floating by boats, and under the sea. As Rachel Kuo argues, hashtags can "be interpreted as signifiers of solidarity, as their collective digital presence becomes a material record of reciprocal commitments."[66] Collectively, the hashtag activism response to the marine life disaster provided a cultural collage of solidarity and publicized a communal commitment to care.

FIGURE 21. Drawing of a cute/kawaii cat mourning fish, April 2016. The marine life kill and responses to it inspired artists to express the affect of the moment. This drawing was reshared more than any other image I found on social media using #ToiChonCa.

On April 27, Deputy Minister of National Resources and Environment Vo Tuan Nhan stated: "As of this time we don't have any evidence to prove a link between Formosa and other plants in the Vung Ang Economic Zone to have been responsible for the mass fish deaths. The investigations by scientists and environmental experts on water samples from the area found no substances that exceeded legal levels."[67] Claiming a "lack of evidence" and calling for more studies is a common delaying tactic when governments or industries want to deflect a public controversy about toxics. Some Vietnamese people were outraged their government wasn't standing up to this foreign-based company, especially when the evidence of trauma was so clearly visible. Yet the state continued to try to minimize the harms. For example, when a reporter asked about the impact of heavy metals on the upcoming tourist season, Deputy Minister Vo Tuan Nhan

FIGURE 22. Art about the marine life kill often was articulated with a range of hashtags on social media, suggesting multiple crises and audiences. Image posted with #vietnam #fish #formosa #pollution #WaterPollution #danger #WeChooseFish #IChooseFish #SaveFish #SaveEnvironment #SaveNatural #SaveOurLife, n.d. *Source:* Dương Thúy Vân (@i.am.yiyi), Instagram. Reprinted with permission of the artist.

interrupted to admonish her: "Don't ask that question. That question damages our country."[68]

As time moved on from the initial surfacing of marine life kills and acute public health impacts, the harms remained. Over 40,000 fisherfolk and 175,000 people involved in the fishing industry were financially impacted.[69] As emphasized on social media and by protestors in the streets, the sea and fishing are vital to this coastal Vietnamese culture as a whole way of life, including recreation, labor, diet, and more.[70] Counterhegemonic friction in response to Formosa Steel affirmed a different way of living, perhaps not new, but one to which many people in Vietnam nonetheless felt attached. As a Taiwanese lawyer shared: "Many victims, upon meeting us, wept profusely and hoped to receive any form of help. As a Taiwanese person, I felt a great sense of shame."[71] The public relations sugarcoating of the disaster did not go unnoticed (see figure 22).

CRIMINALIZATION AND CENSORSHIP

Toward the end of April, human rights activist Truong Minh Tam was arrested for recording video and taking photographs of the ecological harms, and

Facebooker Chu Manh Son was arrested for filming a demonstration.[72] On May 1, 2016, the first large public protest of thousands of people mobilized. As one fisherman observed: "The government tried to hide information and protect Formosa. If Formosa remains in Vietnam and doesn't give us our clean environment back, we will continue to protest."[73] Protests in the streets continued, in addition to online activism, though some participants began to cover their faces with paint or masks or signs, and some remained afraid to voice dissent. Public protest had been limited in Vietnam, which "has very little freedom of speech, and no protected right-to-assemble. A public protest, especially one that goes against the interests of the government, is essentially illegal. For this reason, protests are rare in Vietnam—very rare."[74] However, starting in May 2016 and for months to come, people did protest offline and online, calling for a right to know (or transparency), independent studies of the chemicals, legal accountability, and relief (for lost food and funds) to hold accountable both the broader company of Formosa Plastics and the national government.[75]

One group leading protests was the Roman Catholic Church. Emphasizing government transparency and corporate accountability, Catholic priest and protest leader Anthony Nam said: "We will protest until the government says what caused the spill."[76] Fan and colleagues argue that the Catholic Church in Vietnam functions akin to the United Church of Christ in the United States, that is, as a key voice in the movement for environmental justice. A Vietnamese faith leader explained: "The Earth is created by God and for us all, but people use it selfishly for individual or corporate . . . interests. Anyone who exploits the Earth is committing a crime. FHS [Formosa Ha Tin Steel] does not care about the environment and people's lives. I want to do things to help those victims who lose their jobs and have no choice but to seek jobs in Taiwan; they have been exploited. I want to protect the Earth and seek equity and justice."[77] This articulation of care links the environment with people and people with each other. The moralization of exploitation of Earth and people as a selfish crime articulates a moving intervention. Further, as Fan and colleagues argue, environmental justice reflects broader concerns beyond the church, including (a) the state's inability to protect everyday people from environmental harm; (b) the lack of procedural opportunities for transparency and public participation in the decision-making process; and (c) "challenges of articulating claims to injustice across scales, cultures, and political systems given the transnational nature of the project."[78] These counterpublic critiques included the state's failure to protect its citizens from harm, to include their voices, and to address multiple scales of injustice.

Dissent continued to be policed. In mid-May a newspaper was fined for two articles titled, respectively, "The People Are Always Those Left Behind" and "The Lament of Fish."[79] On a May 29 talk show, *60 Open Minutes*, the theme was "Sharing on social media, what is it for?" On the show, the hosts questioned another reporter, Phan Anh, about his motive in sharing a clip of a dead fish experiment. Phan Anh's call—"Don't be silent"—gained attention and publicity on Facebook.[80]

Central to the ethics of care articulated was marine life. By the end of May, on YouTube videos of whales were uploaded—first, one of authorities trying to save an ill seventeen-ton whale and, two days later, another of a ten-ton whale that was "dead and decomposing."[81] Beyond bearing witness to the loss of life, residents wanted to study the cause of death. Unfortunately the government did not support them, creating more mistrust. As *Loa*, a Vietnamese podcast, reported: "Vietnamese fishermen have long considered whales to be holy creatures. While Hanoi has refused to comment on the mysterious deaths, locals say the discovery of new carcasses casts another spotlight on the authorities' negligence."[82] Again, the impact of toxic pollution was not lost as a sign of the violation of the whale or as an indicator for human well-being.[83] Losing a sacred whale was traumatic, culturally. The juxtaposition between that collective sense of loss and the lack of adequate care from the government mobilized even more acts of counterhegemonic friction.

But then Big Tech began to censor the content of everyday people and activists related to the marine kill. Since the Vietnamese government controls the nation's mainstream news media, dissent often arises online and using indirect critiques; however, in this instance, the government convinced Big Tech to censor and become part of its surveillance apparatus. As Dan Vineburg, who reported from the streets of Vietnam and whose own video of the May 1 protest went viral, observed: "News of the first protest spread over Facebook and Twitter, as it did in the Arab Spring. . . . But these messages could only spread so far. Most of the facts were in Vietnamese, making it difficult for a foreign audience to decipher. The Vietnamese government also has a habit of blocking access to social media. Before the protest on May 15th, Facebook, Instagram and Twitter were blocked nationwide."[84] While social media initially allowed for online organizing of protests, messages began to fail to send through Facebook, and police began to show up at the doors of people who had posted online. Some reported that particular words "such as 'ca chet' [dead fish], 'Formosa', and 'bieu tinh' [protest] were being blocked in their messages by mobile phone service providers."[85] Growing numbers of people also reported being blocked from logging on to Facebook, Instagram, and Twitter, with a rise in the use of VPNs (virtual private networks), presumably to bypass censorship.[86] Access to Facebook also would become intermittent during protests in real time, as protesters tried to share photographs and videos of being beaten and arrested.[87]

Censorship may have been on the rise because then US President Barack Obama visited at the end of May 2016. Often governments try to make a good impression when international press arrives, like many of us do when we host guests. People protesting Formosa Steel's pollution and the government's delay hoped he would help advocate on their behalf.

Amid all of this, Obama was visiting to announce the lifting of a US embargo on selling lethal arms to Vietnam.[88] Gaining a great deal of social media attention—and more in the United States than his speech—was Obama's appearance upon his arrival in Vietnam on celebrity chef Anthony Bourdain's show *Parts Unknown*. With almost twenty-five thousand likes and over twelve thousand retweets, Bourdain's tweet on May 23 made international headlines, including a photograph of the two men smiling over bowls of Bun Cha (pork lemongrass noodles), seafood spring rolls, and beer, with the message: "Low plastic stool, cheap but delicious noodles, cold Hanoi beer." The picture was celebrated for its relatability, with two tall US men sitting on relatively low plastic chairs as a marker of their humble capacity to enjoy street food in an everyday setting for working-class people in Vietnam—cultivating what might appear to be an anti-imperialist ethos, or at least a warm affinity between nations.[89]

The next day, May 24, Obama made a statement affirming improved relations between the two nations and, although he did not use the words *fish* or *Formosa*, his speech recognized that while he had met with some who were trying to expand rights in Vietnam, he was not given access to others. Obama emphasized: "What I've heard consistently from all of them is a recognition that Vietnam has made remarkable strides in many ways—the economy is growing quickly, the Internet is booming, and there's a growing confidence here—but that, as I indicated yesterday, there are still areas of significant concern in terms of freedom of speech, freedom of assembly, accountability with respect to government."[90] Here, Obama also linked principles of public life—speech, assembly, and accountability—with the growing use of the Internet. He also closed by emphasizing the importance of dissent: "By the way, in the United States, where there are all sorts of activists and people who are mobilizing, oftentimes are very critical of me, and don't always make my life comfortable but, ultimately, I think it's a better country and I do a better job as President because I'm subject to that accountability."[91] Rather than pointing fingers or claiming US superiority, Obama emphasized how many in his own country were critical of him, but that he thought being open to those opinions in governing as a national leader was more productive rather than trying to silence them. In contrast to carbon-heavy masculinity, this performance of a global leader who can accept criticism offered a diplomatic way to reimagine dissent as what I'd call "a fellow traveler," someone who is not a member of a community but who empathizes with and tries to be helpful to the community. Despite his not explicitly mentioning the marine kill, a cartoon circulated on Facebook of Obama having dinner with Vietnamese officials saying the words many hoped he might have communicated between the lines or behind the cameras: "I choose fish."[92]

Protests in Vietnam continued. Blogger Nguyen Chi Tuyen (also Anh Chi) began a campaign with the slogan "To knock pans for transparency," which encouraged people to record videos of themselves hitting pans together in their kitchens and then post them online, to call for the public's right to know.[93] A Green Trees protest on World Environment Day centered the message: "Our future in our hands."[94] These ongoing actions showed public pressure was not waning.

On June 30, 2016, the Vietnamese government held a press conference at which Formosa Ha Tinh Steel Corporation and the Formosa Plastics Group admitted responsibility for the toxic chemical leak from its steel plant that had led to the marine life kill and pledged US$500 million (VND 11,500 billion) for cleanup and compensation, an arrangement the government had negotiated in private.[95] The funds would be paid to the government to distribute, which did not invite public participation on how to do so. Tran Hong Ha, the Vietnamese minister of natural resources and environment, recognized fifty-three violations and the finding of phenol and cyanide as the main toxic pollutant. One government official claimed it was "the most serious environmental disaster Vietnam has ever faced." Another government official declared the situation resolved and portrayed the company as the victim in need of sympathy, stating: "One should not hit a man when he is down." This resolution, however, did not satisfy many, which they expressed on social media immediately, declaring the need to have more transparency, criminal proceedings against Formosa Plastics, and the shutting down of the steel plant.[96]

Despite ongoing protests, transnational attention, and the recognition that there was indeed an environmental disaster, the Vietnamese government continued arresting people for dissent. Hundreds of peaceful protesters were beaten or arrested in the streets. The government's extreme response to online and offline advocacy received increased international attention. According to the NGO Human Rights Watch, over one hundred bloggers and social media activists were "serving prison sentences simply for exercising their rights to basic freedoms such as freedom of expression, assembly, association, and religion."[97] NGO Amnesty International declared that this time period "represent[s] an upturn in the use of the criminal justice system in a crackdown against human rights defenders and activists engaged in advocacy relating to the disaster which has included intimidation and harassment, and wide scale surveillance of activists."[98]

The police also targeted online activism, largely mobilized on Facebook, YouTube, and WeChat. In October 2016 the Vietnamese government declared the California-based prodemocracy group Viet Tan a "terrorist organization."[99] And there were many more bloggers or "netizens" (internet activists) blocked, intimidated, and arrested, including Brotherhood for Democracy founder Nguyen Van Dai. In 2017 Catholic priest and protest leader Nguyen Dinh Thuc said he was beaten by police, witnessed arrests, and had his vehicle towed, as well as seeing others' vehicles towed.[100] In 2018 three online bloggers and activists were sentenced to prison, including Le Dinh Luong (twenty years), Nguyen Hoang Bing (fourteen years), and Tran Thi Xuan.[101]

Then, prominent blogger Nguyen Ngoc Nhu Quynh, with the pen name Mẹ Nấm (Mother Mushroom), was arrested and sentenced to ten years in prison. Surprisingly, she was released one year later, after which she promptly flew to the United States with her mother and two children, continuing to write about the "inhuman" prison conditions and injustice in Vietnam.[102] She had been an online environmental activist since 2016, gaining attention for her anti–bauxite mining posts and for cofounding an independent writers' association. She shared news and visuals of cartoon mushrooms (attributed to Vietnamese resilience and a nickname for one of her daughters) or, as Bui recalls, a drawing of her head covered with hundreds of fish. Like most bloggers and social media activists, she was arrested under Article 88 of Vietnam's Penal Code, which prohibits anything considered to be "propaganda against the state." A Facebook post, now removed but published by the BBC upon her arrest, shows her displaying a sign that says (transl. from Vietnamese): "Fish need clean water. Citizens need transparency."[103]

While social media initially was used to bear witness, show solidarity, and publicize public concerns, it also was being used to thwart the right to know and to repress activists. According to John Sifton, Asia advocacy director of Human Rights Watch: "Facebook's users in Vietnam are being jailed simply for using the platform as it was intended: to communicate information and opinions to other users."[104] Big Tech approved many of these actions: "Between January 2016 and June 2019, the Vietnamese government made nearly 250 requests for Google to remove or restrict more than 9,000 items, up from a handful of requests in prior years. Google complied with 77% of these requests . . . [most of which] involved criticism of the government. . . . Facebook restricted 1,600 items in the second half of 2018 [alone], up from 22 items in the same period in 2017."[105] Complying with most requests, Big Tech's dominance exacerbated the already volatile political atmosphere.

Increasingly in Vietnam, Big Tech has become part of the story about the undemocratic regimes and movements globally. A case highlighted by Amnesty International was that of Nguyen Nang Tinh, a music teacher who was arrested in 2019 and given a sentence of eleven years. During the trial his Facebook posts were used as evidence. "They described his use of Facebook as a 'threat to national security' and accused him of using Facebook to 'spread propaganda against the government.'"[106] Amnesty International reports evidence that "prisoners being tortured and otherwise ill-treated, routinely held incommunicado and in solitary confinement, kept in squalid conditions, and denied medical care, clean water, and fresh air."[107] Pellow has argued that "political prisoners" facing "environmental injustices is particularly ironic because they come from social movements that were fighting for environmental justice in the first place."[108]

CONCLUSION

In March 2017 the director of Ly Son Marine Protected Area, Mr. Phung Dinh Toan, and the environmental NGO International Union for Conservation of Nature (IUCN) Viet Nam initiated a project to raise awareness that sea turtles are endangered. They chose An Binh, a small fishing community, to launch their public awareness campaign, asking for volunteers from across the country to paint murals. According to IUCN: "In the beginning, the project team considered using the standard communications tools, such as brochures. However, the traditional way of simply presenting facts seemed to be missing the mark. The team then decided to devise a new strategy to touch local communities on an emotional level. They need a much more effective way to change behaviour, and so came up with a bold solution: why not create huge paintings of sea turtles on the houses of a fishing village? Shortly after, the 'I love the ocean/Born to be wild' project was born."[109] The murals varied in colors, elements, and messages. Considered a success by residents and organizers, the warm reaction to the mural project appeared to become a celebration of community affinities with endangered turtles (see figure 23).

Likewise, the rejection of Formosa Steel's fait accompli framing of a binary choice between steel or fish energized a groundswell of creative interpretations of how to communicate counterhegemonic frictions. A compelling affective network quickly mobilized online, underscoring a profound collective affinity for fish. Following the popularity of #ToiChonCa/#IChooseFish, we can witness how a culture might articulate not only crisis but care through pleasure (the joy of eating fish-shaped food or wearing fish fashion), labor (mourning a loss of work), and solidarity (connecting protests in the streets and firsthand harms). This polyvocal network created a textured and complicated understanding of the cultural meaning of the marine life kill. With such dissent, an opportunity for a just transition can emerge because it troubles extractivism as the taken-for-granted regime, condemns hyperrationalized logics of expendability and expansion of capitalism, and articulates a set of affective relations that constitute a life worth living.

Like most environmental disasters, the toxic impacts of 2016 marine life kill have lingered. According to Turen Van Truong and colleagues,

> People could not eat local fish, dead fish washed ashore for months, and seafood markets collapsed. . . . Out of our 520 respondents, most respondents (over 92 percent) experienced an impact on at least two of their livelihood activities that depended on marine resources and could not work during the disaster. On average people stopped all fishing-related activities for over nine months: this was a period of precarity for most households. . . . It took eighteen months before contamination levels (of oceanic water and in fish) were low enough that it was safe for nearshore fishing to continue.[110]

FIGURE 23. Still from AkzoNobel (@AkzoNobel). "We [heart] our new #HumanCities murals in Vietnam." Marine life continues to be threatened and celebrated throughout Vietnam. In response to endangerment of the species due to fishing practices, a mural campaign was launched in 2017 on the Vietnamese island of An Binh in 2017, led by the NGO IUCN.

In 2021 three workers at the Formosa Steel plant died from workplace hazards.[111] And according to Human Rights Watch, more than 130 people remain imprisoned in Vietnam for blogging and activism.[112]

Resistance, however, also continues. As of 2020, after being dismissed by Vietnamese courts and lower Taiwanese courts, almost seven thousand petitioners had been informed that the Taiwan Supreme Court will have their cases heard for compensation that they have yet to receive.[113] JFFV continues to help victims of the disaster fight for their rights in courts.[114] Bui is producing a film, *Red Sea: Vietnam's Modern Disaster*, directed by independent filmmaker and exiled Vietnamese activist Thuc Pham, to try to raise further awareness and support for the ongoing court cases of people in Vietnam against Formosa Ha Tinh Steel and its parent company, Formosa Plastics, as well as the release of all political prisoners.[115]

International attention turned toward Vietnam when people protesting environmental injustices were imprisoned for their opinions or even for documenting environmental harms and human rights violations. Once they were imprisoned, there has been evidence of poor treatment. These events are alarming for anyone who cares about human rights. As noted, incarceration increasingly warrants the attention of environmental justice studies, as ecological degradation and human sequestration too often become articulated or linked together as disposable.

Networked cultures of transnational care and solidarity provide a way to resist oppressive environmental patterns. Fan, Chiu, and Mabon argue that "in an authoritarian context, access to this science and technology cannot be taken for granted and may itself be restricted or controlled. In such settings, new transnational spaces become vital sites for environmental justice struggles."[116] When communities witness each other's harms, buffer harms for those most at risk, and build capacity for collective action, they build counterhegemonic friction enabled by international law and support.

The global movement for detoxification is a movement for deimperialization. That is, the marine life kill, which harmed the whole ecosystem including fish and people, was more than a devastating material violation—it also was a profoundly deep violation of Vietnamese culture. The corporation's assumption that foreign steel was superior to the traditional way of life in Vietnam was a contemporary expression of imperialism. Therefore the ongoing movement to detoxify requires deimperialist tactics of rejecting steel, holding the corporation accountable, and building grassroots efforts to receive redress.

The collective community response to the marine life kill of 2016 might be described by some as an acute expression of crisis, yet the discourse of #ToiChonCa/#IChooseFish also signified care through a collective affirmation of marine life and human rights. While a range of reactions were expressed, a networked culture of care emerged as witnesses to ecological horrors, human ailments, and indifference from economic and political elites, as well as creators of cultural pride in a whole way of life inspired by marine life affinities and appetites. "The significance of studying affect," according to John Erni, "is that it offers a new and innovative template with which to trace and re-draw the contours and texture of identity and difference in relation to power."[117] In this context, the intensity of the affective art online conjured a sense of collective solidarity to challenge dominant narratives of "progress" through polluting industrialization. An affinity for fish online helped elude some censorship or imprisonment, because it is often hard to pinpoint as a direct critique, when many could use plausible deniability about whether their intent was to draw the attention of, and toward, the state.

Big Tech energies should not be overromanticized—nor should hashtag activism. Big Tech continues to lobby governments to gain more access and make more profits off everyday people. They are not just "the good guys" placed in a difficult position by "bad governments."[118] Further, making connections between recycling waste workers and social media labor, Liboiron and Lepawsky remind us that commercial content moderation (CCM) is "unusually harmful to workers" instructed to censor and report, including self-harm and suicide.[119] They also point out most Facebook CCMs are contracted third-party laborers, facing poor pay and working conditions, as well as being required to sign nondisclosure agreements (NDAs). More than magically "free" services, Big Tech might be better imagined as part of infrastructure, functioning more like other public accommodations and common publicly supported private services, such as railroads and telephone companies, in which all should have a right to equal access in theory.[120] This designation, of course, varies by where we log in.

Barriers to participation do not exist only in Vietnam. It is therefore important not to create a straw man argument about Vietnam's authoritarian government. In the United States, the fossil fuel industry with ALEC has lobbied the US government to decrease freedom of expression.[121] And according to the ICNL (International Center for Not-for-Profit Law), since 2017 more than forty-five states have considered and thirty-six have enacted bills to "restrict the right to peaceful assembly."[122] And violence against environmentalists is on the rise globally: "Between 2002 and 2017, 1,558 people in 50 countries were killed for defending their environments and lands."[123] Similarly, the international watchdog group Global Witness reported, "2019 shows the highest number yet have been murdered in a single year. 212 land and environmental defenders were killed in 2019—an average of more than four people a week."[124]

Likewise, Big Tech's relationship to suppression is not isolated to Vietnam.[125] In 2017 in Italy, for example, "many Twitter posts speak of state repression and police violence and attach links to videos and images, which—as a sousveillance tactic [the counterhegemonic flip of surveillance]—make visible and challenge police violence and abuses" committed by the Italian government, media, and TAP (Trans Adriatic Pipeline) against anti-TAP protestors.[126] Often everyday people, environmental educators, and climate scientists have been censored by Facebook while oil companies have been "getting the green light" for climate denial ads and ads opposing corporate tax hikes. In 2020 Facebook had 431 million views of twenty-five oil company ads, with Exxon leading the spending in advertising.[127] InfluenceMap found that Facebook had accepted money to share advertisements "denying the reality of the climate crisis or the need for action were viewed at least 8 million times in the US in the first half of 2020" alone.[128]

In 2021 Reporters Without Borders filed a lawsuit in Paris "alleging the U.S. giant allows 'the large-scale proliferation of hate speech and false information on its networks.'"[129] Meanwhile, US Republicans since 2020 have started publishing a tsunami of books against Big Tech censorship, feeling that they were robbed of an election due to a liberal bias of the platform, especially after Trump was permanently banned from Twitter and indefinitely from Facebook in 2021 after denying the results of the democratic election.[130] At the baseline of both perspectives is a recognition of Big Tech's dominance over basic freedoms, particularly expression.

Rather than scapegoating Big Tech, I want to emphasize that—while elected officials need to hold Big Tech accountable to its users to improve human rights globally—the issues that arise with Big Tech censorship, surveillance, and disinformation speak to the need to consider how publics, the state, and Big Tech all create or erode the conditions of possibility for healthy public discourse. There is no climate disinformation on Facebook without climate deniers; there are no pages taken down of Vietnamese activists without the pressure of the Vietnamese government on Big Tech. It is too easy to leave all the blame with Big Tech, creating still another straw man argument that claims its founding figures have ruined democracy or human rights; however, both democracy and human rights had problems long before Big Tech was introduced in Vietnam and globally. "Questions about the responsibility of platforms," as Tarleton Gillespie writes, are "part of longstanding debates about the content and character of public discourse" more broadly—which platforms and users have a responsibility to address.[131]

This chapter follows the lead of grassroots advocates who have networked transnational global solidarity among communities impacted by Formosa Plastics offline and online. Throughout the aftermath of the 2016 marine life disaster, digital media platforms have provided spaces for dissent and surveillance. Using Big Tech is an act of impure politics, one that is not without fault or culpability. Some are starting to hold Big Tech accountable to the principles of human rights through not only legislation but also data strikes, data poisoning, and conscious data contributions.[132] Environmental justice studies, then, needs to engage Big Tech as part of not only networks of care but also stories of uneven transnational power relations, spatial geographies of imperialist oppression, and undemocratic regimes, and as a source of climate chaos.

Conclusion

#BreakFree(FromPlastics)

No one seems to care. Not individually. Not collectively. Not politically. No one has ever cared about these populations. Caring, in one sense, is about positionality. Who cares about the "over there" when there's so much to care about here?

—DIANA TAYLOR, FOUNDER AND FORMER DIRECTOR,
HEMISPHERIC INSTITUTE OF PERFORMANCE AND POLITICS

Each of us is precious. We, together, must break every cycle that makes us forget this.

—ADRIENNE MAREE BROWN, WRITER AND ACTIVIST

Despite the popularity of care discourses today, performances of uncaring—both everyday and spectacular—abound. Caring, as Taylor suggests in the epigraph, often depends on our capacity to risk caring for strangers "over there" when caring for ourselves and ecologies "here" often feels overwhelming enough. Plastics have become signs of collective affinities and anxieties, proximities and distances, presences and absences. Ignoring these attachments and detachments misses why plastics have become so important to so many—as well as why environmental policies more broadly remain contested today. Given the economic and political power of the plastics-industrial complex, we should not take for granted that many of the voices amplified in this book care beyond their own self-interest and have cared enough to try to move others to cherish how, as brown (in the epigraph) insists, "each of us is precious."

Plastics, I have argued, have become at this moment in history an *articulator of crisis.* A conjunctural analysis provides a more complicated story about the conditions of possibility in which environmental issues are negotiated through uneven and contested power relations. To address why plastics have proliferated and how we might challenge the hegemony of the plastics-industrial complex, we need to engage the charged affects—angry, shameful, impassioned, and more—that have become entangled with and exceed plastics. *Beyond Straw Men* focuses on hashtag activism not to police tone, but to dwell in and unravel what is being negotiated in the name of plastics. As Sarah Jaquette Ray reminds us: "This is the real challenge—to hold space for both righteous anger and curious compassion."[1]

Environmental advocacy has never been solely led by economically elite whites in the Global North, and the movement to reduce plastics should not be falsely framed as mere distraction. The Global South, including the Global South of the North, has led resistance against the hegemony of the plastics-industrial complex. Plastics have become articulated as a crisis in the Global South as drainage during climate-induced flooding drowns, toxins cause illnesses, and food becomes poisoned. Activists in Kenya identify harm to livestock, aesthetic concerns, infrastructural damage, and public health risks. Hashtag activists in Vietnam underscore how the plastics-industrial complex also has assumed too little about the significance of fish to beloved local cultural values. Those of us living in the Global North, then, would do well to trouble taken for granted canons of knowledge in environmental studies to reorient toward Global South epistemologies of waste management, marine life conservation, disaster response, human rights, global solidarity, and much more. As Erica Nuñez (@Erica__Nunez) of The Ocean Foundation tweeted: "The #PlasticsTreaty process cannot be driven solely by a Global North perspective. It needs to be inclusive and really look to address the root causes of #PlasticPollution."[2]

In 2022 the Rwanda-Peru proposal ambitiously raised the bar for a global plastics treaty, proposing a legally binding agreement to address the whole life/death cycle of plastics as a transboundary crisis.[3] Coalitions to hold corporations and governments accountable to implement and fund aspirational pledges continue to be vital at the level of structural transformation, and inclusion of voices of frontline communities and workers remains a struggle. A promising sign of inclusivity is that global waste pickers were invited to give testimony during these negotiations and were acknowledged for the first time in a UNEP Resolution, celebrated with #RecyclingWithoutWastePickersIsGarbage.[4]

Plastic policies in Bangladesh, Kenya, and more remind us that even with an ambitious global plastics treaty, plastics will remain contested for decades to come, online and offline. "Even with immediate and concerted action," Winnie W. Y. Lau and colleagues argue, with over 700 million metric tons of plastic waste accumulated to date, "coordinated global action is urgently needed to reduce plastic consumption, increase rates of reuse, waste collection, and recycling; expand safe disposal systems; and accelerate innovation in the plastic value chain."[5] This book has focused on the lived, everyday meanings and interactions of successfully negotiating the current hegemony of the plastics-industrial complex, including false solutions and social inequities.

The Global South, of course, is not monolithic. India reportedly was comparatively slower to join the excitement, proposing a last minute resolution, though eventually did come to support the idea of plastics warranting a global diplomatic response.[6] As I write, Haiti has yet to ratify the Basel Convention's plastic waste amendments.[7] And some of the top polymer producers currently are state-owned corporations based in Saudi Arabia, China, and the United Arab Emirates. China, if still considered part of the Global South (which is debatable), has reportedly coerced Uyghurs (an ethnic minority) to create products from plastics, from vinyl to medical supplies.[8] Countries with single-use bans, such as Bangladesh, continue to struggle with transforming their cultures to reflect the laws. And the Global South of the North is not united in voting for candidates who will hold petrochemical corporations accountable. Nevertheless, as the main source of pollution and delay, when the Global North ignores and excludes voices of the Global South in environmental decision-making, unsustainable patterns of environmental injustice become reinforced once again.

The successful networked cultures of care I have described within and beyond the Global South have been far from disembodied, impersonal, or measured. I do not aspire to fetishize regimes of rationalization or imperialist logic that emphasize pacifying the status quo. Recommendations to just talk more to people about plastics (or climate) are an improvement over silences that protect unsustainable beliefs and practices, but not all conversations are equally meaningful or transformative. *Beyond Straw Men* follows affective hashtag activism that has moved people in generative ways to collectively desire, agitate for, and inspire structural change in relation to our intimate entanglements with the plastics crisis. Wakibia was motivated by anger, and his success was predicated on a collective structure of feeling anger with throwaway culture in Kenya, confirmed by Wakhungu and vernacular terms like "African flowers" and "flying toilets." Lonely Whale marketed its animating affect in its name, and its mission has been to try to nudge social change with humor and fun to counter a globally felt sense of loneliness. (And even in its humor, Lonely Whale suggests The Ocean Is Angry.) Lonely Whale also listened to disability backlash accounts of shame. Vietnamese hashtag activists created art in the face of trauma, generating affective networks through a range of creative expressions of collective affinity with fish. The ethics of care, then, need not be normative about communication styles. Instead of recommending politeness, two-siding the issue, civility, or bright-sided positivity as responses to agonistic—too often called "polarized"—politics, those in the preceding pages have shared felt, collective senses of anger, loneliness, shame, and trauma. Such affective social structures do not necessarily indicate failure or crisis about the world's capacity to create a sense of belonging to public, political life. Instead, becoming more attuned to a wider affective spectrum may remind us of what *energizes* concerted desires and actions as we continue addressing plastics.[9]

Networked cultures of care develop offline and online. While hashtag trends go viral quickly, movements for structural transformations often move more slowly. Wakhungu lobbied for years beyond the hashtag trend, and Wakibia approached strangers in the streets to take pictures he would then post online at the end of the day. Lonely Whale created in-person events to pilot its campaign and continued to work with everyday people, corporate partners, and governments long after the headlines switched focus. Disabled people and accessibility advocates followed up offline to support community needs, in addition to providing consultation to nonprofits, corporations, and governments. Survivors of the marine life kill in Vietnam have grown their digital organizing to connect with transnational communities, and they have mourned, protested, and stood on trial offline. In each of these cases, offline and online networks were vital to challenging hegemonic norms about our modes of attachment to—and detachment from—plastics, from how they are made to where they are disposed. Hashtag activism, then, isn't the only advocacy that matters, but it does play vital political and cultural roles in ways that a myopic focus on plastics science or material statistics might miss.

Neoliberal individualism, American exceptionalism, planetary fatalism, and purity politics enact straw man arguments. Instead of owning responsibility for harms caused, having the courage to try to make a difference, or appreciating the pragmatic nuances of advocacy and law, these discourses too often lay all-or-nothing traps. This book was written with the belief that we might be ready for more nuanced perspectives. Placing environmental advocacy on a pedestal is what hegemonic discourses do to environmentalists for a reason—because they know it's a straw man argument of hypocrisy some people enjoy hating—and it's not reflective of the impure, lived praxis of the successful environmental advocates, which I have highlighted in the preceding pages.

As illustrated by the reticulate hashtag publics in this book, contesting power is complicated. While #BanPlasticsKE critiqued the plastics-industrial complex, a key #ISupport ally was a government official attempting to change the system from within. While Lonely Whale's hashtags challenged the habits and assumptions of throwaway culture in partnership with diverse celebrities and established institutions, they also received backlash from conservatives and progressives. While Big Tech enabled hashtagged resistance and solidarity in Vietnam, it also became complicit with state surveillance and carceral politics. Meaningful social change included but exceeded the governance of state and corporate power, involving everyday online and offline reflexivity and interactions.

Despite straw man depictions, impure politics abounds in successful anti-plastics pollution advocacy. Maathai called for a plastic bag ban while using a different type of plastic bag to grow trees. Wakhungu and Wakibia wanted to ban more than bags through changing the culture but settled for what they could achieve at the time, including not only a specific category of bags but incarceration.

Likewise, Lonely Whale aimed to address plastic pollution in the oceans but decided straws would be a useful gateway plastic for the longer struggle. It also quickly pivoted to embrace the critique of disability advocates and began advocating for accessibility accommodations in bans, reminding us that social change is not static, universal, or merely two-sided. While survivors of the marine life kill in Vietnam continue to pursue redress, they have built a global solidarity network to catalyze their capacity through allies who share a common foe. All these stories—and more—serve as reminders that environmental abolition, in a world where environmentalists are not the most powerful actors, requires impure politics. Abolition, in this sense, includes and exceeds the military-industrial complex as what Ray Acheson calls "a political project of promiscuous care."[10]

In contrast to romanticizing the carbon-heavy masculinity of the Global North stands Ahmed's feminist killjoy, which offers a reclamation of a hegemonic insult as a way of thinking and acting in the world. Contemporary political figures might include less ambitious though not necessarily less generative styles of the good-natured activist and a fellow traveler. There probably were more even within the pages of this book. (What might you name the figures of Ives-Rublee and Kane in their dialogue? Or those in Vietnam who created and circulated art?) Likewise, these figures may arise accented differently depending on the cultural context; for example, Natasha Myhal (Sault Tribe of Chippewa Indians) and Clint Carroll (Cherokee) write about "good-natured joking" as a key principle that guides Indigenous Cherokee Medicine Keepers, who emphasize "having a good time and laughing" through storytelling, fellowship, and relationships with the land.[11] While reactionary politics of aggressive masculinity continues to fuel global politics, it seems worthwhile to recognize public or prominent figures who complain, nudge, and cajole others in ways that try to fertilize the world in more inclusive, life-affirming ways, however imperfect. Valuing these counterhegemonic roles—embodied through diverse lived and situated communities globally—is a reminder of the abundant positions of collective resistance available to challenge those who appear to have insatiable appetites for carbon and attention.[12]

Of course, it is important not to romanticize hashtag activism. The responsiveness of a Kenyan politician online who feels a duty to her constituents on an issue she already cared about is quite different than Vietnamese politicians who worked with Big Tech for surveillance and punishment. Few seem as talented as Wong at creating successful hashtag activist trends more than once. Online networks, like globalization, are uneven and partial. As in offline advocacy, inequalities persist. Plastic regulations have included fines and incarceration, which are predicated on punitive relations and systemically enforced in ways that exacerbate discrimination. Celebrities have undeniable cultural influence over hegemonic norms and behaviors but continuing to fetishize celebrity culture can also deflect attention from the carbon-intensive, jet-setting lifestyles of the economic and political elites and can marginalize the voices of those most impacted. Big Tech remains powerful and contested globally on topics such as social media privacy, commercialization, harassment, and censorship. Platforms can become co-opted and rendered obsolete. I do not identify these shortcomings of policy, messengers, and technology to encourage despair or cynicism, but to recognize key barriers to democracy, abolition, justice, and sustainability.

Engaging environmental media, Hautea and colleagues have called for research exceeding textual online analysis to show why affective publics "care (or don't)."[13] I have analyzed a dense archive of official and vernacular social media posts across platforms, as well as interviewed people to help me understand some of the motivations for creating hashtags, the affective reactions to them, and the ways hashtag activism has mattered to environmental policy and practices offline. While studying digital media enabled me to appreciate how collective critique emerges and is negotiated, interviews helped me better understand the motives, missteps, and maintenance of disperse yet networked advocates.

The *Communicating Care* podcast interviews and this book have not been exhaustive. There are plenty more voices resisting plastic pollution worth engaging, including Indigenous-led initiatives and the more than 20 million people who work as waste pickers recovering, sorting, recycling, and reusing globally.[14] There also are related plastic controversies and discussions around regulating other macroplastics—polystyrene food containers, thin packaging, tire abrasion, fishing nets, mylar balloons, cigarette filters, single-use stirrers, and beverage bottles—as well as microplastics and nanoplastics, which are all produced and promoted by the global plastics-industrial complex.[15] Further, there are countless environmental and climate justice advocates with expertise in intersectional praxis that I have not yet interviewed, including some negotiating as I write for a global plastics treaty. My desire is that the shortcomings of the first season of my podcast and this book inspire further analyses of and hashtag activism about plastics, as well as the myriad of critiques that have been provoked by the proliferation of plastics globally.

In the introduction I argued that networked cultures of care can help navigate crises; this was not to indicate that we all are on the same journey on the same part of the globe, but that all our paths would benefit from their lead. By way of conclusion, to catalyze conversations that are far from over, I suggest four hashtag activism trends, drawing on insights from previous chapters that involve resistance worth following.

#Greenwashing

As I emphasize throughout, hashtags provide no political guarantees. The plastics life/death cycle is toxic and contributes to climate chaos. Rather than placing the responsibility for plastic waste on individuals through cleanups and calculators, the multinational corporations that produce and create demands for plastics must be held accountable—as they have benefited the most and have financial resources to profoundly transform infrastructure. While some companies are becoming circular economy leaders (such as Patagonia), astroturf groups abound—it is not a coincidence that the colloquial term for organizations that appear to have grassroots but do not is named after a plastic product brand.

Consider the ACC, which launched the America's Plastic Makers® in 2021, claiming to support a #CircularEconomy to #EndPlasticWaste. Members include BASF, ChevronPhillips, Dow, Dupont, ExxonMobil Chemical, and Shell Chemicals. While they propose "at least 30% recycled plastic by 2030" and national recycling standards for plastics, they also call for more studies and "an American-designed producer responsibility system," which has served as a signal for business-driven US neoliberal solutions that are far behind the rest of the world.[16] Their advertising promoting false solutions often mobilizes not only greenwashing and carewashing but also "woke-washing," to make them appear inclusive by co-opting social causes and diverse spokespeople while not holding themselves accountable to the Global South.[17]

Some companies are starting to be held accountable for their disinformation campaigns. A Reuter's report found Unilever was receiving affirmation as a green leader for calls to regulate plastic sachets while it had lobbied to prevent those same regulations.[18] An increasing number of lawsuits challenging greenwashing are being filed globally.[19] The Instagram (@bluetritonbrands) of BlueTriton, which owns Poland Spring and many other popular plastic bottled water brands, is filled with "green" hashtags, such as #ThisIsHowWePlanet, #MadeForABetterTomorrow, and #CommunityFirst. Earth Island Institute, for example, filed suit against BlueTriton. As Sharon Lerner reports, in the company's motion to dismiss, it admits: "Many of the statements at issue here constitute non-actionable puffery." General counsel at the Earth Island Institute Sumona Majumdar emphasizes how the company's social media has been key to its puffery spin: "When you look at their Instagram feeds and their statement about sustainability, it seems like a fait accompli. But in this brief they field, they're admitting that they use these sustainability commitments just as marketing tools."[20]

Resistance to greenwashing has also begun focusing on advertisers. Clean Creatives was established in 2020 to call on PR and ad agencies to stop supporting the fossil fuel industry.[21] Climate and environmental justice organizations have identified Chevron as one of the worst actors in the Global South (particularly the Amazon) and the South of the North (as in Richmond, California). Paul Paz y Miño, associate director of Amazon Watch, claims that part of Chevron's crime has been a lack of remorse and an ethos of bullying: "Not only did it set the terrible precedent of deliberate toxic dumping and destruction in the Amazon rainforest, but Chevron has never shown even an ounce of remorse or good faith in any of its negotiations or in the face of communities virtually destroyed by its acts. On the contrary, it has created a brand for itself as the most ruthless corporate bully on the planet."[22] Further, Megan Zapanta, the Richmond organizing director for the Asian Pacific Environmental Network, makes linkages between Chevron's poor environmental and political records in relation to their false advertising:

> Working with Asian immigrant and refugee communities who live at the fenceline of the Chevron Refinery in Richmond, [California] we've seen firsthand how Chevron not only pollutes Richmond's air but also our politics. Through mailers, billboards, and even a Chevron-owned local news site, Chevron tries to paint itself as a good employer and a necessary pillar of the community. We know better. The Chevron refinery visibly spews pollutants, while continually going out of their way to avoid paying their fair share of taxes and hiring very few local residents. Richmond residents are fighting for a Just Transition away from big corporate polluters towards a clean and locally governed economy. We're done with Chevron's lies.[23]

Talented people in advertising and PR increasingly agree with this and are refusing jobs that bolster unsustainable corporations. As I write, these campaigns have led to a partnership with Clean Creatives to pressure the advertising company Wavemaker (and its parent company WPP) to drop Chevron as a client.[24]

As more people become aware of the myth of recycling, there are also signs that the chasing "recycling" arrows on plastics might become more accurately regulated.[25] Laws—locally, nationally, regionally, and globally—can help catalyze structural transformations to create the conditions for substituting plastics and, when there is no replacement yet, enforce extended producer responsibility (EPR). Such measures can require corporate accountability to remove the recycling symbol from any plastic that is not recyclable in most communities (counties, wards, etc.) in a country, which would help address wishcycling. These policies will require legally binding conversations about infrastructure, as well as what is meant by "recyclable" (readily broken down as reusable feedstock or . . .). To date, EPR already has been passed into law in more than sixty countries, with Europe and Latin America leading the trend.[26] US state governments such as California, Colorado, Maine, and Oregon also have passed EPR laws—and more are being considered.[27] Global policymakers also are talking about a #CircularEconomy and #Upstream solutions more often, though they range in their aspirations from promoting technocratic fantasies in which plastics remain flowing at accelerating rates and striving to achieve a detoxified and just transformation.[28] Legal trends to stop #greenwashing continue, which is important for corporate accountability.

#BrandAudit

I have argued that bans are imperfect policy solutions that are meaningless without the infrastructure of enforcement for economic and political elites, as well as the will of the people. Bans make headlines, but they don't necessarily entail holding elites accountable upstream to prevent production to "turn off the tap" of plastics in the first place. While the plastics-industrial complex has framed littering narrowly as the problem and cleanups as the downstream solution, the anti–plastic pollution movement has begun to hold producers instead of consumers primarily accountable.

Since the privilege of proxemics allows economic and political elites of the Global North to evade the negative impacts of plastics consumption, these audits offer a powerful data tool to count what has been discounted, and how. Vital to addressing the plastics crisis is public reporting and monitoring of unjust environmental patterns, such as environmental racism, pollution colonialism, and waste imperialism. The aforementioned #BreakFreeFromPlastic coalition was founded in 2016 with more than eleven thousand organizations and individual supporters worldwide and organizes a global #BrandAudit campaign. The coalition ranks and publicizes results. As noted, each year since 2018, it has organized people to collect plastic waste and identify which brands are found the most—and to date, Coca-Cola always tops the list.[29] In addition to shifting blame from consumers to producers and from cleanups to source reduction, the campaign confronts false solutions for the plastics crisis. For example, Abigail Aguilar, Plastics Campaign regional coordinator of Greenpeace Southeast Asia, links the corporate responsibility campaign to greenwashing and the climate crisis: "It's not surprising to see the same big brands on the podium as the world's top plastic polluters for three years in a row. These companies claim to be addressing the plastic crisis yet they continue to invest in false solutions while teaming up with oil companies to produce even more plastic. To stop this mess and combat climate change, multinationals like Coca-Cola, PepsiCo, and Nestlé must end their addiction to single-use plastic packaging and move away from fossil fuels."[30] Auditing the plastics-industrial complex, then, creates an opportunity to challenge unsustainable discourses of plastics, advocating for a reduction of production, as well as, for example, redesigns that ban toxic chemicals from plastics. As this book goes to press, the campaign is tagging companies online for accountability. In calling out plastic producers, it also emphasizes that "the first essential step is to establish a new commitment focused entirely on reuse-based product delivery systems and packaging-as-a-service business models."[31]

Lakshmi Narayan, cofounder of SWaCH Waste Picker Cooperative in Pune, India, underscores the degraded labor conditions of waste pickers in the Global

South as part of what this campaign highlights: "Corporations rely on informal waste workers to collect their packaging, allowing them to meet sustainability commitments and justify their use of high quantities of single-use plastic packaging. Yet the current shift to lower value plastic packaging is threatening the livelihoods of the waste pickers, who cannot resell such low-grade items. The systems that waste pickers operate in must change."[32] While anti-environmental backlash often frames environmental policies against labor (e.g., as "jobs vs. environment"), once again, environmental advocates and labor advocates are in alliance, making linkages between crises and working together toward a just transition through public participation in plastics decision-making.[33]

By publicly shaming corporations such as Coca-Cola instead of individual consumers, the broader #BreakFreeFromPlastic coalition calls for systemic transformation that prioritizes life over profits without waiting solely on governmental reform. In 2022, for example, the coalition created a culture-jamming animated short of the song "You've Got the Whole World in Your Hands" showing the iconic company's plastic bottles in scenes of oil pumping and fracking, public health impacts, and boardrooms planning to triple production. Instead of ending solely on the critique, the video ends with an uplifting message that the beverage industry could begin supporting refill stations.[34]

To address environmental privilege, The Big Plastic Count is an ongoing campaign in Britain to encourage government to ban plastic exports, organized by Greenpeace and Everyday Plastic, connecting personal awareness of plastic footprints with structural transformation.[35] If adopted, the principle of proxemics moves the costs and benefits of plastics closer together.

Big Tech can assist in solutions also. Wakibia recently was part of a single-use plastic bottle brand audit in Nakuru Park. Using an online app (wastebase.org), the auditors were able to scan more than three thousand bottles.[36] Technology, in this context, might be used to reverse surveillance patterns from everyday people to mapping the harms of transnational corporations—to ask them to become more accountable to the entire lifecycle of their products. Likewise, the Basel Action Network (BAN), a core member of the Break Free From Plastic Movement, also is working with researchers at Columbia and Yale Universities to create an *Atlas of Plastic Waste* to map plastic waste globally, which highlights private and public sites of concentrated plastic waste.[37] This global crowd-sourcing campaign promises to trouble the idea that plastics just magically "go away." Likewise, Yelp, a platform that features crowd-sourced reviews of restaurants and businesses, partnered with the Plastic Pollution Coalition to allow searchable sustainability attributes related to plastics, such as "Plastic-free packaging," "Provides reusable tableware," "Bring your own container allowed," and "Compostable containers available."[38] Big Tech, then, can support initiatives such as these, as well as reduce its own use of plastics and production of greenhouse gases.

#BeyondPlastics/#ÉtéSansPlastique/#SinPlástico

Plastic production continues to explode in ways that defy imagination. This proliferation impacts habits of our everyday lives, from being asked at the grocery store if we want "paper or plastic" to how we spend money itself: "Will you be paying with paper or plastic?"[39] And those are the moments when we're asked our preference. Caring about plastic pollution does implicate refusing plastics we can live without—even if our answers are different.

In response, a growing global movement has argued—invoking the eloquent play on words by Wakibia—that "less plastic is fantastic." New policies and practices require, and ideally enable, a shift in cultural attitudes, beliefs, and behaviors, which requires a shift in imagination in reducing waste, reassessing our (de)attachments, and holding elite corporations accountable for unsustainable practices. Plastics and climate, as noted, are linked to living on a finite planet with seemingly infinite urges to consume, making them relevant to each other in practice and policy.[40] All plastics cause harm to someone in their production, let alone considerations of exposure through consumption and waste. Currently, some of us use some plastics to live, but, none of us need all plastics all the time. The fact that some of us have forgotten how to live without plastic conveniences in such a short period of time conversely may offer inspiration for how quickly we may be able to reverse course. As I wrote in chapter 5, *every person is necessary, every plastic is not.*

Plastic-free discourse imagines a different world where plastics are no longer necessary. Reducing production requires a systemic critique of the habits and attitudes that perpetuate throwaway culture and eliminating nonessential use within the next decade. Key, however, is the collective work of care, not scapegoating ("at least I'm not as bad as") or fulfilling neoliberal fantasies of individualism ("I can just focus on me"). If world governments reallocated subsidies for fossil fuels (and therefore plastics) or shut them down, what cultures of care should we prioritize? Just as all plastics are not the same, finding ways out of this mess will vary by culture.

In places that have banned certain types of single-use plastics, some of the solutions involve remembering traditional ways of living, such as the noted return to jute in Bangladesh. In Mexico City, people are beginning to return to *ayate* or mesh bags. As the city's director of environmental awareness, Claudia Hernández, has said: "We have a very rich history in ways to wrap things. We are finding people are returning to baskets, to *cucuruchos* (cone-shaped paper)." She emphasizes that while all the details have not been worked out about which plastics are important for health and which can be replaced, the ban "is an invitation, a provocation to rethink the way we consume."[41]

Humans cannot avoid making waste completely, as Liboiron and Lepawsky argue, but we can imagine within our specific contexts of impurity how to "discard well," including discarding less.[42] With the encouragement of customers and government regulations, more companies, industries, and NGOs may start eliminating unnecessary plastics when possible. Related hashtag activism (#SkipTheStuff, #SkipTheStraw, #SayNoToStyro, #ZeroWaste) is becoming more common, sharing how a world "beyond plastic" might manifest.[43] The Story of Stuff Project has an ongoing campaign about Keurig single-use plastic pods for coffee, which it estimates already "could wrap around the planet more than 10 times" and can be replaced with paper or reusable filters.[44] The Italian pasta company Barilla has begun phasing out the thin plastic film on its boxes in the United Kingdom with the message: "No more plastic window. CHANGING OUR WORLD ONE PACK AT A TIME."[45] The tourist industry also has begun to reduce the large amount of single-use plastics offered to visitors.[46] The NGO Practice Greenhealth is helping health-care systems to reduce waste, including single-use plastics.[47] Greenpeace emphasizes that initiatives to support reuse and refill also should come from the food industry, since affordable, durable, nontoxic, convenient, and simple designs for a just transition will vary across food types and geography.[48] As this book goes to press, Bloomberg Philanthropies also is launching a #BeyondPetroChemicals campaign to fund community-based organizations resisting industrial expansion in the Gulf Coast and Appalachia, spotlighting RISE St. James's victory over Formosa Plastics as a success story they hope to help repeat.[49]

One way our imaginations are enclosed or emancipated is through storytelling.[50] For colonized people more broadly, de Onís insists: "Fluid storytelling efforts are necessary for deep contextualization and to acknowledge the resiliency and entwinement of energy coloniality and energy privilege to demonstrate their iterative and obstinate qualities."[51] To resist the mythic stories promoted by the plastics-industrial complex, it is necessary to critically interrupt hegemonic narratives to create a sense of presence about how power inequities create toxic patterns and how they are inspired to advocate for collective action, as well as to reimagine and to retell stories of everyday life with a different relationship to plastics and the planet (and all that entails).[52] Anti-colonial approaches to plastics centering consent, for example, provide a way of rearticulating relations.[53] Likewise, recognizing resistance against petrochemical industries as part of a broader struggle for abolition appears to be emerging.[54]

In addition to podcasts, there are a growing number of documentaries about the plastics crisis drawing on inspirational discourses such as these.[55] In under ten minutes, *The Story of Plastic* amplifies a range of stories about plastics that feature workers. *Glass, Metal, Plastic: The Story of New York's Canners* features two stories of Black people living in New York and making a living as recyclers. One of the storytellers, Pierre Simmons, tells his own life history of employment, living with cancer, canning (recycling), and advocating to transform laws. At one point, he states: "Caring about what people think about you is bondage, so the first thing that has to be suspended is ego."[56] What a compelling way to articulate resistance to stigma about garbage and the damage ego can do to our sense of self, as well as to the health of Earth. I cannot do his story justice here, but I want to at least gesture to how storytelling can link contextual and personal plastic entanglements. I hope more stories are documented at the intersection of plastics and labor, including how, at least in the United States, prison labor is exploited.

To reduce the distance, literally and figuratively, between those who have been benefiting the most and those bearing the greatest costs, GAIA (the Global Alliance for Incinerator Alternatives) is organizing toxic tours to include not only sharing harms with those not experiencing them daily but also stories of "treasures" of local zero waste initiatives, including "community gardens, repair shops, and secondhand stores."[57] These creative and pragmatic treasures underscore the harms of environmentally unjust patterns, as well as practices of reattachment to the communities and systems we all depend upon to live, work, and play.[58]

#Tortuga

Like the polar bear for the climate crisis, the turtle with a plastic straw has become the icon of the anti–plastic pollution movement—or maybe it's the songs of the lonely whale? Or a seal slap in Chilé? Or fish dying in Vietnam? Or a seahorse holding a cotton swab?[59] The point is: as alarms about plastics have been sounded by marine life endangerment, more members of our own species have been moved to care about our interdependence within broader ecologies.

Drawing on Indigenous epistemologies, including the Zapatistas, Escobar argues for a counterhegemonic value system called "radical relationality," a notion "that all entities that make up the world are so deeply interrelated that they have no intrinsic, separate existence by themselves."[60] Banerjee emphasizes the contextuality of Indigenous interspecies justice eloquently: "What works for one place and in a particular culture may not work for another place and in another culture. What makes biodiversity conservation so beautiful is that it is a pluriverse—so many ideas, so many practices, so many forms of human-nonhuman kinship that exist around the world."[61] Likewise, expanding beyond Indigenous epistemologies, Pellow emphasizes that the work of critical environmental justice requires "articulating a viewpoint that all humans and more-than-human actors are *indispensable* to the present and for building sustainable and just resilient futures."[62]

There are principles that guide planetary research commitments, such as interdependence, but each articulation within each ecosystem warrants genuine listening—in the transformational sense—to nonhuman kin. Consider, for example, Gumbs's call in *Undrowned: Black Feminist Lessons from Marine Mammals*: "Instead of continuing the trajectory of slavery, entrapment, separation and domination and making our atmosphere unbreathable, we might instead practice another way to breathe. I don't know what that will look like, but I do know that our marine mammal kindred are amazing at not drowning."[63] Each culture is informed by different traditions and affinities related to marine life, but humans would do well not to ignore the global pattern of aquatic alarms and lessons of care if we too wish to remain unbowed and undrowned.

In *Virtual Menageries: Animals as Mediators in Network Cultures*, Jody Berland has warned that the popularity of nonhuman animals online in an age of digital capitalism may be part of a longer history of colonial objectification and distorted pleasure that risks ignoring environmental destruction offscreen.[64] Likewise, the presence of a particular animal trending online does not guarantee improved human-nonhuman relations. Popularity, though fetishized in and through media, signals attention, but not necessarily the quality of engagement. A meaningful global plastics treaty requires a broader understanding of plastics' entire lifecycle beyond marine life, as Global South environmental leaders emphasize not only water but also land connections and human rights.

Calls for #PlasticFreeOceans, #OceanPlastic, #OceanJustice, and more marine-related hashtags, however, have animated anti–plastic pollution advocacy globally. Since communicating for the environment, whether it is the ocean or a whale, is an interpretive act that requires expanding our species' imagination, art often plays a significant role on and off screens. The preceding chapters noted the importance of photographs, videos, and drawings for advocates to circulate statements on everyday life, mobilize broader public support, and inventively affirm cultural self-pride to detoxify their lives and dignify their way of life. Creative expression, then, remains one of the vital ways to muster desire-based research in a world on fire. As D. Soyini Madison writes: "Water as politics needs beauty."[65]

When I was talking with climate justice organizer and artist Michelle Gabrieloff-Parish, she underscored the importance of reconnecting with nature and poetry:

> How can we be in relationship with that part of nature as well, not just take from it, not just extract knowledge from it, but actually start being in relationship with it? . . . This is one of the gifts that the environmental justice movement has really given us is: the recognition that everywhere is nature. . . .

And this culture that we're in has really created a fantastical divide between people, creatures, Earth. . . . People are disconnected. They don't know: where does the water come from that's coming out of your faucet? . . .

And I just love the ways in which poetry can help us connect that we don't have to speak in complete sentences, and we don't have to just speak to the parts of us that reside in the intellect and, you know, the three pounds of mass that sit on top of our neck . . . I have a poem somewhere or another about some revolutionary jellyfish that actually went into a nuclear power plant and stopped it up--and I'll have to go to find the story, because I wrote it a long time ago.

Gabrieloff-Parish's insights speak to holding ourselves accountable to a broader sense of scale in terms of belonging across geographies and temporalities, calling on humans to consider how we can affectively reconnect with life beyond humans and engage other humans to pursue justice. Poetry and other artistic acts remain vital, as she notes, to unsettle the oppressive logics that confine ways of thinking and knowing. Reconnecting with nonhuman kin through creative expression in this age of seemingly constant crisis can become a practice that can nurture communal healing and celebrate inclusive joy.

ON AWARENESS, ACTION, AND ACCOUNTABILITY

Plastics remain heatedly contested online and offline. I offer these four hashtag activism trends about plastics to encourage further awareness of and action in response to the consequences of our attachments and detachments. While the plastics-industrial complex may promise infinite transformation, anti–plastic pollution hashtag activists who critically interrupt this narrative have insisted on recognizing ecological, cultural, and biological harms. Through photography, videos, hashtags, and art, networked cultures of care have cultivated relations in ways that appear far safer and healthier than status quo predictions of our species' future.

Plastics remain a dramatic articulator of crisis—and care—charging a wide range of controversies and conversations in ways that implicate and exceed plastics. Hegemonic straw men arguments about anti–plastic pollution advocacy pervade dominant culture with daily reminders to not question the popularity of plastics, to not listen to the Global South or marine life, and to not be foolish enough to believe social and ecological justice is possible. Better to mock people who are questioning these norms, economic and political elites too often suggest. And yet *Beyond Straw Men* has shown that plastic hashtag activism has become a courageous way for some to mobilize affective publics to rehearse and to constitute a wide range of counterhegemonic beliefs, practices, and policies—even as it can be used for surveillance, carceral politics, and ableism. Through and beyond hashtag activism, environmental and disability justice networks have succeeded in transforming policy and cultural norms across uneven power relations and disproportionate impacts, despite the odds. Plastics have become conductors through which we are negotiating nothing short of culture: who matters—and who doesn't—as well as the world we are making—and breaking.

In conclusion, I want to note an encouraging story unfolding about the movement resisting plastic pollution. Promoting #OceanOptimism, the US-based NGO Ocean Conservancy is known for partnering with plastic polluters, such as Coca-Cola, for international beach cleanups for decades.[66] More recently, however, Ocean Conservancy hired its first vice president of conservation, justice, and equity and successfully advocated for California to pass a new plastics reduction act in 2022 with a broad coalition, which involved gaining the support of environmental justice organizations.[67] Then, in July 2022, Ocean Conservancy issued a statement apologizing for a report it had published in 2015, *Stemming the Tide*, which erroneously supported incineration and scapegoated or placed blame for plastics in the wrong direction: "By focusing so narrowly on one region of the world (East and Southeast Asia), we created a narrative about who is responsible for the plastic ocean crisis—one that failed to acknowledge the outsized role that developed countries, especially the United States, have played and continue to play in generating and exporting plastic waste to this very region. This too was wrong. We apologize for the framing of this report and unequivocally rescind any direct or indirect endorsement of incineration as a solution to ocean plastic." The statement then notes Ocean Conservancy has moved to correct the harm by removing the report from its website and promotional materials, as well as to recommit "to a healthier ocean protected by a more just world." This public apology was significant, as the 2015 report led to the creation of the Philippines-based #BreakFreeFromPlastic coalition, in part, to counter Ocean Conservancy's problematic narrative about plastics.[68]

Ocean Conservancy's statement prompted GAIA and the Break Free From Plastic Movement to start "engaging in a repair and transformative justice process with OC to identify ways to mitigate the harm caused." Froilan Grate, the regional director of GAIA-Asia Pacific stated: "The apology is an invitation to hear the voices and concerns of communities and groups in the Asia Pacific region who have been disproportionately impacted by this framing, and for whom this issue is very personal. This is a time for the rest of the world to listen and to follow their lead."[69] He followed up: "This unprecedented report retraction is an opportunity to interrupt decades of waste colonialism."[70]

This apology offers a reminder that to #BreakFreeFromPlastic we all need to break free of the global and intersectional inequalities that have created the conditions of possibility for the unsustainable conjuncture we find ourselves in today. It also provides another exemplar of how the environmental movement can continue to become more inclusive not by ignoring errors and exclusions, but through engaging processes of transformative justice, including self-reflection, naming harm, apologizing, and acting to do better through changed behavior. "Ultimately," as Farhana Sultana argues, "there is no singular or ideal climate justice for all, no singular arrival point, but a becoming, of doing more and better, unlearning to relearn, and continually taking stock in various collectives for equitable and transformative praxis."[71]

Bans, as I have argued, are imperfect policy solutions that strive to counter the weight of the plastics-industrial complex, but require a cultural transformation of our everyday lives to matter. In the moments when the world will—once again—inevitably fumble, hesitate, compromise, and falter in addressing the most pressing crises of our current conjuncture, we need to collectively resist uncaring. Naming uneven operations of power and harm while recognizing those who have successfully transformed the world for the better are critical moves for ushering in a more viable tomorrow. Although we undoubtedly will continue to act imperfectly, insisting that we all are precious and indispensable requires resisting ecological and social detachments through nourishing networked cultures of care. If more of us risk stepping into the fray—as angry, lonely, shamed, and traumatized as many of us feel—we might foster relations that are more egalitarian, biodiverse, and dignified.

NOTES

INTRODUCTION: CARE AMID OCEANS OF TROUBLE

1. Freinkel, *Plastic*. Relatedly, Executive Director José T. Bravo of Just Transition Alliance invites students and staff to conduct a "toxic romance evaluation." Pezzullo, "José Toscano Bravo."

2. Berlant, *Queen of America*.

3. Geyer, Jambeck, and Law, "Production, Use, and Fate of Plastics." On plastics pre-1950 see Altman, "Five Myths about Plastics."

4. World Economic Forum, "New Plastics Economy," 17.

5. Alaimo, *Exposed*, 131.

6. Just Transition Alliance, "Lifecycle of Plastics." Chapter 1 elaborates.

7. Leslie et al., "Plastic Particle Pollution in Human Blood"; Jenner et al., "Microplastics in Human Lung Tissue"; and Ragusa et al., "Raman Microspectroscopy Detection."

8. De Wit and Bigaud, "No Plastic in Nature." This is ironic in countries where "plastic" has become a colloquial expression for financial transactions using credit.

9. Geyer, Jambeck, and Law, "Production, Use, and Fate of Plastics"; and Sullivan, "How Big Oil Misled the Public." Chapter 2 elaborates.

10. Geyer, Jambeck, and Law, "Production, Use, and Fate of Plastics."

11. US Energy Information Administration, "How Much Oil Is Used?"; CIEL, "Fueling Plastics"; and Lavers et al., "Far from a Distraction."

12. Zhu, "Plastic Cycle."

13. Enck, foreword to *The New Coal*.

14. World Economic Forum, "New Plastics Economy"; Lerner, "Africa's Exploding Plastic Nightmare"; and Gardiner, "Plastics Pipeline."

15. Quoted in *PBS Frontline*, "Plastic Wars."

16. Wang et al., "Early Warning Signals of Critical Transition."

17. Haraway, *Staying with the Trouble*; and Wiens, "Virtual Dwelling."

18. "Strawman arguments" are named as such due to the androcentrism of the English language and argumentation; following the publisher's norms of style, "straw man" is written as two words throughout this book. In this book, I (she/her) reference "men" and "women" when noting research based on those categories, terms people use to self-identify, or the cultural norm of patriarchy (e.g., "ad men," "strawman"). I don't distinguish between "women-identified" and "women" because trans women are women; likewise, trans men are men. When people self-identify as "they," I follow their lead for accuracy and dignity.

19. Articulation theory identifies connections creating consequential unities, which exceed the elements independently. For Stuart Hall's argument see Grossberg, "On Postmodernism and Articulation." See also Laclau and Mouffe, *Hegemony and Socialist Strategy*; Pezzullo, *Toxic Tourism*; and Pezzullo, "Anti-Toxic Activism on Screen." On voice see Watts, "'Voice' and 'Voicelessness'"; chapter 5 elaborates.

20. It appears to be in vogue for Global North critics to claim that single-use plastics distract. In "Plastics Straws Aren't the Problem," Mintor calls it a "fashionable global protest movement" that might "sound virtuous" and "hip" but really is "ineffective" and "could distract" while "likely to make matters worse." In "Capitalism Is Killing the Planet," Monbiot defines "micro consumerist bollocks" as "tiny issues such as plastic straws and coffee cups," claiming we "are obsessed with plastic bags" and think "we're doing the world a favour by buying tote bags"; Monboit and I agree that overall reduction of waste is ideal and "small things . . . should not matter to the exclusion of things that matter more"—but there's a good deal of derision before he makes that point. In contrast see Lavers et al., "Far from a Distraction."

21. On the climate emergency as predicated on coloniality see Agarwal and Narain, *Global Warming in Unequal World*.

22. Shome, "Thinking Culture and Cultural Studies," 215.

23. Thaker, "Environmentalism of the Poor," 200.

24. Dutta and Pal, "Theorizing from the Global South," 367.

25. Okune, "Open Ethnographic Archiving," 25–26.

26. Pellow, *Resisting Global Toxics*, 3.

27. Antonio Gramsci articulated "hegemony" as predominant by consent, that is, a dominant way of life that everyday people have consciously or not acquiesced in as acceptable or the way things are. As he sat in jail with Potts disease, a World War II political prisoner of Benito Mussolini's fascist regime, Gramsci wrote what is now called *The Prison Notebooks*. He focused on two hegemonic conjunctures: (1) how Italians could democratically elect a fascist leader (who jailed him) and (2) how northern Italians positioned southern Italians as subaltern (*subalterno*) or inferior to themselves culturally (which he resisted as a Sardinian). Culture, from a Gramscian perspective, is not the ultimate culmination of a society but a struggle; this idea deeply influenced Stuart Hall, Gayatri Spivak, and more, including myself (also of southern Italian heritage). Studying hegemonic struggles considers why people come to believe what we believe and the consequences of those beliefs. Focusing not on violence or coercion, hegemonic analysis considers how everyday people might come to identify and to act (vote, imprison, etc.) in alignment with positions of political and economic power even when they are counterproductive to one's own liberation or self-love.

28. Madison, *Acts of Activism*, 20.

29. Other terms (First/Third, Developed/Least) imply a globally agreed upon, universal hierarchy of achievement, which has been disastrous for people and the planet. Some use "poorer" and "darker" nations, including Prashad, *Darker Nations* and *Poorer Nations*.

30. My understanding of *reorienting* is shaped by my position in the Global North, as well as: (1) Said's critique of "Orientialism," which he uses to name how the West (especially Western European English and French cultures) tends to consolidate power by portraying the East (i.e., the Orient or countries in the Middle East, North Africa, and Asia) as inferior and derivative; and (2) Ahmed's phenomenological orientation insofar as our positioning of ourselves "affect(s) what we do, and how we inhabit space." Said, *Orientalism*; Said, "Orientalism Reconsidered"; and Ahmed, *Queer Phenomenology*, 28.

31. Despite the long-standing value of *intersectionality* in environmental justice praxis (Di Chiro, "Mobilizing 'Intersectionality'"), the current popularization is attributable to critical race theorist Crenshaw in "Mapping the Margins," who argues that US legal systems are inadequate for addressing how specific lived experiences are not reducible to isolated oppressions. In short, oppression of Blacks and of women cannot fully account for how Black women are oppressed.

32. Haraway, "Situated Knowledges." "Positioning and not centering" was articulated to me by my friend and series coeditor, Salma Monani. Raised in the Global North, I am a European American with dual citizenship (US and Italia/the European Union). Since 2015 I have lived on the traditional lands of Ute, Cheyenne, and Arapaho nations. Throughout, I elaborate on my position(s) that appear most relevant.

33. On disposability see Pezzullo, *Toxic Tourism*. On indispensability see Pellow, *Critical Environmental Justice?* On "discard studies" see Liboiron, "'Matter out of Place.'"

34. Banerjee, "Resisting the War on Alaska's Arctic." In *Total Liberation*, Pellow's concept of "socioecological justice" resonates with this inclusion of nonhuman kin in environmental justice studies.

35. Hall et al., *Policing the Crisis*, 1–2 (emphasis added).

36. Hauser, *Vernacular Voices*, xix.

37. In *Things Worth Keeping*, Harold emphasizes the importance of designers reassessing the value of attachment to consumerism of human-made objects in relation to disposability.

38. I am grateful to personal correspondence with John Clarke (June 21, 2021) as I tried to make sense of the mugging event that animated *Policing the Crisis*. Their book examines public discourse about three youth of mixed ethnic immigrant backgrounds robbing and injuring an Irish man in a working-class, predominantly Asian and African Caribbean immigrant neighborhood in Britain. The youth (Paul Storey, James Duignan, and Mustafa Fuat) robbed Robert Keenan of 30 pence, keys, and cigarettes; they also attacked him multiple times over a period of hours. Fuat, who was noted for Cypriot roots and "oriental rugs" in his home, and Duignan, identified as "multi-ethnic," called an ambulance, leading to their arrest and more lenient sentences (ten years for both fifteen-year-olds); sixteen-year-old Storey was framed as the lead instigator from a home of abuse and as having a West Indian father (Hall et al., *Policing the Crisis*, 83). On the immigrant, working-class neighborhood see Cottle, "Stigmatizing Handsworth."

39. Liboiron, *Pollution Is Colonialism*, 27. Throughout, I note Indigenous identities in parentheses after names of people who provide their Indigenous identities in their published work; I do not include identities after those who don't.

40. *OED Online,* "Single-use" and "Throwaway."

41. Williams, *Marxism and Literature.*

42. Loeffelholz, "First Giant Step?"

43. Stouffer, "Plastics Packaging." For awareness of this anecdote, I am indebted to Liboiron, Altman, and more plastics experts.

44. For example, see Chellel and Dontoh, "West Africa Is Drowning in Plastic." On resistance to this post–World War II promotion of throwaway culture see Strasser, *Waste and Want.*

45. Lerner, "Africa's Exploding Plastic Nightmare"; and Jambeck et. al., "Plastic Waste Inputs."

46. Sze, *Fantasy Islands,* 26; dominant US imaginaries of China resonate with the utopian and exploitative Western frontier myth (35). On China and US environmental communication see Pezzullo, "Introduction."

47. China's policy follows tensions with the United States, the increased difficulty of sorting plastic waste, and Wang's documentary *Plastic China,* which publicized the hidden abode of plastic "recycling."

48. Lerner, "Africa's Exploding Plastic Nightmare."

49. Lerner, "Africa's Exploding Plastic Nightmare."

50. The Human Rights Watch report *They're Poisoning Us,* details how the China ban also shifted European Union plastics to Turkey, with devastating impacts.

51. Office of the Deputy Prime Minister (UK), "EEA Glossary": "The principle of proximity implies that waste should generally be managed as near as possible to its place of production, mainly because transporting waste has a significant environmental [and, I would add, social] impact."

52. California established the first single-use plastic ban in the United States in 2014. For a list of US bans see Rapoza, "China Quits Recycling U.S. Trash."

53. Plasticbaglaws.Org, home page. On US plastic bag legislation see NCSL, "State Plastic Bag Legislation."

54. Kidwell, "States Reversing Bans on Plastics."

55. *Collins Dictionary,* "Why Single-Use?"

56. Amid industry backlash and enforcement concerns, India's ban launched on July 1, 2022, including nineteen single-use plastics but not others, such as chip bags and beverage bottles. Ghosal, "Cups, Straws, Spoons"; and Dasgupta, "India's Single-Use Plastic Ban."

57. The European Union, "Circular Economy."

58. UNEP, "Blue Awakening."

59. UNEP, "Historic Day in the Campaign"; and McVeigh, "'Historic' Treaty on Plastic Waste." The treaty must be passed by 2024 to establish legal accountability. On ways the oil, gas, and petrochemical industry, including plastics, exploited the pandemic, see Feit and Muffet, "Pandemic Crisis, Systemic Decline."

60. Briggs, "Plastic Pollution."

61. On the origin of the environmental justice movement see Pezzullo, "Performing Critical Interruptions"; and Pezzullo, "Environmental Justice and Climate Justice." Currently, the Warren County Environmental Action Team has a Facebook page and a Twitter account (@WCEAT1).

62. Bullard, "Interview"; and Bullard et al., *Toxic Wastes and Race.*

63. Bullard, *Unequal Protection*, 11.

64. Chakraborty, "Proximity to Extremely Hazardous Substances"; and Chakraborty, "Unequal Proximity to Environmental Pollution."

65. UN, "Disability-Inclusive Services." Listening to US-based disability justice activists, I use first-person language ("disabled people"), except when engaging people or research that uses people-first language ("people with disabilities"). Environmental justice studies has underengaged disability studies (Vasquez, "Environmental Injustice and Disability"); disability studies has been critiqued for colonial assumptions (Meekosha, "Decolonising Disability").

66. Pellow, *Resisting Global Toxics*, 141.

67. Park and Pellow, *Slums of Aspen*, 5.

68. Pellow, *Resisting Global Toxics*, 8–9. On the plastics waste trade see "Truth behind Trash."

69. Betasamosake Simpson, *Dancing on Our Turtle's Back*, 17.

70. Quoted in Boyd, introduction to *Month and a Day*, xi (emphasis added). On colonialism and plastics in the nineteenth century see Altman, "Five Myths about Plastics."

71. Calil et al., *Neglected*, 24. Chapter 1 elaborates.

72. Higgins, "Race, Pollution, and Mastery of Nature," 252.

73. Liboiron, *Pollution Is Colonialism*, 9, 41–42 (emphasis added).

74. Nishime and Williams, "Afterword," 251.

75. On extractivism see LeBrón, *Policing Life and Death*, esp. 236; and Gómez-Barris, *Extractive Zone*, esp. xvii.

76. De Onís, *Energy Islands*, 5.

77. Gumbs, *Undrowned*. In *On Black Media Philosophy*, Towns also writes about water as freedom and a guide for runaway enslaved people (67–70).

78. Climate Justice Alliance, "About." See also Klein, *This Changes Everything*, 424; and Simpson, *As We Have Always Done*, 3.

79. Gingrich-Philbrook, "Autoethnography's Family Values."

80. Tuck, "Suspending Damage."

81. Following turns in cancer biology and conservation biology, in "Nature's 'Crisis Disciplines'" Cox argues that studying crises ethically requires environmental communication research to not only identify problems but also make judgments and, ideally, intervene. Buell has asked if "crisis discourse [is] still necessary or even . . . useful and helpful at all?" (*Apocalypse to Way of Life*, 199). In *Plastics Unlimited*, Mah provides a synthesis of literature on the limits of crisis as a frame, though she ultimately insists we address plastics as such (20–22). In *Beyond Straw Men*, I am interested in when people imagine plastics as a crisis and to what ends.

82. Drawing on examples from airline taglines to platform-based markets for "everyday care needs" (pet care, babysitting), the London-based Care Collective coined *carewashing* as when dominant institutions "capitalize on the very care crisis they have helped to create." Chatzidakis et al., *Care Manifesto*, 11. In "Care Not Growth," Di Chiro writes of "how the deracinated idea of 'caring for climate' so effortlessly slips into gratuitous greenwashing" (304). Simmonds articulates "care as an affective relation whose leading ethic is

to create attachments within infrastructures of inequity." Recounted in Liboiron, *Pollution Is Colonialism*, 114–55.

83. Di Chiro, "Mobilizing 'Intersectionality'."

84. Kenner, *Breathtaking*, 8.

85. De Onís, *Energy Islands*; Guitiérrez Aguilar, *Horizontes Comunitaros-Populares*; Quizar, "Logic of Care"; Sun and Dutta, "Meanings of Care"; and Whyte and Cuomo, *Ethics of Caring*.

86. Pezzullo, "Environment," "Between Crisis and Care," "Introduction," "Environmental Justice and Climate Justice," and "Bats, Breathing, and Bella Vita Verde."

87. Chatzidakis et al., *Care Manifesto*, 10. On neoliberalism, Asen summarizes three key characteristics: (1) imagining people as individuals rather than as having subjectivities constituted through relationships, (2) privileging self-interest as the primary motivation over public engagement or the cultivation of relationships to address collective concern, and (3) dismissing structural conditions instead of grappling with their impacts on agency. "Introduction," 1–5.

88. Gumbs, *Undrowned*, 56. I am grateful to Dr. Amber Johnson for recommending *Undrowned*; on their care praxis see www.justicefleet.com.

89. Di Chiro, "Care Not Growth," 306.

90. Ghani, "We Need a Culture of Care."

91. Douglas's *Purity and Danger* continues to influence my thinking and, more recently, Liboiron and Lepawsky, *Discard Studies*.

92. Quoted in Petsko, "Q&A."

93. Altman, (@rebecca_altman). "This Wk, I Have a Teen Febrile w/ COVID" (shared with permission). An exceptional author, Altman is writing a memoir-sociological US history of plastics: *The Song of Styrene*.

94. GAIA Asia Pacific @ZeroWasteAsia, "#BreakFreeFromPlasticFriday."

95. Haraway, *Simians, Cyborgs, and Women*.

96. Lugones, "Purity, Impurity, and Separation," 464.

97. Kothari et al., *Pluriverse*.

98. Escobar, "Global Doesn't Exist."

99. Fraser, "Transnationalizing the Public Sphere," 28; Asen, "Introduction"; and Taylor, *Disappearing Acts*.

100. Pezzullo and Cox, *Environmental Communication*, ch. 6.

101. Scheible, *Digital Shift*, 106.

102. On democracy involving stranger relations see Allen, *Talking to Strangers*.

103. Piatek, "#sandiegofire—the First Successful Hashtag." Hashtags were used previously on other social media platforms; Messina notes Jaiku. On digital information during crises see Sutton, Palen, and Shklovski, "Backchannels on the Front Lines"; Starbird and Palen, "'Voluntweeters'"; and Takahashi, Tandoc, and Carmichael, "Twitter during a Disaster."

104. Messina, "Hashtag Turns 13."

105. Wilz, *Resisting Rape Culture*, 11.

106. Jackson, Bailey, and Welles, *#HashtagActivism*, xxxv, xxvii.

107. Jackson, Bailey, and Welles, *#HashtagActivism*, 185–86. See also Papacharissi, "Affective Publics," 4; and Poole and Giraud, "Right-Wing Populism," 4.

108. Kuo, "Reflections on #Solidarity," 187.

109. Alaoui, "The Arab Spring between the Streets and the Tweets," 36.

110. Jackson and Welles, "Hijacking #myNYPD"; Hodges and Stocking, "Pipeline of Tweets." "Counterpublics" are not homogenous or binary; see Pezzullo, "Resisting 'Breast Cancer Awareness Month.'"

111. Tufekci, *Twitter and Tear Gas*, ix; Wired News Report, "Rebel Movement's Life on the Web." On interpellation see Butler, *Bodies That Matter*.

112. Duarte, "Connected Activism," 21. Duarte focuses on EZLN, Idle No More, and the Rio Yaqui water rights.

113. Seigworth, "Capaciousness," i. Grossberg defines affect as "the nature of the concern (caring, passion) in the investment." *We Gotta Get Out*, 82, 83, 87.

114. Escobar, *Pluriversal Politics*, xi.

115. Seigworth and Gregg, "Inventory of Shimmers," 1.

116. Cram, *Violent Inheritance*, 7, 33.

117. Garcia-Rojas, "(Un)Disciplined Futures," 255. In *Toxic Tourism*, I cite James Baldwin, *Fire Next Time*, to theorize the sensuous, which I imagine as akin to affect (12).

118. Vats, "Cooking Up Hashtag Activism."

119. Papacharissi and de Fatima Oliveira, "Affective News and Networked Publics," 280.

120. Greene and Kuswa, "'From the Arab Spring to Athens,'" 283.

121. "If we continue looking for public 'participation' or 'engagement' according to standards inherited from antiquated notions of what these behaviors involve, we will neglect . . . the rich and flourishing diversity of ways people . . . are expressing their concerns . . . within the vernacular context of the everyday." Ingraham, *Gestures of Concern*, 190–91.

122. Hautea et al., "Showing They Care (Or Don't)," 12.

123. Pezzullo, *Toxic Tourism*. I started teaching and publishing about anti–plastic pollution advocacy in the 2016 edition of Pezzullo and Cox, *Environmental Communication*.

124. On reticulate publics see Hauser, *Vernacular Voices*.

125. Na'puti and Cruz, "Mapping Interventions," 15.

126. On the vernacular: Hauser, *Vernacular Voices*; Ono and Sloop, "Critique of Vernacular Discourse"; and Aguayo, "Bodies That Push Buttons Matter."

127. Fiesler and Proferes, "'Participant' Perceptions of Research Ethics."

128. Following the Association of University Presses' best practices, I tried to maximize benefits (e.g., donating the book's proceeds and providing compensation for figure permissions and interviews) and minimize harms (e.g., anonymizing, asking for permission, and leaving most sources out). It's complicated; see Clark-Parsons and Lingel, "Margins as Methods."

129. My first publication (Pezzullo, "Performing Critical Interruptions") was based on Warren County, North Carolina, known as the birthplace of the environmental justice movement, where *environmental racism* was coined; after years of participant observation, I lobbied with the community, creating the lobbying pamphlet used when securing funds to detoxify the landfill. Since then my praxis has involved protesting, petitioning, organizing, volunteering services (e.g., publicity, coauthoring inclusive policies, community teachins), reallocating resources through grants and donations, codesigning and cofacilitating trainings, consulting (NGOs, governments, and corporations), blogging, cocreating public art exhibits, speaking at public events, providing testimony, ongoing relations, and more.

130. Pezzullo, "Afterword," 188. De Onís subsequently wrote a similar alliteration, calling for "decarbonizing, decentralizing, democratizing, and decolonizing." De Onís, *Energy Islands*. Together, we have published: de Onís and Pezzullo, "Ethics of Embodied Engagement"; Pezzullo and de Onís, "Rethinking Rhetorical Field Methods." More recently: Liboiron and Lepawksy have encouraged research practices of "defamiliarization, denaturalization, decentering, and depurifying." *Discard Studies*, 132.

131. I am inspired by Angela Aguayo's Rural Civil Rights Project, summarized in "Bodies That Push Buttons Matter," and Karma R. Chávez's *Palestine on the Air*. Despite my limitations, my methodological touchstone remains Madison, *Critical Ethnography*.

132. On broader trends, see Eldridge and Ferrucci, *Institutions Changing Journalism*.

133. Kinkaid, Emard, and Senanayake, "Podcast-as-Method?"

134. The Just Transition Alliance, for example, has tweeted my podcast episode with their executive director: "#EarthDay."

1. #ThereIsNoAway: CARBON-HEAVY MASCULINITY AND THE LIFE/DEATH CYCLE OF PLASTICS

Epigraph: Pezzullo, "José Toscano Bravo."

1. Glennon, "Unfolding Tragedy in Bangladesh"; and Ahmed and Gotoh, "Impact of Banning Polythene Bags."

2. Pezzullo, "Sharir Hossain."

3. Environmental Justice Foundation, "Climate Displacement in Bangladesh."

4. Pezzullo, "Sharir Hossain."

5. *UN News*, "Climate Change as 'Threat Multiplier.'"

6. Chakrabarty, *Climate of History in a Planetary Age*.

7. Mahmud, "Cash and Carry On."

8. Clapp and Swanston, "Doing Away with Plastic Bags."

9. Sultana, "Polybag Ban Fails."

10. Antara, "The Plastic Pandemic."

11. Werft, "Eliminating Plastic Bags in Rwanda."

12. Dsilva, "Going Surgical on Plastics."

13. DDT is still used globally, though the phaseout continues. Banned in the United States in 1996 and globally in 2002, the last leaded gasoline was used in Algeria as late as 2021. Domonoske, "World Has Stopped Using Leaded Gasoline." The Montreal Protocol stipulates regulation of almost one hundred substances. UNEP, "About Montreal Protocol."

14. Geyer, Jambeck, and Law, "Production, Use, and Fate of Plastics."

15. Thompson, "Amazing History of Bendy Straw."

16. Chalmin, "History of Plastics."

17. Skager, "Plastic Grocery Bags Invented"; and Altman, "American Beauties."

18. UNEP, "Visual Feature."

19. World Economic Forum, "New Plastics Economy," 6.

20. Barthes, *Mythologies*, 97. Barthes was influenced by plastic's roots linguistically in Greek as *plastikos*.

21. Barthes, *Mythologies*, 99.

22. America's Plastic Makers & American Chemistry Council, "5 Actions for Sustainable Change." For critical analysis of ACC's campaigns see Fortun, "Essential2life"; and Pezzullo, "Contaminated Children."

23. Freinkel, *Plastic*. On turtles see Kearney, "When Plastics Saved Turtles."

24. Science History Institute, "Science Matters."

25. A pilot study funded by the Plastic Pollution Coalition, which is a project of Earth Island Institute, identified the frequency of single-use items in sixty-four episodes of thirty-two popular television shows. Weinstein, Rogers, and Rosenthal, *Flip the Script*.

26. Paystrup, "Plastics as a 'Natural Resource,'" 176.

27. Kana, "'World Biggest Plastic Pyramid.'"

28. On the love/hate relationship with toxics in US culture see Pezzullo, *Toxic Tourism*.

29. Bullard, *Unequal Protection*, 11.

30. International Institute for Sustainable Development, "Doubling Back and Doubling Down."

31. Charles, Kimman, and Saran, "Plastic Waste Makers Index." Formosa Plastics (see chapter 6) is listed in the top twenty. ExxonMobil and Dow also are counted among the top companies responsible for greenhouse gasses. Two research centers offer rankings: Climate Accountability Institute and the UMass Amherst Political Economy Research Institute.

32. These founding narratives operate as one genre of what Kuhn ("Negotiating the Micro-Marco Divide") calls "authoritative texts," devices providing insights into corporate authority or how organizational complexity about identity and appropriateness of aspirational or unfolding strategies are condensed into a common organizational sense of belonging. For more on organizational communication theories of narratives see Kuhn, "Communicatively Constituting Organizational Unfolding."

33. ExxonMobil, "Press Release."

34. ExxonMobil, "Our History."

35. Dow Chemical Company, "Compelling Investment."

36. Dow Chemical Company, "Golden Age of Inorganics."

37. Sinopec Holdings, "Our Story."

38. Strong, "Taiwanese Tycoon Needs 90 Minutes."

39. Sinopec Holdings, "Monthly Net Profits."

40. Alaimo, *Exposed*, 108.

41. Sze, *Environmental Justice*, 7.

42. Berlant, *Queen of America*, 2. Challenging "the invisibility of dominant masculinities" ideally enables analysis of "the socially constructed, institutionalized yet shifting form of masculinist identity that systematically dominates femininities and alternative masculinities." Previously, in *Homegrown*, hooks and Mesa-Bains argued: "We have to constantly critique imperialist white supremacist patriarchal culture because it is normalized by mass media and rendered unproblematic" (61). This is also reaffirmed in Ashcraft and Flores, "'Slaves with White Collars,'" 3.

43. Oxfam, *Confronting Carbon Inequality*, 8.

44. Alaimo, *Exposed*, 108.

45. On necropolitics see Mbembe, "Necropolitics"; and Fleetwood, "Failing Narratives."

46. Liboiron, "Plastics in the Gut."

47. A book launched at the same time that mine was going to press illustrates these relations clearly and with nuance: Liboiron and Lepawsky, *Discard Studies*. On the circuit of culture as a heuristic for studying environmental advocacy see Pezzullo, "Contextualizing Boycotts and Buycotts."

48. UN, "Drowning in Plastics," 34.

49. Ragusa et al., "Plasticenta"; Freinkel, *Plastic*; and Pezzullo, "Contaminated Children."

50. Geneva Environment Network, "Plastic Pollution Dialogues"; and Orellana, *Report on Implication for Human Rights*. Microplastics have been found in amounts ten times greater in baby feces than in adults'; see Zhang et al., "Polyethylene Terephthalate and Polycarbonate Microplastics."

51. Calil et al., *Neglected*, 24.

52. Odonkor and Gilchrist, "Transforming the Plastics Value Chain."

53. Global Alliance for Incinerator Alternatives (GAIA), "Plastic Pollution and Waste Incineration"; and Thompson et al., "Our Plastic Age."

54. Ribeiro-Broomhead and Tangri, "Zero Waste and Economic Recovery."

55. These impacts are well documented; see Union of Concerned Scientists, "Hidden Costs of Fossil Fuels." On US oil extraction and attachments see LeMenager, *Living Oil*.

56. Just Transition Alliance, "Lifecycle of Plastics."

57. Muttitt and Kartha, "Equity, Climate Justice."

58. Global Newswire, "Shale Gas Creating Renaissance."

59. Carpenter, "Consequences of America's Plastics Boom."

60. *Cracker plants* refer to facilities that heat ethane (part of gas) until it "cracks" into ethylene, which is used to make plastic. Southwest Pennsylvania Environmental Health Project, "Health behind Plastics Cracker Plants."

61. Cirino, "Plastic Pollution."

62. Just Transition Alliance, "Lifecycle of Plastics."

63. Altman, "Five Myths about Plastics."

64. Gribkoff, "In Depth."

65. Lusher et al., "Microplastics in Arctic Polar Waters."

66. For an app attempting to identify marine debris see "Marine Debris Tracker." This has been written about by research collaborators Jambeck and Johnsen, "Debris Data Collection and Mapping"; and Duncan et al., "Message in a Bottle."

67. UNEP, "Single-Use Plastics"; Wetherbee, Baldwin, and Ranville, "It Is Raining Plastic"; and Carrington, "Microplastic Pollution Revealed."

68. Allen et al., "Atmospheric Microplastics." Scientists have found that land-based disease-carrying microorganisms can attach to microplastics to transmit through seawater. Zhang et al., "Zoonotic Protozoan Parasites with Microplastics."

69. Helvarg, *Blue Frontier*, 3. It's regrettable that *Blue Frontier* uses a white settler frame ("frontier," "wilderness") to elevate the ocean's status. Helvarg himself has reported on killing buffalos and Indigenous peoples as part of the founding of the US frontier in *War Against the Greens*, 46.

70. "Dr. Vandana Shiva an Interview by Andy Opel," 498–99 (emphasis added).

71. For graphics summarizing global data on plastic marine litter see UNEP, *Drowning in Plastics*.

72. LeMenager, *Living Oil*, 191.

73. Na'puti, "Possibilities for Communication Studies," 95; see also Na'puti, "Archipelagic Rhetoric."

74. Quoted in Van Gelder and Shiva, "Earth Democracy."

75. For more about ongoing campaigns to protect water as sacred to life on Turtle Island, the Indigenous Environmental Network is a worthwhile starting place: www.ien earth.org.

76. Carrozza and Fantini, "Italian Water Movement," 100.

77. The Story of Stuff Project, "Story of Bottled Water."

2. HAVE A COKE AND A #FootprintCalculator: THE MYTH OF RECYCLING AND TRANSNATIONAL GREENWASHING

1. On the plastics-industrial complex's greenwashing, see Lerner, "Waste Only." On public relations and greenwashing, see Stauber and Rampton, *Toxic Sludge Is Good for You!*

2. Pezzullo, "Resisting 'Breast Cancer Awareness Month,'" 346; and Pezzullo and Cox, *Environmental Communication*, 93.

3. Reich, *Toxic Politics*, 235–51. These patterns persist. In 2021 ExxonMobil lobbyist Keith McCoy explained how plastics are handled by the company like climate, providing stories to justify delay and false solutions, arguing: "You can't ban plastics." Quoted in Carter, "Inside Exxon's Playbook."

4. TRAC, "Greenhouse Gangsters vs. Climate Justice."

5. Loepp, "Hashtags, Real Problems, Symbols."

6. Loepp, "Hashtags, Real Problems, Symbols."

7. Wilkins, "Recycling Won't Solve Plastic Pollution."

8. Wilkins, "Recycling Won't Solve Plastic Pollution."

9. Keep America Beautiful, "Mission & History." On W. Howard Chase, who led these efforts, see "Manipulating the Masses." On ongoing efforts to block bottle bills see Lerner, "How Coca-Cola Undermines Plastic Recycling."

10. *BBC News*, "#Trashtag"; and Greenspan, "Man Who Popularized #Trashtag-Challenge."

11. Tsiaoussidis, "MrBeast Smashes Team Seas Goal."

12. Toner, "He's Doing the 'Dirty Work'"; and Bhatia, "Dia Mirza and Afroz Shah."

13. Environmental advocates suggest Rs might also include Refuse, Repair, Rot, Regift and Recover. In *Things Worth Keeping*, Harold calls for fostering "sensuous enchantment" with commodities to "manage the material excesses of consumer capitalism by building on, rather than repudiating, our attraction and attachment to objects" (18).

14. Geyer, Jambeck, and Law, "Production, Use, and Fate of Plastics"; Calil et al., *Neglected*, 15; and USEPA, "Plastics."

15. Econie and Dougherty, "Contingent Work in US Recycling." Bill Keegan, president of a waste and recycling facility in Minnesota, is commonly credited for coining *wishcycling*. Altman cautions that the term may appear to blame individuals rather than systems; "On Wishcycling."

16. Freinkel, *Plastic*, 162.

17. Dunaway, *Seeing Green*, 98–101; and Wilson, "Student Designed Recycling Logo."

18. Dunaway, *Seeing Green*, 98, 101.

19. Venhoeven, Bolderdijk, and Steg, "Why Going Green Feels Good."

20. Dunaway, *Seeing Green*, 101; and van Doorn and Kurz, "Warm Glow of Recycling."

21. Ministry of Economy, Trade, and Industry, "Containers and Packaging Recycling Law"; and Package Recovery Organization Europe, "Green Dot Trademark."

22. *PBS Frontline*, "Plastic Wars."

23. Dunaway, *Seeing Green*, 247–48.

24. Sanchez, "History of Plastic."

25. Disinformation campaigns are "strategic courses of action enacted through communication that are undertaken deliberately to spread false data"; "Misinformation is the sharing of false ideas, whether or not intended." Pezzullo and Cox, *Environmental Communication*, 224, 223.

26. On the rhetorical significance of myths see Hernandez et al., "U.S. Offshore Wind Energy Communication"; Rushing, "Evolution of 'The New Frontier'"; and Moore, "Rhetorical Criticism of Political Myth."

27. Green consumerism is not simply a transaction—the purchase of a certain product— but involves a *discourse* about the identity of individual consumers in which purchasing is imagined as "an act of faith": "it is based on a belief about the way the world works." Smith, *Myth of Green Marketing*, 89.

28. Liboiron and Lepawsky, *Discard Studies*, 12.

29. Calma, "Coca-Cola Says People Want Its Plastic."

30. Multiple advertising companies claim credit. One designer claimed to have created four of the seven symbols. Jarvis, "Coca-Cola Live Positively." I note Ogilvy's and MCN's work below. Another designer claims POSSIBLE. Abbey, "Coca-Cola Live Positively Redesign." Some credit the Brazilian campaign to the PR company DPZ (Dualib, Petit & Zaragoza Propaganda). Ripoll, "'Live Positively'."

31. Massive Change Network, "Case Study"; and Ran, "CSR Special."

32. Ripoll, "'Live Positively.'"

33. Ripoll, "'Live Positively,'"108.

34. Cloud, *Control and Consolation*; Ahmed, *Promise of Happiness*; and Ehlers and Krupar, *Deadly Biocultures*.

35. Ehrenreich, *Bright-Sided*, 57–58.

36. Berlant, *Cruel Optimism*.

37. Infamously rearticulated by comedian Eddie Murphy in 1987, the slogan also can be interpreted to mean looking out for one's self-interest and keeping your judgments to yourself, or "Have a Coke and a smile and shut the f*ck up." Murphy's punchline recounts what comedian Richard Pryor said when asked what he thought of Coca-Cola spokesperson and comedian Bill Cosby calling to tell Murphy his comedy was too raunchy. Pryor's implication is that getting paid to sell soda is not uplifting the community either, so as a hypocrite, Cosby shouldn't criticize. Townsend, *Eddie Murphy: Raw*. More recently, some have noted the irony of Cosby encouraging "civility" at a time when he allegedly was drugging, sexually assaulting, or raping approximately sixty women. Lovett, "Murphy Fans Furious at Bill Cosby."

38. Coca-Cola's role in water scarcity and poor health habits is well-documented; see Schmidt et al., "Corporations, Obesity and Planetary Health"; Gómez, "Coca-Cola's Political and Policy Influence"; and Ciafone, "If '*Thanda Matlab* Coca-Cola.'"

39. Messe Frankfurt, "Coca-Cola Introduces World-First."

40. Massive Change Network, "Case Study."

41. A compelling study based on interviews with Indigenous plastic pollution decision makers. Fuller et al., "Plastics Pollution as Waste Colonialism."

42. Break Free From Plastic Movement, "Top Plastic Polluters."

43. Greenpeace, "Climate Emergency Unpacked," 3.

44. Break Free From Plastic Movement, *Missing the Mark*, 9. This ambitious report ranks seven of the most polluting corporations according to which most commonly offer false narratives of the plastics crisis, including "beach clean ups are a solution" and "packaging made from plastic collected from the ocean is solving pollution." Since then Coca-Cola has created a video about plastic bottle recycling featuring Bill Nye, the popular children's science educator. The Coca-Cola Co., "Coca-Cola Company and Bill Nye."

45. Lerner, "Africa's Exploding Plastic Nightmare."

46. Lerner, "Africa's Exploding Plastic Nightmare."

47. Lerner, "Africa's Exploding Plastic Nightmare."

48. For example, see Hilary, "Criticized for Plastic Pollution."

49. Ahmed, *Living a Feminist Life*.

50. This agreement is cited widely, including in Bazvand, "Importance of Foreign Investment Attraction."

51. BP, "Early History—1909–1924."

52. Pharr, *Time of the Right*, 11.

53. BP, "Early History—1909–1924." According to BP, Cambridge University Press is publishing four volumes on its history. The British government sold its shares in 1987. *Los Angeles Times*, "Government to Sell 31.5% BP Stake."

54. Meredith, "BP Beats First-Quarter Estimates."

55. BP, "Late Century—1971–1999."

56. Ogilvy and Mather's story of origin began in London in 1850 with a British man, Edmund Charles Mather. In 1921 his son hired a Scottish Irish man, Frances Ogilvy, who convinced the firm to hire his younger brother, David Ogilvy. David became known as the father of advertising; since then the firm has represented many fossil fuel companies. About a 1958 contract with Standard Oil that ended with Shell, Ogilvy claims: "The biggest account I have ever got was Shell." Ogilvy, *Confessions of Advertising Man*, 10. Ogilvy's puffery ignores his having been hired decades before through nepotism, declaring: "I had gone to New York and started an advertising agency. . . . My agency was an *immediate* and *meteoric* success" (emphasis in the original, 15). He died in 1999, the year his firm started contracting with BP. Brown, "BP Vote of Confidence."

57. Baar, "Global Role on Ogilvy's BP."

58. *PR Week*, "PR Week Awards 2001."

59. Holmes, "Taking BP Beyond Petroleum." Ogilvy Berlin also claims to have created "Coca-Cola's first major sustainability campaign in the UK"; "Ogilvy." This story became linked to Ogilvy's Coca-Cola "LGBTQ" campaign: "Love Cans—Coca-Cola."

60. Carpenter, "Abandoned 'Beyond Petroleum' Re-Brand"; and Meiners, "Ten Years Later."

61. On Rees's eureka moments, such as observing his new computer having a smaller "footprint" on his desk, see https://williamrees.org/footprint/ .

62. BP, "BP Carbon Footprint Calculator."

63. On public relations and oil in the twentieth century, see Ewen, *PR!*. For more recent research see Franta, "Early Oil Industry Disinformation"; Brulle, "Networks of Opposition"; and Supran and Oreskes, "Rhetoric and Frame Analysis."

64. "BP Ad: Carbon Footprint."

65. Learmouth, "How 'Carbon Footprint' Originated."

66. "Corporate Responsibility Report," 19.

67. Doyle, "Where Has All the Oil Gone?," 218.

68. Yoder, "Footprint Fantasy."

69. Kaufman, "Carbon Footprint Sham."

70. Schneider et al., *Under Pressure*, 107, 108; and Schwarz et al., "Hypocrite's Trap."

71. BP, "Drive down Carbon Footprint."

72. Pezzullo and Cox, *Environmental Communication*, 131.

73. There are gamification apps that claim to track successful impact; see Cook, *Cranky Uncle vs. Climate Change*.

74. Gilbert, "Against Commodification of Everything," 557, 551. This quote reminds me of Michael Pollan's "Cornucopia of Choices" segment in a documentary that was released the same year, Kenner, "Food, Inc.," during which he discusses the "illusion of diversity" at US grocery stores versus the fact that much of the processed food is "the clever rearrangement of corn."

75. Funding for disinformation campaigns is also excluded. Brulle, Aronczyk, and Carmichael, "Corporate Promotion and Climate Change."

76. Doyle, "Where Has All the Oil Gone?," 225.

77. Pattee, "Forget about Carbon Footprint"; and Pattee, "Leading Climate Scientist Katharine Hayhoe." Another model considers investments, organizational choices, and actions; see Nielson, et al., "Role of High-Socioeconomic-Status People."

78. McKibben, "Multiplication Saves the Day." Brownstein, Kelly, and Madva agree we should not frame individual and systemic choices as mutually exclusive or oppositional; "Individualism, Structuralism, Climate Change."

79. Nielsen et. al, "Role of High-Socioeconomic-Status People."

80. In "World's Richest People Driving Global Warming," Roston, et al. report:

> The single-most polluting asset, a superyacht, saw a 77% surge in sales last year. An 11-minute ride to space, like the one taken by Amazon founder Jeff Bezos, is responsible for more carbon per passenger than the lifetime emissions of any one of the world's poorest billion people. . . . One-tenth of all flights departing from France in 2019 were on private aircraft. In just four hours, those individually-owned planes generate as much carbon dioxide as an average person in the European Union emits all year. Fourth-fifths of the people on the planet never get on an airplane in their entire lifetime.

81. *PBS Frontline*, "Plastic Wars."

82. Douglas, *Purity and Danger*.

83. Sontag, *Under the Sign of Saturn*, 88.

84. Grossberg, *We Gotta Get Out*.

85. Pezzullo, "Contextualizing Boycotts and Buycotts," 126.

86. Pezzullo, "Contextualizing Boycotts and Buycotts," 132. Bsumek et al. identify four ways impure politics can play a pragmatic role as meaningful gestures, through "promoting

articulation and solidarity, interrupting dominant discourses, enacting alternative futures, and applying leverage at sites of decision making." "Strategic Gestures in Climate Change Rhetoric," 40.

87. West, *Transforming Citizenships*, 192.

88. Renegar and Sowards, "Contradiction as Agency," 16.

3. FROM #BanPlasticsKE TO #ISupportBanPlasticsKE: PISSED OFF ONLINE, PICTURING PARTICIPATION, AND POLICING POLLUTION IN KENYA

Epigraph: Institut Open Diplomacy, *Stepping Up Climate Action*; Pezzullo, "Her Excellency Prof. Judi Wakhungu."

1. During that time I received intensive language training in Kiswahili, though my aptitude has rusted over time. The archive and interviews in this chapter originally appeared in English because that is the predominant language of social media and government in Kenya, a legacy of British colonialism (1895–1963).

2. As I write, thirty-eight of fifty-four countries in Africa have legislated some form of a national single-use plastic ban, including Benin (2017), Botswana (2017), Burkina Faso (2014), Burundi (2018), Cameroon (2014), Cape Verde (2016), Chad (2005), Cote de'Ivoire (2014), Djibouti (2016), Egypt (2017), Eritrea (2005), Ethiopia (2016), Gabon (2010), Gambia (2015), Ghana (2015), Guinea-Bissau (2016), Ivory Coast (2013), Kenya (2017), Madagascar (2015), Malawi (2015), Mali (2013), Mauritania (2013), Mauritius (2016), Morocco (2015), Niger (2014), Nigeria (2014), Republic of the Congo (2012), Rwanda (2008), Senegal (2015), Seychelles (2017), Somalia (2015), South Africa (2004), Togo (2018), Tunisia (2017), Uganda (2007), United Republic of Tanzania (2006, 2019), Zambia (2019), and Zimbabwe (2010). More African nations have regional bans, charges/deposits, or bans forthcoming. Sources were cross-referenced in Nyathi and Togo, "Overview of Framework Approaches."

3. For a digitally sophisticated interactive media archive of plastic bags, see Tsing et al., "Plastics Saturate US."

4. To read the ban, see NEMA (@NemaKenya), "Kenya Gazette."

5. During her tenure (2013–2018) as Kenyan cabinet secretary for environment and natural resources, Wakhungu enacted a wide range of landmark legislative measures regarding water, climate, and waste. Her first legislative victory was the Wildlife Conservation and Management Act 2013 (No. 47 of 2013).

6. Smith, "Kenya Burns Ivory Stockpile"; and Craig, "Africa's Elephant Ivory."

7. "Kenya Ivory Amnesty." Wakhungu emphasizes the international pressure for unsustainable poaching, not Indigenous hunting. There is a difference between the legacy of foreign white conservation in Kenya and local Indigenous practices. On the long history of the dangers of the myth that whites have been conservation saviors and Blacks "poachers" or unable to manage wildlife, see Mbaria and Ogada, *Big Conservation Lie*, 240. On how this colonial pattern continues to impact conservation, see Asher, "Next Big Green Lie." On this pattern in the United States see Jacoby, *Crimes against Nature*.

8. Indigenous tribes in the region had lived alongside wildlife for generations. The marginalization and regulation of Maasai particularly is controversial; see Byaruhanga and Onyiego, "Nashulai."

9. Osman, "Kenya's War on Plastic."

10. Behuria, "Ban the (Plastic) Bag?," 8.

11. For more on the Greenbelt Movement, see "Who We Are."

12. Ochieng, *Groundwork for Good Life*, 180.

13. Chirindo, "Bantu Sociolinguistics," 456.

14. Lantern Books, "Wangari Maathai Talks."

15. Quote from Lantern Books, "Wangari Maathai Talks."

16. Behuria, "Ban the (Plastic) Bag?," 9.

17. Behuria, "Ban the (Plastic) Bag?," 10. Wakhungu also emphasized the power of KAM's lobby in the podcast. Pezzullo, "Her Excellency Prof. Judi Wakhungu."

18. Ochieng, *Groundwork for Good Life*, 179.

19. UNEP, "Meet James Wakibia"; McCarthy, "Fed Up with Plastic"; and Carroll, "Helping to Solve Kenya's Waste Problem."

20. Tabuchi, Corkery, and Mureithi, "Big Oil Is in Trouble."

21. Behuria, "Ban the (Plastic) Bag?," 11, 12.

22. Pezzullo, "James Wakibia."

23. Pezzullo, "Her Excellency Prof. Judi Wakhungu."

24. Maathai, *Unbowed*, 73, 515.

25. Maathai, *Unbowed*, 515. On the Japanese term *mottaini*, Maathai expanded the initial intent of stopping wastefulness through reducing, reusing, recycling, and respecting to include connections to her broader values of "human rights, social justice, and overall peacebuilding." Kinefuchi, "Chapter 8," 153.

26. Gorsevski, "Wangari Maathai's Emplaced Rhetoric," 12.

27. Meiu, "Panics over Plastics," 233. See the introduction on how academics invoke "moral panics."

28. Maathai, *Unbowed*, 272.

29. Quoted in Lacey, "Flower of Africa."

30. Calling plastic bags "African flowers" appears to have been a vernacular saying in Kenya long before anyone cited it in print. Burrows, "Plastic Bag."

31. Quoted in Lacey, "Flower of Africa."

32. Lacey, "Flower of Africa."

33. This is not just a pattern in Kenya. In "American Beauties," Altman writes: "Now there are Instagram accounts about bags in the wild, for example, @ISpyABag, and hashtags to track bags' migrations, e.g., #HookedPlastic, #BagsInTrees, and #WitchesKnickers." Hashtags changed to camel case for accessibility.

34. "Flying Toilets in Every Direction." No longer online, a photo-essay with first-person testimony was published: *BBC News*, "In Pictures: Flying Toilets."

35. Watts, "Eight Months on, Is the World's Most Drastic Plastic Bag Ban Working?"

36. UNEP, "Economic Instruments," 23.

37. UNEP, "Economic Instruments,"15.

38. UNEP, "Economic Instruments," 16.

39. UNEP, "Economic Instruments," 16.

40. Njuguna, "Efficacy of the Ban," 104.

41. "Ban on Polythene Bags?"

42. UNEP, "Economic Instruments," 56.

43. Pezzullo, "Her Excellency Prof. Judi Wakhungu."

44. "Meet James Wakibia."

45. Pezzullo, "James Wakibia."

46. Pang and Law, "Retweeting #WorldEnvironmentDay," 55.

47. Sebeelo, "Hashtag Activism," 96.

48. Silver and Johnson, "Majorities Own Mobile Phones."

49. Silver and Johnson, "Majorities Own Mobile Phones."

50. "Kenya has enjoyed at least two decades of democratic transition since the election of Mwai Kibaki in 2002." Bosch, Admire, and Ncube, "Facebook and Politics in Africa," 352.

51. Bosch, Admire, and Ncube, "Facebook and Politics in Africa," 354.

52. Okune, "Open Ethnographic Archiving," 7.

53. Pezzullo, "Her Excellency Prof. Judi Wakhungu."

54. Quantitative Twitter data verified through paid services from Track My Hashtag.

55. Rentschler, "Bystander Intervention"; Skoric et al., "Social Media and Citizen Engagement."

56. Papacharissi, *Affective Publics*, 27; and Papacharissi, "Structures of Storytelling."

57. Papacharissi, *Affective Publics*, 311.

58. Papacharissi and de Fatima Oliveira, "Affective News and Networked Publics," 14.

59. Papacharissi, "Structures of Storytelling," 5.

60. Pezzullo, "James Wakibia."

61. Pezzullo, "James Wakibia."

62. Pezzullo, "Her Excellency Prof. Judi Wakhungu" (emphasis in original).

63. Pezzullo, "James Wakibia."

64. Pezzullo, "James Wakibia."

65. Pezzullo, "Her Excellency Prof. Judi Wakhungu."

66. Pezzullo, "James Wakibia."

67. Pezzullo, "James Wakibia."

68. Avaaz.org petition that Wakibia started on October 1, 2015, titled "Kenya Should Ban Single Use Plastic Bags," to the Government of Kenya Cabinet Secretary of Environment, National Environment Management Authority, Board of Governors, Plastic Bag Manufactures, Supermarkets, and UNEP. It closed after the ban was announced in March 2017.

69. Pezzullo, "James Wakibia."

70. Pezzullo, "James Wakibia."

71. For example, see Bollen, Pepe, and Mao, "Modeling Public Mood and Emotion"; Römpke, Fritsche, and Reese, "Get, Feel, Act Together"; and Postmes and Brunsting, "Collective Action in Age of Internet."

72. As I write, there are sixty-seven images on the Facebook page for IsupportbanplasticsKE (@banplasticsKE). They do not include statements; however, they have shaped my understanding of the range of images used during the campaign, which is consistent with Wakibia's Twitter feed and online research on the hashtag.

73. Seigworth and Gregg, "Inventory of Shimmers," 14.

74. Pezzullo, "James Wakibia."

75. *BBC News*, "Kenya Plastic Bag Ban."

76. Pezzullo, "Her Excellency Prof. Judi Wakhungu."

77. Njuguna, "Efficacy of the Ban," 101.

78. Houreld and Ndiso, "World's Toughest Law against Plastic."

79. National Environment Management Authority (NEMA), "Court Upholds Plastic Bag Ban."

80. Wakibia, "Officially Closing My #BanPlasticsKE Campaign."

81. Pezzullo, "James Wakibia."

82. National Environment Management Authority (NEMA), "2 Years On."

83. Watts, "Is Most Drastic Ban Working?"

84. Numbers derived from Statistica.com, "Extreme Poverty Rate in Kenya 2017–2021"; and Valle, "Exxon, Chevron Paid Their CEOs."

85. For these reports and more from Pellow's research team, see Global Environmental Justice Project, UC Santa Barbara, "What Is the GEJP?" Turning to incarceration is part of a broader "critical environmental justice studies" shift. Pellow and Brulle, *Power, Justice, Environment*, 3.

86. Global Environmental Justice Project, UC Santa Barbara, *Environmental Justice Struggles*, 57. The report cites Ondieki, "Kenya's Criminal Justice System."

87. Davis, *Are Prisons Obsolete?*, 107.

88. Quoted in brown, *We Will Not Cancel Us*, 1; see also Gilmore, Bhandar, and Toscano, *Abolition Geography*.

89. Rentschler, "Bystander Intervention."

90. National Environment Management Authority (NEMA), "2 Years On."

91. Boniface Mwangi, "Poor Traders Will Be Jailed."

92. Wathuti, "The poor have a voice too."

93. NEMA Kenya (@NemaKenya), "Nema Arrested Three People Using the Banned Plastic Bags"; and NEMA Kenya (@NemaKenya), "Two People Were Arrested in Syokimau, Machakos."

94. News Moto, "NEMA Arrest a Major Distributor"; and NEMA Kenya (@NemaKenya), "Today Morning @NemaKenya Together with DCI Officers."

95. Kahiu, *Pumzi*.

96. Wachira, "Maathai's Environmental Afrofuturist Imaginary."

97. One Planet Network, "Reducing Plastic Pollution," 79.

98. Pezzullo, "James Wakibia."

99. Clean Up Kenya, "Statement from James Wakibia."

100. UNEP, "Kenya Bans Single-Use Plastics."

101. More on this effort may be found at Kenya Plastics Pact.

102. Ochieng, *Groundwork for Good Life*, 201.

103. Büscher, *Truth about Nature*.

104. "Meet James Wakibia."

4. ENGAGING #StrawlessInSeattle AND #StopSucking: THE LONELIEST WHALE, SPORTING FUN, AND AMERICAN EXCEPTIONALISM

Epigraph: Kimmerer, *Braiding Sweetgrass*, 58.

1. Figginer's video has over 84 million views: The Leatherback Trust, "Plastic Straw Removed from Sea Turtle's Nostril."

2. While Moore's naming raised awareness, it contributed to imagining plastics as inert. Alaimo, *Exposed*, 135–38.

3. *Business Insider*, "Adrian Grenier Asks #StopSucking On Plastic Straws."

4. "Lonely Whale (Philanthropy)."

5. Lonely Whale, mission.

6. #SkipTheStraw sometimes was used with #StopSucking, suggesting synergy. The conclusion elaborates on the Ocean Conservancy.

7. Wu, "Q&A"; and Seattle Public Utilities, "Straws & Utensils."

8. Abidin et al., "Tropes of Celebrity Environmentalism."

9. Hall, "Deconstructing 'the Popular,'" 453.

10. The initial conjecture was that 52 might be a new species. The current educated guess is that 52 is a hybrid of two whale species, and there might be more than one. Zeman, *Loneliest Whale*.

11. This chapter focuses on dominant US culture to provide the most relevant context for Lonely Whale. For an Indigenous perspective see Na'puti, "Archipelagic Rhetoric" and "Ocean Possibilities for Communication Studies."

12. Whiting, "How Humans Have Affected Whale Populations."

13. International Whaling Commission, "Welcome to IWC Web Archive."

14. On blue ecology see Alaimo, *Exposed*; and Ayana Elizabeth Johnson, "Projects."

15. McQuay and Joyce, "It Took a Musician's Ear."

16. O'Dell, "Songs of the Humpback Whale."

17. Rothenberg, "Nature's Greatest Hit."

18. "Declaration of Rights for Cetaceans."

19. On whale tourism and marine life communication see Milstein, "Performer Metaphor"; Schutten and Burford, "'Killer' Metaphors"; and Burford and Schutten, "Internatural Activists.'"

20. Galoustian, "Beluga Whales Form Social Networks."

21. The story opened with the line "Humans are social animals, and the lifeblood of society is conversation." Revkin, "Song of Solitude."

22. Revkin's story noted that some Deaf people wondered if the whale might be Deaf, which inspired the children's book Kelly, *Song for a Whale*.

23. Jendukie, "BTS (방탄소년단)—Whalien 52."; PotatoMcFry. "Doreen the Whale"; and Sofar Sounds. "52 Hertz Whale."

24. Many express loneliness for a reason (like a breakup). When I reference "affect" as a structure of feeling, I'm signaling a broader transnational felt sense of identification with 52 that shapes social arrangements.

25. Zeman experienced "a particularly difficult breakup" when he read Revkin's story. Pritchett, "Leonardo DiCaprio's New Documentary." A book should be written about when breakups spark social movements.

26. Approximately fifty-eight thousand whales were killed to create oil for Britain and its allies in World War I. Imperial War Museums, "Why Whales Were Vital."

27. Zeman, *Loneliest Whale*.

28. Lonely Whale, "Against the Current with Adrian Grenier." This idea has been studied; see Nuojua, Pahl, and Thompson, "Ocean Connectedness and Consumer Responses."

29. Ives, "Gateway Plastic."

30. These historical landmarks are repeated often; see Footprint, "History of Straws."

31. Thompson, "Amazing History aof Bendy Straw." Bernard's patent application is publicly available at https://patents.google.com/patent/US2094268A/en.

32. Quito, "Bendy Plastic Straw Used in Hospitals." The idea of "universal design" wasn't coined until 1985 by Ronald Mace. Hamraie, *Building Access*, 6.

33. Lonely Whale, "What We've Achieved So Far."

34. For Ocean Conservancy CleanUp Reports see https://oceanconservancy.org/trash-free-seas/international-coastal-cleanup/annual-data-release/. For a campaign raising awareness about cigarette butts as plastic pollution see https://uofttrashteam.ca/cigarettebutts/.

35. Lonely Whale cites Cress often, including in "Understanding Plastic Pollution." This repeated statistic of 500 million straws was estimated by Cress in 2011 when he was nine years old for a school report, leading to his launch of the Be Straw Free Campaign. The statistic is greatly debated. One estimate is as low as 170 million per day and another 390 million per day. Cress has emphasized that it is difficult to estimate with corporate secrecy, but the campaign is less about precision than about recognizing that single-use consumption adds up. Chokshi, "Statistic Shaped Debate on Straws."

36. Brueck, "Cities and Businesses Banning Plastic Straws."

37. Doucette, "Plastic Bag Wars."

38. Ives, "Gateway Plastic."

39. Brueck, "Cities and Businesses Banning Plastic Straws."

40. *Business Insider*, "Adrian Grenier Asks #StopSucking on Plastic Straws."

41. Kandel, "Stages in Adolescent Drug Use."

42. One helpful summary of gateway theory is Schwandt et al., "Integrated Gateway Model."

43. Jordan, "Do Plastic Straws Make a Difference?" Richard Stafford and Peter J. S. Jones argue that climate and overfishing pose more of a crisis than plastic pollution in marine environments: "Ocean plastic can provide a convenient truth that distracts us from the need for more radical changes to our behavioural, political and economic systems . . . as well as the cause of plastic pollution, i.e. over-consumption." Stafford and Jones, "Viewpoint," 187.

44. In this pilot study, plastic bag bans decreased this behavior, whereas voluntary reuse of bags was more likely to lead to someone buying a spontaneous treat. Karmarkar and Bollinger, "BYOB." If this spillover (of candy or bag of chips) is packaged in single-use plastic, the study suggests another way single-use plastics are reduced through plastic bag bans.

45. Nudges may reduce support for carbon taxes; see Hagmann, Ho, and Loewenstein, "Support for a Carbon Tax"; and Maki, "Potential Cost of Nudges." These essays do not engage the climate justice movement critique of carbon taxes.

46. Truelove et al., "Positive and Negative Spillover"; Carrico, "Climate Change and Spillover"; and Truelove et al., "Plastic Bottle Recycling to Policy Support."

47. Ives, "Gateway Plastic."

48. Freinkel wrote about psychology and nudging as it relates to plastics (*Plastic*, 168–69). On peer influence see Graziano and Gillingham, "Spatial Patterns of Solar Photovoltaic System Adoption."

49. Liebe, Gewinner, and Diekmann, "Effects of Green Energy Defaults."

50. Lonely Whale, mission.

51. Carpenter, "Consequences of America's Plastics Boom."

52. Auxier and Anderson, "Social Media Use in 2021."

53. Cook, "Seattle a Top Social Media City."

54. Each of these pledges was touted as a success by Lonely Whale.

55. Market-based solutions and environmental advocacy have long collaborated. DeLuca, "Trains in the Wilderness."

56. Asen, "Neoliberalism and Public Good" and Pezzullo and Cox, *Environmental Communication*; Pezzullo, "Contextualizing Boycotts and Buycotts"; Mukherjee and Banet-Weiser, *Commodity Activism*; Luxon, "Economics-Oriented Discourse Strategies"; Miller and Lellis, "Audience Response to Marketplace Advocacy"; Littler, *Radical Consumption*; Jennings, Allen, and Phuong, "More Plastic Than Fish"; and Mukherjee and Banet-Weiser, *Commodity Activism*.

57. Pezzullo and Cox, *Environmental Communication*, xx.

58. Tombleson and Wolf, "Rethinking the Circuit of Culture," 15. In "Making Fame Ordinary," Littler argues that celebrity culture also is fueled by desires that arise from economic inequities.

59. ALS is a degenerative disease that impacts nerve cells in the brain and the spinal cord to prohibit nourishment of muscles. It also is known as Lou Gehrig's disease, after the professional baseball player who had been very successful retired at age thirty-nine with ALS.

60. ALS Association, *Evaluation of ALS Grant Programs*.

61. Doyle, Farrell, and Goodman, *Celebrities and Climate Change*.

62. Pezzullo, "Hello from the Other Side." On defining "sexy" and the risks of popular culture eclipsing structural transformation, see Pezzullo, "Articulating 'Sexy' Anti-Toxic Activism on Screen."

63. Gössling, "Celebrities, Air Travel, Social Norms."

64. "Most Innovative Companies."

65. Shorty Awards, "#STOPSUCKING Silver Distinction."

66. Lonely Whale, *Celebrating 5 Years*.

67. The organization seems to bend toward London/United Kingdom, with the Sucker Punch campaign more toward New York and Los Angeles audiences. Shorty Awards, "#STOPSUCKING Silver Distinction."

68. Kane, "Lonely Whale."

69. Dittoe PR, "Dittoe PR Takes Over Seattle."

70. Pezzullo, "CU Boulder Alumna Emy Kane." As an alumna of the campus where I work, Kane kindly spoke with me during the pandemic. I have not eaten meat since I was nine nor fish since I was eighteen (when I moved out of my parents' home), and I have always loved the ocean; however, I have lived in landlocked states to date and have identified as an environmental or climate justice advocate more than as an ocean conservationist.

71. Lonely Whale Foundation, "Show Me the Work."

72. AdCouncil praise is noted by Mitchell, "Lonely Whale Launch #StopSucking Campaign."

73. Lonely Whale, "What We've Achieved So Far.

74. Lonely Whale, "Sucker Punch."

75. Pezzullo, "CU Boulder Alumna Emy Kane."

76. The PSA was created by a creative agency: "Lonely Whale Foundation: #Stop-Sucking by POSSIBLE."

77. Shorty Awards, "#STOPSUCKING Silver Distinction."

78. Lonely Whale, "#StopSucking for a Strawless Ocean."

79. Ives, "Hey, Could You #StopSucking?" The most liked Instagram post of 2016 was a Coca-Cola ad featuring singer Selena Gomez with a plastic straw and soda bottle. Baxter-Wright, "Most Liked Instagram Photos 2016."

80. Ray Page, quoted in Mitchell, "Lonely Whale Launch #StopSucking Campaign."

81. Consider the prosopopoeial short from Conservation International, "Nature Is Speaking, or Carson's argument in *Sea Around Us*: "The sea, though changed in a sinister way, will continue to exist; the threat is rather to life itself" (xiii).

82. Pezzullo, "CU Boulder Alumna Emy Kane."

83. Pezzullo and Cox, *Environmental Communication*, 227–28.

84. Brouwer, "Risibility Politics," 232, 232, 234.

85. Lonely Whale, "#StopSucking for a Strawless Ocean."

86. Ministerio Del Medio Ambiente, "Campaña Ciudadana Chao Bombillas."

87. Lonely Whale, "#HydrateLike."

88. Lee, "Last Straw?"

89. Pompeo, "@theellenshow Did You Know Humans Use 500 Million Plastic Straws a Day?"

90. Grenier, "Kicking off #StrawlessInSeattle."

91. LEW, "Lonely Whale and Grenier Partner."

92. *Valeur Magazine*, "Strawless In Seattle."

93. Lonely Whale, "Strawless In Seattle."

94. Oprah, "Oprah's Favorite Things 2018." Known for gifting and providing reusable straws in her interviews, Oprah's line riffs off the popular 1980s public safety PSA, "Friends don't let friends drive drunk."

95. "Walking the Talk in Vanuatu."

96. McCarthy, "Taiwan Announces Ban"; Everington, "KFC Taiwan Stops Serving Plastic Straws"; and Morgan, "Taiwan's Plastic Straw Ban."

97. Knorr, "7 Countries, States, and Cities"; and Picchi, "Ikea."

98. Stossel, "John Stossel."

99. Carlson, "Tucker Carlson."

100. Carlson and I disagree on most matters, as my environmental justice commitments are antithetical to his embrace of white nationalism and authoritarian leadership, summarized in Duignan, "Tucker Carlson."

101. The 2017 Ocean Conservancy report has been retracted and removed, for reasons noted in the conclusion. The damaging impression was global; see Dunning, "Refusing a Straw with Your Cocktail."

102. Charles, Kimman, and Saran, "Plastic Waste Makers Index."

103. Charles, Kimman, and Saran, "Plastic Waste Makers Index."

104. Jambeck et al., "Plastic Waste Inputs"; Law et al., "United States' Contribution of Plastic Waste"; and Holden, "US Produces Far More Waste."

105. As Jason Hinkel established in "Quantifying Responsibility for Climate Break-down": "The Global North was responsible for 92%. By contrast, most countries in the Global South were within their boundary fair shares, including India and China (although China will overshoot soon)."

106. Ashcraft, *Wronged and Dangerous*, 4. The argument ("while the US pollutes, China is worse, so the US is not the problem") also exemplifies the commonplace fallacy of relative privation ("while x might appear bad, y is worse, so x isn't really bad").

107. Mercieca, "Guide to Trump's Dangerous Rhetoric."

108. Johnson, *I the People*, 178.

109. On Trump's hashtagged homonationalist affects see Hatfield, "Toxic Identification."

110. Edwards, "Make America Great Again," 190.

111. Ott and Dickinson, *Twitter Presidency*. Trump was banned from the platform after January 6, 2021, for "risk of further incitement of violence." Twitter, "Permanent Suspension of @realDonaldTrump."

112. A classic study of the red-baiting, homophobia, and more insults of environmental backlash discourse is Helvarg, *War against the Greens*.

113. For some, "Make America Great Again" offers a nostalgic slogan for when the United States became a leading global superpower after World War II, a time when white supremacy, heterosexuality, and patriarchy were imagined as less inhibited by, well, the rights of everyone else.

114. Parscale, "I'm so over Paper Straws. #LiberalProgress."

115. Schaltegger, "Trump Sold $200,000 of Plastic Straws"

116. Rosenberg, "Turtle May End Up."

117. I am sharing this and the next two posts anonymously without platform or exact date. These posts are emblematic of the vernacular of those who bought Trump straws in 2019. They are quoted verbatim instead of fabricated (which some social media researchers do to further obscure sources) because, given the Republican Party's concerns over "fake news," it seems important not to create fake posts. Hashtags were edited into camel case for accessibility.

118. *Business Insider*, "Responding to Trump's Plastic Straws."

119. *CNN*, "Climate Crisis Town Hall"; Locker, "Why Straw Bans Are Straw Dogs"; and Bradner, "Plastic Straws Subject of Culture War."

120. Justice, "Joe Biden in Favor of Banning Straws."

121. Curated by Patrick Chandler and the Arbor Institute, the exhibit, held in Boulder, Colorado, in 2021. included artwork from Chandler, David Oonk, Kelsi Nagy, Chris Jordan, and Marcus Eriksen.

122. Eriksen, "Opinion"; and Eriksen et al., "Camels Eating Plastic Waste."

123. Pezzullo, "Hello from the Other Side," 805.

124. Lugones, "Playfulness."

125. Osnes, Boykoff, and Chandler, "Good-Natured Comedy," 2.

126. Ewert, "Moving beyond the Obsession."

127. Pezzullo, "CU Boulder Alumna Emy Kane."

128. Eli Pariser coined *filter bubble*. He served as executive director of MoveOn.org and cofounded Avaaz (the platform Wakibia used to circulate his petition), both initiatives that helped define digital engagement for contemporary social movements. Though filter bubble

is useful, Pariser has argued that we need to move our media approach beyond exposure to "how we come into contact." Schiffer, "'Filter Bubble' Author Eli Pariser." On Pariser see "Eli Pariser." For a longer historical perspective see Striphas, *Algorithmic Culture*.

129. On the problem of both-siding climate see Boykoff and Boykoff, "Balance as Bias."

130. Justice, "Joe Biden in Favor of Banning Straws."

5. #SuckItAbleism INTERVENES: ECO-NORMATIVE SHAMING, VOICING JUSTICE, AND PLANETARY FATALISM

Epigraph: Wong, "The Rise and Fall of the Plastic Straw," 5 (1): 3.

1. On medical uses of plastic straws see Danovich and Godoy, "Why People Want Bans to Be Flexible."

2. Ferdman, "Things Learned about Plastic Straw Bans."

3. Wong was born with spinal muscular atrophy: Wong, "Disability Visibility Project." Wong has written, consulted, and advocated extensively; she founded the Disability Visibility Project in 2014, which partners with the US Library of Congress StoryCorps to collect oral histories of disabled people. Wong explains the name of her organization thus: "The usage of the word 'visibility' in the project name is metaphorical. It is not meant to privilege one sensory experience over others" (Disability Visibility Project, "About"). I have written about ocularcentrism in Pezzullo, *Toxic Tourism*. For more on "visibility" as a disability argument see Johnson and Kennedy, "Introduction," 161.

4. As noted in the introduction, following the US disability justice movement, I favor first-person language ("disabled people") rather than people-first language ("people with disabilities"); I use people-first language when engaging people or research that does.

5. Fraser identifies this type of discourse as vital to counterpublics: "where members of subordinated social groups invent and circulate counterdiscourses to formulate oppositional interpretations of their identities, interests, and needs." Fraser, "Rethinking the Public Sphere," 123. In "'One Tweet Make So Much Noise,'" Ellcessor argues that digital media platforms can allow people with disabilities to "build communities, challenge assumptions, and provide more resonant and complex representations of disability," 259.

6. Hanson, "Straw Wars."

7. A longer engagement with the relationships between the disability movements and environmental movements exceeds this book; while writing, I coauthored an essay to think through some of those tensions and possibilities with disabled colleagues. Cram, Law, and Pezzullo, "Cripping Environmental Communication." Currently, I do not identify as having permanent intellectual disabilities, mental illness, or sensory disabilities, though I have lived with poor eyesight, allergies, respiratory challenges, and other health complications. I pass as and benefit from the privilege of ableism.

8. Di Chiro, "Polluted Politics."

9. Charlton, *Nothing about Us without Us*, 3.

10. Available widely, including "Principles of Environmental Justice."

11. Alston and Panos Institute, *We Speak for Ourselves*.

12. "Jemez Principles for Democratic Organizing."

13. "Principles of Working Together." Full disclosure: I was one of the few people in the room who helped draft these principles.

14. Watts, "'Voice' and 'Voicelessness' in Rhetorical Studies," 192 (emphasis added). Although Watts initially did not intend to think about voice beyond the human, his work also speaks to the value of listening to nonhuman voices: Watts, "Coda."

15. Sowards, ¡Sí, Ella Puede!, 156.

16. Heumann, *Being Heumann*, ix. She also has written a young adult version: Heumann and Joiner, *Rolling Warrior*.

17. A worthwhile documentary on this movement showing values of community, pleasure, and coalitions is Lebrecht and Newnham, *Crip Camp*.

18. Heumann, *Being Heumann*, 23.

19. Piepzna-Samarasinha argues in *Care Work*: "Disability justice is to the disability rights movement what the environmental justice movement is to the mainstream environmental movement" (22). On the relationship between the environmental and environmental justice movement see Sandler and Pezzullo, *Environmental Justice and Environmentalism*.

20. Sins Invalid, "Principles of Disability Justice."

21. Blackpast, "(1982) Audre Lorde."

22. Hamraie, *Building Access*, 5.

23. *Care & Climate*, 9.

24. Chen, "Toxic Animacies, Inanimate Affections," 274.

25. Piepzna-Samarasinha, *Care Work*, 134.

26. Wong, "Rise and Fall of Plastic Straw," 2.

27. On the disability activist history of smashing sidewalks in protest, which leads to curb cuts, see Hamraie, *Building Access*, 95–130. Hamraie argues that "curb cuts are politically, materially, and epistemologically adaptive technologies around which two distinct approaches to disability inclusion—liberal, assimilationist positions and crip, anti-assimilationist positions—have cohered" (99).

28. Pezzullo, "Joe Andenmatten."

29. Escobar, *Designs for the Pluriverse*.

30. Hautea et al., "Showing They Care (Or Don't)."

31. Watts, "Climatologist Michael E. Mann.'"

32. Quoted in Freinkel, *Plastic*, 165.

33. UNEP, "Latin America Wakes Up."

34. I am quoting this anonymously since it is a common joke. Also consider this headline from the satirical online magazine *The Onion* in 2010: "'How Bad for the Environment Can Throwing Away One Plastic Bottle Be' 30 Million People Wonder."

35. In the context of climate, Heglar claims disavowal of personal consumption in the United States suggests "moral bankruptcy of the highest order. "I Work in the Environmental Movement."

36. Other single-use plastics for medical uses are also under scrutiny. For example, on average, a hysterectomy in the United States involves twenty pounds of waste, mostly plastic. Thiel et al., "Environmental Impacts of Surgical Procedures."

37. This comparison arose again on social media when I was finalizing my book, illustrating the endurance of plastic straws as an articulator of crisis. The following tweet extends Chico Mendes's oft-quoted critique about class struggle and gardening: Jason Hickel (@jasonhickel), "Environmentalism without class struggle is using paper straws

while the rich take 9 minute flights in their private jets." He then was retweeted by Ives-Rublee. See Mia Paddle (@SeeMiaRoll), "As @SFdirewolf says, #SuckItAbleism." Many criticized celebrity Taylor Swift during this weekend; these two are illustrative discourse from verified accounts.

38. Heglar, "Home Is Always Worth It."

39. On disability hashtags see Ellis, "#Socialconversations"; Ellis and Goggin, "Disability and Media Activism"; and Ellis and Goggin, "Disability Media Participation."

40. Walker, "#CripTheVote," 149.

41. Ellcessor, "'One Tweet Make So Much Noise,'" 259.

42. Walker, "#CripTheVote," 167.

43. The website is https://millionsmissing.meaction.net/.

44. Pezzullo and Cox, *Environmental Communication*, 288.

45. Johnson, "Breaking Down."

46. Rodriguez, "Alice Wong Says #suckitableism"; and Hanson, "Straw Wars."

47. Wong, "Last Straw." Although Wong does not name Lonely Whale here, she cites Ives in another article, and her hashtag is clearly a play on their message. Wong lists herself as a consultant for Starbucks. Disability Visibility Project, "Hire Me!" Throughout her advocacy, Wong calls for listening and making more space for disabled people as not just tolerable but valued.

48. Wong, "Rise and Fall of Plastic Straw," 4.

49. Soto, "Plastic Activists Need to Know."

50. DeFrane, "Single-Use Activism." The writing bio is from The New Twenties, "Writers' Co-Op."

51. These quotes are shared anonymously; both are frequently used lines of argument in person and online. Although they were ignored in most (all?) news articles on the topic, restaurant and other food industry workers repeatedly found themselves on the frontlines of single-use plastic controversies, and it would be worthwhile to interview them on these exchanges and their agency or role. #StrawShaming appears to have been started by people who are not disabled. While some overlap occurred with conservatives, bots, in jest, and more, in this chapter I focus only on perspectives that trended among multiple disabled people and accessibility advocates.

52. Ochieng, *Groundwork forGood Life*, 104.

53. Jacquet, *Is Shame Necessary?*, 18–19. As an exemplar of greenwashing and pinkwashing, I have written about how the founders of Susan G. Komen profit from the entire cancer life cycle: producing cancer (through toxic pesticides), detecting cancer (through creating mammogram machines), and treating cancer (through creating a leading treatment drug). Pezzullo, "Resisting 'Breast Cancer Awareness Month'"; and Pezzullo, *Toxic Tourism*.

54. Jacquet, *Is Shame Necessary?*, 109–110, 115–116, 143. On Greenpeace's use of image events see De Luca, *Image Politics*.

55. Jacquet, *Is Shame Necessary?*, 106.

56. Jacquet, *Is Shame Necessary?*, 174.

57. Kämmerer, "Scientific Underpinnings of Shame."

58. Brown, "Shame vs. Guilt."

59. If a person passes as abled and the one shaming didn't realize she/he/they were disabled, being asked to share the specifics of one's health history in a public space violates

one's privacy. The consubstantiality of shame through the mutual constitution of bodies and space is well addressed in West, "PISSAR's Queer and Disabled Politics"; and West, *Transforming Citizenships*.

60. Bennett argues that medical discourse tends to turn into a management discourse that individualizes disability instead of placing responsibility systemically. Bennett, "Troubled Interventions." Bennett also notes how the individualized approach of self-management is often manifested in the phrase "take care"; Bennett, *Managing Diabetes*, 4.

61. *Discovering Alternative Straw Use.*

62. A comical and informative video explaining this critique is Squirmy and Grubs, "Should We Ban Straws?"

63. Hell on Wheels, "Podcast."

64. For example, Bibipins sells "Suck It Ableism" pins with a cartoon of Bubble Tea with a plastic straw (https://bibipins.com/products/suck-it-ableism-pin). "STRAWS ARE ACCESS" T-shirts were designed and sold by Dissent Clothing, showing fists up in the air clenching plastic straws and #SuckItAbleism (with credit given to Alice Wong): "Straws Are Access! T-Shirt" (https://www.teepublic.com/t-shirt/4914900-straws-are-access). "ABLEISM IS TRASH" products were created by Mia Mingus and Amita Swadhin with artwork by Jee Hei Park; proceeds go to Wong's Disability Visibility Project: "Featured Products" (https://ableismistrash.creator-spring.com/?). Kaalyn's straw chart can be purchased on everything from mouse pads to shirts (https://society6.com/hellonwheels).

65. Ho, "'People Need Them.'"

66. Consider disability justice advocate Lydia Brown (@autistichoya) tweeting: "Eco-Ableism is insisting your restaurant will never provide straws, even if asked by disabled people who need them (or risk aspiration/choking/not being able to drink), because they're 'thinking of the fishes.' When the u.s. miliary is the single biggest polluter on the planet."

67. Brown, *We Will Not Cancel Us*, 52, 55, 54. The entire book is relevant, particularly "Unthinkable Thoughts: Call-Out Culture in the Age of Covid-19."

68. "Mia Ives-Rublee, Disability Policy Director and Public Speaker." In addition to offline advocacy, Ives-Rublee also engages in related #SuckItAbleism hashtag activism.

69. Beyond Plastics, "Parks Should Go Plastic-Free."

70. Lonely Whale's accessibility talking points are available for download at https://www.dropbox.com/s/s9wn3tw23qjmyqo/Championing%20Accessibility.pdf?dl=0. For these talking points in action see Ives, "Hey, Could You #StopSucking?"

71. "Supported policy change in Seattle which will become the largest metropolitan city to ban the single-use plastic straw in July 2018." Lee, "Last Straw?"; and Lonely Whale, "What We've Achieved So Far."

72. Mingus, "Four Parts of Accountability."

73. Pezzullo, "CU Boulder Alumna Emy Kane."

74. Starbucks, "Follow Up to Starbucks News."

75. Starbucks, "Starbucks Commitment to Access." For more on backlash against Starbucks see Blake, "Starbucks' Decision Lets Me Down"; and Quito, "Bendy Plastic Straw Used in Hospitals."

76. AB-1884 Food Facilities: Single-Use Plastic Straws.

77. Brueck, "California First US State to Ban Plastic Straws"; Romero, "Want a Plastic Straw in California?"; and Filloon, "California Bans Giving Out Plastic Straws."

78. Bratskeir, "Milo Cress."
79. On critical interruptions see Pezzullo, "Performing Critical Interruptions."
80. Two phenomenal pieces on the work that needs to be done are Mingus, "Access Intimacy"; and Rage, "Access Intimacy and Institutional Ableism."
81. Wakibia, "Deaf Women Community in Ethiopia."
82. Teki Paper Bags, "Why?."
83. Pezzullo, "José Toscano Bravo."
84. Meekosha, "Decolonising Disability"
85. Grech and Soldatic, "Disability and Colonialism," 1.
86. Grech and Soldatic, "Disability and Colonialism," 4.

6. CREATING #ToiChonCa (#IChooseFish): TRAUMA, AFFECTIVE ART, AND BIG TECH DOMINANCE

Epigraph: *Radical Philosophy*, "Interview: Forgetting Vietnam."
1. For a map of the disaster see "Formosa Toxic Waste Spill."
2. De Onís likewise imagines "oceans as large, interconnected regions that are not constrained by the edge of landscapes. Observing and honoring the blurring of land, sky, and water requires illuminating how island inhabitants work toward creating good lives with sea and oceanic ti(d)es." *Energy Islands*, 16.
3. The trending hashtag tended to be all lowercase; it's converted to camel case for accessibility.
4. Regrettably I am not fluent in Vietnamese, nor have I ever visited Vietnam. Throughout, I rely on tonal marks and translations of others. Writing in English, I use "Vietnam" and "Vietnamese," though I do not alter quotes using the country's own name: "Việt Nam." The "Vietnam War" is a US term, whereas the Vietnamese tend to call it the "American War." I do not change either use in quotes, as these language choices reflect meaningful cultural perspectives.
5. For a timeline see Trang, "Timeline."
6. Green Trees, *Overview of Marine Life Disaster*. Green Trees is a nonprofit NGO in Vietnam, which has had an established Facebook group (Vi Mot Ha Noi Xanh/For a Green Hanoi) since 2015 to protect urban trees.
7. Altman, "Who Sings from the Resins?" This phrasing emerges out of a dialogue. Ansel Adams's photography moved Altman to write: "I began to understand how metals, petrochemicals, and plastics work as their own kind of trinity." Artist and photojournalist Raymond Thompson Jr., in turn, named them a "holy trinity." Altman attributes her thinking to essayist Catherine Venable Moore, as well as to Ken Geiser. Personal correspondence with author, August 21, 2021.
8. On American imperialism and communication technologies, see Towns, *On Black Media Philosophy*.
9. Chen, *Asia as Method*, vii.
10. Tsing, *Friction*, 5.
11. Tsing, *Friction*, 6.
12. Balcom and Balcom, *Indigenous Writers of Taiwan*.

13. "History: Taiwan."

14. On January 29, 1955, then US president Dwight D. Eisenhower passed a law with the backing of the US Congress called the Formosa Resolution, requesting Congress to preapprove military force in Taiwan if deemed necessary to counter China. Office of the Historian, "56. Joint Resolution by the Congress." I cannot do justice to explaining complicated US-Taiwan-China relations here; for a more detailed and nuanced account of this contested history see Hartnett, *World of Turmoil*.

15. For another account see Funding Universe, "Formosa Plastics Corporation History."

16. Tanzer, "Y. C. Wang Gets up Early," 88.

17. As noted in chapter 2, the trope of meritocracy propagates the fallacy that those who work the hardest will achieve the most success. Pharr, *Time of the Right*.

18. Tanzer, "Y. C. Wang Gets up Early."

19. *New York Times*, "Y. C. Wang."

20. Formosa Plastics Group, "Introduction," 4, 16.

21. Formosa Plastics Corporation and Subsidiaries, "Consolidated Financial Statements," 8.

22. *Forbes*, "Global 2000."

23. Formosa Plastics Group, "Introduction," 1–2.

24. *Forbes*, "Global 500."

25. Collier, "Retired Texas Shrimper Wins Settlement"; and *Dirty Money*, "Point Comfort."

26. The fines were based on Clean Air Act amendments to include hazards after the 1984 Dow Chemical disaster in Bhopal, India. Ramirez, "Formosa Plastics Will Pay $2.8 Million."

27. Laughland, "Louisiana Plastics Plant Put on Pause." On RISE St. James's campaign against Formosa, see Cirino, *Thicker Than Water*, ch. 8.

28. Earth Justice and RISE St. James, "Decision is a Major Victory." In 2021, after hearing testimony of residents from southern Louisiana, UN human rights experts stated: "This form of environmental racism poses serious and disproportionate threats to the enjoyment of several human rights of its largely African American residents, including the right to equality and non-discrimination, the right to life, the right to health, right to an adequate standard of living and cultural rights." UN, "USA."

29. Disaster STS Network, "Formosa Plastics Global Archive"; Tim Schütz, Shan-Ya Su, Hung-Yang Lin, Chia-Liang Shih, Wen-Ling Tu, Paul Jobin and Kim Fortun at Disaster STS Network, home page." For the full video see www.stopformosa.org/post/formosa-plastics-toxic-tour.

30. For example, see Vuong, "In Wake of Formosa Plastics."

31. Companies Market Cap, "Largest Taiwanese Companies."

32. Oxfam, "Ten Richest Men Double Fortunes." For an affective theorization of the profit versus connection frame of social media considering public discourses of Zuckerberg/Facebook in comparison to Julian Assange/WikiLeaks, see Seigworth and Tiessen, "Mobile Affects."

33. "Big Tech and Climate Policy." There has been an unofficial Big Tech competition to become net zero since at least 2016, though this still seems supplemented by carbon offset purchasing, which climate justice activists emphasize is a false solution. Hern, "Facebook Has Reached Net Zero."

34. Ma, "Facebook and Google Balance Booming Business."

35. Consider this regional example: in a study of Android in the Philippines of advertising and data-driven services, researchers found a lack of "transparency, attribution, and accountability." Gamba et al., "Pre-Installed Android Software."

36. Amnesty International, *Surveillance Giants*, 5.

37. White, "Vietnamese Netizens Defiant."

38. Consider Lopez's analysis of the optimistic reading of YouTube as a space where "self-representations would counter the preponderance of troublesome racial stereotypes that have long saturated" mainstream US media of Asian Americans; she cautions that in US mediascapes "the success of Asian American YouTubers has been located at the level of the talented few." Lopez, "Asian America Gone Viral," 159.

39. White, "Vietnamese Netizens Defiant."

40. Lee, "Invisible Networked Publics."

41. Green Trees, *Overview of Marine Life Disaster*, 2.

42. Nguyen, "From Cyberspace to Streets."

43. Nguyen, "From Cyberspace to Streets."

44. I analyzed 133 posts tagged with #toichonca, 120 posts with #tôichọncá, and 178 posts with #ichoosefish. Since the more popular hashtag was the first, I have favored it.

45. A fluent reader could find a rich archive on other social media platforms; I focused on these for reasons I hope will become apparent.

46. Mrs. Bui escaped from her homeland by boat with her two young children in 1979. In 1988 she published *Bot Bien* (*Sea Foam*), a novel describing the hardships of people who have fled Vietnam. Now based in Austin, Texas, she is also the founder and president of the Vietnamese American Heritage Foundation (VAHF), as well as an award-winning producer of documentary films.

47. Fan, Chiu, and Mabon, "Environmental Justice," 2.

48. Fan, Chiu, and Mabon, "Environmental Justice," 5.

49. Green Trees, *Overview of Marine Life Disaster*, 56

50. *Viêt Nam News*, "Ministry Opposes Steel Region Plans."

51. Tran, "Vietnamese Diasporic Films," 213.

52. *Viêt Nam News*, "PM Hails Efforts at Construction."

53. Thuy, "Chinese Laborers Working without Permit."

54. Ngoc, "Scaffolding Collapse in Formosa."

55. Charlier, "Taiwan Company under Fire."

56. Trang, "Timeline."

57. Dinh, "Minister of Natural Resources and Environment Tran Hong Ha"; Tong, "Vietnam Fish Deaths"; and Tinh, "Exec Makes Shocking Remarks" (emphasis added).

58. Pezzullo, "Nancy Bui."

59. Food and Agriculture Organization of the United Nations, "Fishery and Aquaculture Country Profiles."

60. Hung and Anh, "Steel Firm to Sack Executive"; Nguyen and Ming, "Formosa Steel Goes on Offensive"; and Duy, "Formosa Representative Nets Termination."

61. Green Trees, *Overview of Marine Life Disaster*, 25.

62. Social media hashtags appeared biased toward English; nevertheless, there was a trending Vietnamese hashtag before the English version. In a study of hashtag activism

related to the Amazon fires of 2019, Skill and colleagues argue that English hashtags are a way to call in a transnational audience, which I believe rings true in the #IChooseFish trend. Skill, Passero, and Francisco, "Assembling Amazon Fires."

63. Dong, "Vietnamese Demand White House Probe."

64. Memes in cultures with limited free speech particularly should be recognized as consequential: Mina, "Batman, Pandaman and the Blind Man."

65. I do not provide citations, as the stakes of dissent with international attention are high.

66. Kuo, "Reflections on #Solidarity," 186.

67. "Panoramic View of Fish Death."

68. Green Trees, *Overview of Marine Life Disaster*, 26.

69. Hung and Anh, "Taiwanese Steel Firm to Sack Executive." Some claim more than 200,000 were impacted, including Davis, "Vietnam's Fishermen Stranded."

70. This language of culture as a "whole way of life" borrows from Williams, *Keywords*.

71. Nguyen, "Victims of 2016 Marine Life Disaster."

72. Green Trees, *Overview of Marine Life Disaster*, 27–28.

73. Ives, "Outrage over Fish Kill in Vietnam."

74. Vineberg, "We Choose Fish."

75. Green Trees, *Overview of Marine Life Disaster*, 73.

76. Paddock, "Toxic Fish in Vietnam."

77. Fan, Chiu, and Mabon, "Environmental Justice," 12.

78. Fan, Chiu, and Mabon, "Environmental Justice," 4.

79. Green Trees, *Overview of Marine Life Disaster*, 30.

80. Green Trees, *Overview of Marine Life Disaster*, 32.

81. Lã Trường Sơn, "Cá Voi Mắc Cạn Tại Biển Nam Định"; and Đặng Văn Tường Channel.

82. Nguyên, "Tons of Fish Are Dying."

83. Residents also hosted a memorial to a dead dolphin that washed up on shore that December. Hung, "Dead Dolphin Washes Up."

84. Vineberg, "We Choose Fish."

85. Nguyen and Ming, "Formosa Steel Goes on Offensive," 33.

86. Perez, "Facebook Blocked in Vietnam."

87. Nguyen and Datzberger, *Environmental Movement in Vietnam*. Their compelling report details much more.

88. Editorial Board, "In Vietnam, Truth Is Criminal 'Propaganda.'" For those unfamiliar with the Vietnam War, here is a short summary. Once Vietnam ended France's colonial rule, the divided nation became pressured to take sides in the Cold War, a period beginning in the mid-twentieth century when the United States (advocating for ideals of capitalism and democracy) fought the USSR and China (advocating for ideals of communism and/or socialism) globally. The Vietnam/American War was a twenty-year regional struggle, including a civil war. Depending on one's perspective, US involvement was part of its imperialist agenda globally or of its mission to help save the South Vietnamese from a repressive governmental regime. North Vietnam won—although the United States claimed victory in the broader Cold War a little over a decade later. The Vietnam/American War was considered the first major US military defeat since becoming a global superpower.

Despite pivoting on the selling and presumed use of lethal arms, the US embargo lift was framed as a sign of healing of the relationship between the two nations. For more, see Chen, *Asia as Method*, 8–9.

89. Bourdain, "Low Plastic Stool." I do not have the space to analyze Bourdain's gender politics or a US-based food show that claims cuisine in Hanoi is "unknown" (to whom?) or its trope of "relatability" (as Bourdain emphasized that he was taking the president to a working-class place and paying the bill, even though the meal was part of a food show). Further, Bourdain subsequently died by suicide, at which time Obama retweeted this tweet. Hopefully, someone else will write about these dynamics. For examples of the celebratory tone of global media coverage of this meal as "delighted," "cool," and "proud," see *BBC News*, "Six Things about Bourdain-Obama Meal"; *Atlas Obscura*, "Bourdain and Obama's Dinner Table"; and Hendrickson, "Bourdain's Meal with Obama."

90. Obama, "Remarks by President Obama."

91. Obama, "Remarks by President Obama."

92. I am sharing this anonymously. This cartoon is also mentioned in Ives, "Outrage Over Fish Kill."

93. Green Trees, *Overview of Marine Life Disaster*, 33.

94. Green Trees, *Overview of Marine Life Disaster*, 33.

95. Minh, "Formosa Spill Roils Public Opinion."

96. Summarized from Trang, "Timeline"; and Sun, "Fish Death Crisis." I am not convinced that the US military spraying Agent Orange (a toxic herbicide) for a decade over the country during the Vietnam/American War wasn't a more serious environmental disaster. Green Trees notes that Radio Free Asia published a report on July 15 finding cyanide, phenol, arsenic, cadmium, and lead: Green Trees, *Overview of Marine Life Disaster*, 39.

97. Human Rights Watch, "New Wave of Arrests"; and Ha and Boliek, "Vietnam Cracks Down on Dissenters."

98. Ha and Boliek, "Vietnam Cracks Down on Dissenters."

99. *The Guardian*, "Vietnam Declares Activist Group Terrorist Organization."

100. Ha, "Vietnamese Authorities Send 'Thugs.'"

101. Bui, "Slide."

102. *The Guardian*, "Mother Mushroom"; and *The Peninsula*, "Vietnamese Blogger Arrested."

103. *BBC News*, "'Mother Mushroom'"; and Bemma, "Vietnam's Mother Mushroom."

104. Human Rights Watch, "Activist Facing Prison Term."

105. Ma, "Facebook and Google Balance Booming Business."

106. Amnesty International, "Let Us Breathe!," 43.

107. Amnesty International, "Surge in Number of Prisoners."

108. Pellow, "Political Prisoners," 15.

109. "Saving Sea Turtles in Viet Nam."

110. Van Truong et al., "Household Recovery from Disaster," 5–6.

111. Hung, "Suspected Gas Leakage."

112. Human Rights Watch, "Free Vietnam's Political Prisoners!"

113. For the lawsuit see "Formosa Plastics Lawsuit."

114. Pezzullo, "Nancy Bui."

115. Justice For Formosa's Victims Presents *Red Sea: Vietnam's Modern Disaster* is available at the website for JFFV. Justice for Formosa's Victims, "About JFFV."

116. Fan, Chiu, and Mabon, "Environmental Justice," 15.

117. Erni, "Introduction," 1.

118. For example, see Cadwalladr and Campbell, "Revealed."

119. Liboiron and Lepawsky, *Discard Studies*, 65. On censoring hashtags across platforms see Gillespie, *Custodians of the Internet*.

120. Dickinson, "Big Tech's Tightening Grip."

121. Noor, "Exxon's Secret Assist"; see also Funes, "Criminalizing Protests a Dangerous Idea."

122. An updated resource is International Center for Not-for-Profit Law, "US Protest Law Tracker."

123. Butt et al., "Supply Chain of Violence," 742.

124. For more, see Global Witness, "About Us."

125. *EJ Atlas*, "Environmental Justice Atlas Database"; and Scheidel et al., "Environmental Conflicts and Defenders."

126. Di Ronco, Allen-Robertson, and South, "Environmental Harm and Resistance," 157. Steve Mann coined *sousveillance*, according to Browne, *Dark Matters*, 18–19.

127. Noor, "Exxon's Secret Assist."

128. Carrington, "Climate Denial Ads on Facebook."

129. Bodoni, "Facebook Faces Complaint."

130. Tannenbaum, "Every Platform Trump Is Banned From." This book was written prior to Musk purchasing Twitter, which occurred as this book went to press; the impacts of that shift remain unclear. Hashtag activism and repression, however, continue to matter across social media platforms.

131. Gillespie, *Custodians of the Internet*, 212–13.

132. Hao, "How to Poison Data."

CONCLUSION: #BreakFree(FromPlastics)

Diana Taylor epigraph: Taylor, *¡Presente!*, 122.
adrienne maree brown epigraph: brown, *We Will Not Cancel Us*, 77.

1. Ray, *Field Guide to Climate Anxiety*, 109.

2. March 15, 2022, 7:20 a.m. Shared with permission.

3. UNEP, "Draft Resolution on Legally Binding Instrument."

4. Global Alliance of Waste Pickers (@globalrec_org), "[BREAKING] It Is a Historical Moment for the #WastePickers Movement!"

5. Lau et al., "Scenarios toward Zero Plastic Pollution."

6. Tabuchi, "World Is Awash in Plastic"; and Aldred and White, "UN Meeting Sets Sights on Treaty."

7. *U.S. Plastic Waste Exports*.

8. Hvistendahl, "Toxic Tiles."

9. Again, my articulation of *energize* is indebted to de Onís, *Energy Islands*, as well as LeMenager, *Living Oil*. As I have written elsewhere: "Although dialogue that allows *only*

space for happiness and optimism can feel oppressive, the opposite also rings true: Creating spaces that enable only sadness and cynicism can feel oppressive as well." Pezzullo and Cox, *Environmental Communication*, 10.

10. Acheson, "Abolition."

11. Myhal and Carroll, "Indigenous Optimism in Colonialcene."

12. My use of *abundant* is indebted to analysis of antidisplacement and food justice organizing in Gordon, "Troubling 'Access,'" as well as a critique of higher education discourse of racialized scarcity in Flores, "Between Abundance and Marginalization."

13. "Future studies can move beyond our textual and discursive analysis to tease out creators' motivations through methods such as ethnography, interviewing, and focus groups." Hautea et al., "Showing They Care (Or Don't)," 12.

14. Boudreau, "Fighting for Global Recognition"; and Morais et al., "Global Review of Waste-Picking."

15. Policy instruments include subsidies and tax breaks, capture of waste, cash deposit for return, education, labels, and limiting or banning plastics. Karasik et al., "20 Years of Government Responses."

16. America's Plastic Makers & American Chemistry Council, "Actions for Sustainable Change."

17. Mahdawi, "Woke-Washing Brands"; Westervelt, "Big Oil's 'Wokewashing.'" I have published on pinkwashing, cause-related marketing for breast cancer that is disingenuous: Pezzullo, "Resisting 'Breast Cancer Awareness Month.'"

18. Brock and Geddie, "Unilever's Plastic Playbook."

19. Pezzullo and Cox, *Environmental Communication*, 93–95.

20. Lerner, "Bottled Water Giant Admits 'Puffery.'"

21. Clean Creatives was founded by Jamie Henn, a well-known climate activist who cofounded 350.org. "Jamie Henn Profile."

22. Henn, "Clean Creatives Pressure Wavemaker."

23. Henn, "Clean Creatives Pressure Wavemaker."

24. Henn, "Clean Creatives Pressure Wavemaker."

25. For example, see Ivanova, "States Take Aim at Symbol."

26. "By 2018, sixty-three countries had EPR regulations for single-use plastics, considering product take-back, deposit-refunds, and recycling objectives. Europe had thirty-eight countries, followed by LAC with nine." Montalvo and Olivares, "Policy Mechanisms Reduce Single-Use Plastic," 10. On EPR debates in Kenya see Betterman, "Why Law Is Likely to Fail."

27. Winters, "California Passes Plastic Reduction Bill." On national efforts in the United States see Break Free From Plastic Movement, "#BreakFreeFromPlastic Pollution Act."

28. In "Typology of Circular Economy Discourses," Calisto Friant, Vermeulen, and Salomone have identified four: Reformist Circular Society, Transformational Circular Society, Technocratic Circular Economy, and Fortress Circular Economy.

29. The global coordinator is Von Hernandez, a Goldman Environmental Prize winner in 2003 for his advocacy of an incinerator ban in the Philippines. He previously was global development director of Greenpeace International, which was the first organization to pilot a brand audit. For more information see Break Free From Plastic Movement,

"Brand Audit Toolkit." They also created global toxic tours online; see #breakfreefrom-plastic, "#breakfreefromplastic Toxic Tours."

30. Eonnet, "Top Plastic Polluters."

31. Break Free From Plastic Movement, *Missing the Mark*, 35.

32. Eonnet, "Top Plastic Polluters."

33. For their Just Transition Principles, see Break Free From Plastic Movement, *Missing the Mark*.

34. GreenpeaceUSA, "Do you know what's REALLY behind the bottles of @CocaCola and @Pepsi?" The Ellen MacArthur Foundation (https://ellenmacarthurfoundation.org) tracks corporate performance in relation to global commitments.

35. To join, watch a video, and/or sign The Big Plastic Count UK petition, see https://thebigplasticcount.com/ .

36. For more, see https://wastebase.org/#/dashboard .

37. Basel Action Network, "Atlas of Plastic Waste."

38. PlasticPollutionCoalition, "Yelp Adds 'Plastic-Free Packaging.'"

39. On credit from orality to plastics see Steele, "History of Credit Cards."

40. Ford et al., "Fundamental Links."

41. *NBC News*, "Mexico City's Ban Takes Effect." Mexico City's ban was supported by over two hundred organizations, which created a coalition. La Alianza México Sin Plástico (AMSP), "About Us?" These campaigns will vary by location; compare, for example the Peruvian NGO L.O.O.P. (Life Out Of Plastic).

42. Liboiron and Lepawsky, *Discard Studies*, esp. 129.

43. For example, see Garcia, "La Nature"; Lauren, "Zero Waste Guide"; and Beyond Plastics, "Beyond Plastics Bill." On how the United Farm Workers imagined "more egalitarian social orders" through visual culture, see Marez, *Farmworker Futurism*, 4.

44. This statistic comes from The Story of Stuff Project petition; see https://action.storyofstuff.org/sign/amount-k-cups-have-been-thrown-landfills-could-wrap-around-planet-over-11-times .

45. Barilla, "Sustainable Packaging"; and Barilla, "2021 Sustainability Report."

46. UNEP, "Reframing Tourism."

47. Kaiser, "What Comes after Pandemic?"

48. Greenpeace, "Throwing Away the Future."

49. Bloomberg Philanthropies, "Bloomberg Launches Campaign."

50. In *Affective Publics*, Papacharissi goes so far as to argue: "'Technologies network us but it is narratives that connect us to each other" (5).

51. De Onís, *Energy Islands*, 153–54.

52. Pezzullo, "Performing Critical Interruptions."

53. Ngata, "Unconquerable Tide."

54. UN, "USA."

55. Examples provided at the link listed in the dedication.

56. Story of Stuff Project, *Glass, Metal, Plastic*. The project has produced many short films on plastics worth watching.

57. Global Alliance for Incinerator Alternatives, "Zero Waste Master Plan," 10. This source also provides a helpful "how to" guide, "Planning a Toxics and Treasures Tour."

58. Future tours might also find ways to tell more inspiring stories about public infra-structure, such as sanitation to deter plastic bag use for human waste (noted in chapter 3), water fountains with bottle refills to encourage reuse, and recycling centers to highlight labor conditions of waste workers.

59. One of my undergraduate students shared this with me; see Hofman, "Photo I Wish Didn't Exist."

60. Escobar, *Pluriversal Politics*, xiii.

61. Banerjee, "Resisting War on Alaska's Arctic." On becoming-with relational world-ing and kin in a technoculture, see also Haraway, *Staying with the Trouble*.

62. Pellow, *What Is Critical Environmental Justice?*, 317.

63. Gumbs, *Undrowned*, 2.

64. Berland, *Virtual Menageries*.

65. Madison, *Acts of Activism*, 117. On how one Louisiana community has created per-formances to mourn and to reconnect with water culture, see McGuffey, "Copious Dwelling in Sinking Landscape."

66. Ocean Conservancy thanks Coca-Cola, Pepsi, and more from the plastics-industrial complex on its website: https://oceanconservancy.org/trash-free-seas/international-coastal-cleanup/partners/.

67. Phengsitthy, "Plastic Reduction, Responsibility Bill."

68. Ocean Conservancy, "Stemming the Tide Statement." The 2015 report had led to more than seven hundred organizations protesting its findings: "Open Letter to Ocean Conservancy." On the organization's more recent commitment to ocean justice, see https://oceanconservancy.org/advancing-ocean-justice/.

69. GAIA and Break Free From Plastic, "Turning Back the Tide."

70. Quoted in McVeigh, "NGO Retracts 'Waste Colonialism' Report."

71. Sultana, "Critical Climate Justice," 122.

BIBLIOGRAPHY

AB-1884 Food Facilities: Single-Use Plastic Straws. Assembly Bill No. 1884. Chapter 576. (2018). https://leginfo.legislature.ca.gov/faces/billTextClient.xhtml?bill_id=201720180 AB1884.

Abbey, Erin. "Coca-Cola Live Positively Redesign." n.d. www.erinabbeyux.com/live positively.

Abidin, Crystal, Dan Brockington, Michael K. Goodman, Mary Mostafanezhad, and Lisa Ann Richey. "The Tropes of Celebrity Environmentalism." *Annual Review of Environment and Resources* 45, no. 1 (October 17, 2020): 387–410. https://doi.org/10.1146 /annurev-environ-012320-081703.

Acheson, Ray. "Abolition." *Thoughts for Change* (blog), Women's International League for Peace & Freedom, 2022. www.wilpf.org/thoughts-for-change/what-we-mean-when-we -talk-about-abolition/.

Adler-Bell, Sam. "Why White Supremacists Are Hooked on Green Living." *New Republic*, September 24, 2019. https://newrepublic.com/article/154971/rise-ecofascism-history -white-nationalism-environmental-preservation-immigration.

Agarwal, Anil, and Sunita Narain. *Global Warming in an Unequal World: A Case of Environmental Colonialism*. New Delhi: Centre for Science and Environment, 1991.

Aguayo, Angela J. "The Bodies That Push the Buttons Matter: Vernacular Digital Rhetoric as a Form of Communicative Agency." *Enculturation* 23 (2016). www.enculturation.net /the-bodies-that-push-the-buttons-matter.

Ahmed, Sara. *Living a Feminist Life*. Durham, Nc: Duke University Press, 2017.

———. *The Promise of Happiness*. Durham, NC: Duke University Press, 2010.

———. *Queer Phenomenology: Orientations, Objects, Others*. Durham, NC: Duke University Press, 2006.

Ahmed, Sarwar Uddin, and Keinosuke Gotoh. "Impact of Banning Polythene Bags on Floods of Dhaka City by Applying CVM and Remote Sensing." In *Proceedings: 2005*

IEEE International Geoscience and Remote Sensing Symposium, 2005, 2:1471–74. Seoul: IEEE, 2005. https://doi.org/10.1109/IGARSS.2005.1525403.

AkzoNobel (@AkzoNobel). "We our new #HumanCities murals in Vietnam." Twitter. June 14, 2017. https://twitter.com/AkzoNobel/status/875012008902631425.

Alaimo, Stacy. *Bodily Natures: Science, Environment, and the Material Self.* Bloomington: Indiana University Press, 2010.

———. *Exposed: Environmental Politics and Pleasures in Posthuman Times.* Minneapolis: University of Minnesota Press, 2016.

Alaoui, Fatima Zahrae Chrifi. "The Arab Spring between the Streets and the Tweets: Examining the Embodied (e)Resistance through the Feminist Revolutionary Body." In *Women of Color and Social Media Multitasking: Blogs, Timelines, Feeds, and Community,* edited by Sonja M. Brown Givens and Keisha Edwards Tassie, 35–67. Lanham, MD: Lexington Books, 2015.

Aldred, Jessica, and Aron White. "UN Meeting Sets Sights on Global Plastics Treaty (China Dialogue Ocean)." *Maritime Executive,* February 13, 2022. www.maritime-executive .com/index.php/editorials/un-meeting-sets-sights-on-global-plastics-treaty.

Allen, Danielle S. *Talking to Strangers: Anxieties of Citizenship since Brown v. Board of Education.* Chicago: University of Chicago Press, 2006.

Allen, Garland E. "'Culling the Herd': Eugenics and the Conservation Movement in the United States, 1900–1940." *Journal of the History of Biology* 46, no. 1 (February 2013): 31–72. https://doi.org/10.1007/s10739-011-9317-1.

Allen, Steve, Deonie Allen, Kerry Moss, Gaël Le Roux, Vernon R. Phoenix, and Jeroen E. Sonke. "Examination of the Ocean as a Source for Atmospheric Microplastics." Edited by Amitava Mukherjee. *PLOS ONE* 15, no. 5 (May 12, 2020): e0232746. https://doi.org /10.1371/journal.pone.0232746.

ALS Association. *Evaluation of the ALS Association Grant Programs: Executive Summary Report.* May 2019. www.als.org/sites/default/files/2020-06/RTI-Report-FINAL.pdf.

Alston, Dana A., and Panos Institute. *We Speak for Ourselves: Social Justice, Race and the Environment.* Washington, DC: Panos Institute, 1990.

Altman, Rebecca. "American Beauties." Photographs by Jan Staller. August 1, 2018. Topic .com. www.topic.com/american-beauties.

———. "Five Myths about Plastics." *Washington Post,* January 14, 2022. www.washingtonpost .com/outlook/2022/01/14/five-myths-plastics/.

———. "On Wishcycling." *Discard Studies,* February 15, 2021. https://discardstudies.com /2021/02/15/on-wishcycling/.

———. (@rebecca_altman). "This Wk, I Have a Teen Febrile w/ COVID." Twitter. January 20, 2022. Shared with permission. https://twitter.com/rebecca_altman/status/1484 187477019668480.

———. *The Song of Styrene: An Intimate History of Plastics.* New York: Scribner, forthcoming.

———. "Who Sings from the Resins? Raymond Thompson Jr. and the Appalachian Ghosts." *Orion Magazine,* 2021. https://orionmagazine.org/article/who-sings-from-the-resins/.

American Chemistry Council (ACC). "Plastics." n.d. www.americanchemistry.com/chemistry -in-america/chemistry-in-everyday-products/plastics.

America's Plastic Makers & American Chemistry Council. "5 Actions for Sustainable Change: A Plan for Congress to Accelerate a Circular Economy for Plastics." 2021. www.plastic makers.org/files/d6b3a34b9a88b1a6ee4da0a73b24562d740f80e4.pdf.

Amnesty International. *Surveillance Giants: How the Business Model of Google and Facebook Threatens Human Rights.* 2019. www.amnesty.org/en/documents/pol30/1404/2019/en/.

———. "Viet Nam: Let Us Breathe! Censorship and Criminalization of Online Expression in Viet Nam." November 30, 2020. www.amnesty.org/en/documents/asa41/3243/2020/en/.

———. "Viet Nam: Surge in Number of Prisoners of Conscience, New Research Shows." May 13, 2019. www.amnesty.org/en/latest/news/2019/05/viet-nam-surge-number -prisoners-conscience-new-research-shows/.

Antara, Nawaz Farhin. "The Plastic Pandemic: Single-Use Plastics Pile up as Covid-19 Delays Ban." *Dhaka Tribune*, April 21, 2021. https://archive.dhakatribune.com/bangladesh/2021 /04/21/the-plastic-pandemic-single-use-plastics-pile-up-as-covid-19-delays-ban.

Aratai, Lauren. "US Ranks Last in Healthcare among 11 Wealthiest Countries despite Spending Most." *The Guardian*, August 5, 2021. www.theguardian.com/us-news/2021 /aug/05/us-healthcare-system-ranks-last-11-wealthiest-countries.

Aronoff, Kate. "Care Work Is Climate Work." *New Republic*, April 9, 2021. https://new republic.com/article/161998/care-work-climate-work.

Asen, Robert. "Introduction: Neoliberalism and the Public Sphere." *Communication and the Public* 3, no. 3 (September 2018): 171–75. https://doi.org/10.1177/2057047318794687.

———. "Neoliberalism, the Public Sphere, and a Public Good." *Quarterly Journal of Speech* 103, no. 4 (October 2, 2017): 329–49. https://doi.org/10.1080/00335630.2017.1360507.

Asen, Robert, and Daniel C. Brouwer, eds. *Counterpublics and the State.* SUNY Series in Communication Studies. Albany: State University of New York Press, 2001.

Ashcraft, Karen Lee. *Wronged and Dangerous: Viral Masculinity and the Populist Pandemic.* Bristol, UK: Bristol University Press, 2022.

Ashcraft, Karen Lee, and Lisa A. Flores. "'Slaves with White Collars': Persistent Performances of Masculinity in Crisis." *Text and Performance Quarterly* 23, no. 1 (2003): 1–29. https://doi.org/10.1080/10462930310001602020.

Asher, Michael. "The Next Big Green Lie: 30–30 Plan Displaces to Conserve." *The Star*, May 10, 2021. www.the-star.co.ke/news/big-read/2021-05-10-the-next-big-green-lie-30 -30-plan-displaces-to-conserve/.

Atlas Obscura. "Anthony Bourdain and Barack Obama's Dinner Table." n.d. www .atlasobscura.com/places/anthony-bourdain-and-barack-obama-dinner-table.

Auxier, Brooke, and Monica Anderson. "Social Media Use in 2021." Pew Research Center. April 7, 2021. www.pewresearch.org/internet/2021/04/07/social-media-use-in-2021/.

Baar, Aaron. "Collins Gets Global Role on Ogilvy's BP." *AdWeek*, December 7, 2004. www .adweek.com/brand-marketing/collins-gets-global-role-ogilvys-bp-76582/.

Balcom, John, and Yingtsih Balcom, eds. *Indigenous Writers of Taiwan: An Anthology of Stories, Essays, & Poems.* Modern Chinese Literature from Taiwan. New York: Columbia University Press, 2005.

Baldwin, James. *The Fire Next Time.* New York: Laurel, 1962.

"The Ban on Polythene Bags—Are Kenyans Ready?" *GBM Blog*, June 25, 2017. www.green beltmovement.org/node/822.

Banerjee, Subhankar. "Resisting the War on Alaska's Arctic with Multispecies Justice." *Social Text*, June 7, 2018. https://socialtextjournal.org/periscope_article/resisting-the -war-on-alaskas-arctic-with-multispecies-justice/.

Barilla. "Sustainable Packaging." n.d. www.barillagroup.com/en/who-we-are/position /sustainable-packaging/.

———. "2021 Sustainability Report: Our Commitment to the Wellbeing of People and the Planet through the Quality of Our Food." July 23, 2021. www.barillagroup.com/en /stories/stories-list/2021-sustainability-report/.

Barthes, Roland. *Mythologies*. Translated Annettee Levers. Paris: Les Lettres Nouvelles, 1957.

Basel Action Network. "Atlas of Plastic Waste Launched." *Recycling Magazine*, November 30, 2021. www.recycling-magazine.com/2021/11/30/atlas-of-plastic-waste-launched/.

Baxter-Wright, Dusty. "These Are the 10 Most Liked Instagram Photos of 2016." *Cosmopolitan*, December 1, 2016.www.cosmopolitan.com/uk/entertainment/news/g4771/most -popular-instagram-photos-of-2016/.

Bazvand, Aboutaleb. "The Importance of Foreign Investment Attraction in Oil and Gas Industry." *IMPACT: International Journal of Research in Humanities, Arts and Literature* 5, no. 5 (May 2017): 47–54.

BBC News. "In Pictures: Flying Toilets." n.d. http://news.bbc.co.uk/2/shared/spl/hi/picture _gallery/07/africa_flying_toilets/html/1.stm.

———. "Kenya Plastic Bag Ban Comes into Force after Years of Delays." August 28, 2017. www.bbc.com/news/world-africa-41069853.

———. "'Mother Mushroom': Top Vietnamese Blogger Jailed for 10 Years." June 29, 2017. www.bbc.com/news/world-asia-40439837?S.

———. "Six Things about the $6 Bourdain-Obama Meal." May 24, 2016. www.bbc.com /news/world-asia-36365988.

———. "#Trashtag: The Online Challenge Cleaning Places Up." March 12, 2019. www.bbc .com/news/world-47536861.

Behuria, Pritish. "Ban the (Plastic) Bag? Explaining Variation in the Implementation of Plastic Bag Bans in Rwanda, Kenya and Uganda." *Environment and Planning C: Politics and Space* 39, no. 8 (December 2021): 1791–1808. https://doi.org/10.1177/2399654421994836.

Bemma, Adam. "Vietnam's Mother Mushroom: 'If I Don't Speak about the Future, Who Will?'" *Medium*, January 9, 2019. https://medium.com/@adambemma/vietnams-mother -mushroom-if-i-don-t-speak-about-the-future-who-will-1e5a19aeb568.

Bennett, Jeffrey A. *Managing Diabetes: The Cultural Politics of Disease*. New York: New York University Press, 2019.

———. "Troubled Interventions: Public Policy, Vectors of Disease, and the Rhetoric of Diabetes Management." *Journal of Medical Humanities* 34, no. 1 (March 2013): 15–32. https://doi.org/10.1007/s10912-012-9198-0.

Beraza, Suzan, dir. *Bag It*. Paramount Vantage, 2011.

Berland, Jody. *Virtual Menageries: Animals as Mediators in Network Cultures*. Cambridge, MA: The MIT Press, 2019.

Berlant, Lauren. *Cruel Optimism*. Durham, NC: Duke University Press, 2011.

———. *The Queen of America Goes to Washington City: Essays on Sex and Citizenship*. Series Q. Durham, NC: Duke University Press, 1997.

Betasamosake Simpson, Leanne. *Dancing on Our Turtle's Back: Stories of Nishnaabeg Re-creation, Resurgence and a New Emergence*. Winnipeg, MB: ARP Books, 2011.

Betterman, Simidi M. "Here Is Why the Proposed Kenya Extended Producer Responsibility Law Is Likely to Fail." Clean Up Kenya. September 24, 2021. https://cleanupkenya .org/here-is-why-the-proposed-kenya-extended-producer-responsibility-law-is-likely -to-fail/.

Beyond Plastics. "The Beyond Plastics Bill." n.d. www.beyondplastics.org/beyond-plastics -bill.

———. "The Break Free From Plastic Pollution Act." 2021. www.beyondplastics.org /legislation.

———. "National Parks Should Go Plastic-Free." June 23, 2021. www.beyondplastics.org /press-releases/national-parks-water-bottles

Bhatia, Anisha. "Dia Mirza and Afroz Shah Urge Plastic Companies to Come Forward and Play Their Part to #BeatPlasticPollution." *Benega Swasth India*, July 16, 2018. https:// swachhindia.ndtv.com/dia-mirza-afroz-shah-urge-plastic-companies-to-beat-plastic -pollution-22877/.

"Big Tech and Climate Policy." InfluenceMap. January 2021. https://influencemap.org /report/Big-Tech-and-Climate-Policy-afb476c56f217ea0ab351d79096df04a.

Blackpast. "(1982) Audre Lorde, 'Learning From the 60s.'" August 12, 2012. www.blackpast .org/african-american-history/1982-audre-lorde-learning-60s/.

Blake, Melissa. "How Starbucks' Straw Decision Lets Me Down." *CNN*, July 11, 2018. www .cnn.com/2018/07/11/opinions/starbucks-plastic-draw-hurts-disabled-like-me-blake /index.html.

Bloomberg Philanthropies. "Michael R. Bloomberg Launches New $85 Million Campaign to Stop Rapid Rise of Pollution From the Petrochemical Industry in the United States." Press Release, September 21, 2022. www.bloomberg.org/press/michael-r-bloomberg -launches-new-85-million-campaign-to-stop-rapid-rise-of-pollution-from-the -petrochemical-industry-in-the-united-states/.

Bodoni, Stephanie. "Facebook Faces Complaint over Hate Speech." *Bloomberg*, March 23, 2021. www.bloomberg.com/news/articles/2021-03-23/facebook-faces-complaint-over -hate-speech-disinformation-claims.

Bollen, Johan, Alberto Pepe, and Huina Mao. "Modeling Public Mood and Emotion: Twitter Sentiment and Socio-Economic Phenomena." November 9, 2009. https://doi.org/10 .48550/ARXIV.0911.1583.

Boniface Mwangi (@bonifacemwangi). "Poor Traders Will Be Jailed Up to 4 Years for Using Plastic Bags." Verified. Twitter. February 17, 2020. https://twitter.com/bonifacemwangi /status/1229649390127779840?lang=en.

Bosch, Tanja Estella, Mare Admire, and Meli Ncube. "Facebook and Politics in Africa: Zimbabwe and Kenya." *Media, Culture & Society* 42, no. 3 (April 2020): 349–64. https:// doi.org/10.1177/0163443719895194.

Boudreau, Catherine. "The Waste Picker Fighting for Global Recognition." *Politico*, April 13, 2022. www.politico.com/newsletters/the-long-game/2022/04/13/the-waste -picker-fighting-for-global-recognition-00024944.

Bourdain, Anthony (@Bourdain). Verified. "Low Plastic Stool, Cheap But." Twitter. May 23, 2016. https://twitter.com/Bourdain/status/734772293830737920.

Boyd, William. Introduction to *A Month and a Day: A Detention Diary*, by Ken Saro-Wiwa. New York: Penguin Books, 1995.

Boykoff, Maxwell T., and Jules M. Boykoff, "Balance as Bias: Global Warming and the US Prestige Press." *Global Environmental Change* 14 (2004): 125–36. https://doi.org/10.1016/j.gloenvcha.2003.10.001.

BP. "Drive down Your Carbon Footprint." 1996–2022. www.bp.com/en_gb/target-neutral/home/calculate-and-offset-travel-emissions.html#/.

———. "Early History—1909–1924." 1996–2022. www.bp.com/en/global/corporate/who-we-are/our-history/early-history.html.

———. "Late Century—1971–1999." 1996–2022. www.bp.com/en/global/corporate/who-we-are/our-history/late-century.html.

"BP Ad: Carbon Footprint." December 11, 2008. YouTube video. www.youtube.com/watch?v=ywrZPypqSB4.

Bradner, Eric. "Plastic Straws Are the Subject of the Latest 2020 Culture War." *CNN*, September 5, 2019. www.cnn.com/2019/09/05/politics/plastic-straws-2020-culture-war.

Bratskeir, Kate. "Milo Cress, the Kid Who Started the Straw Ban Movement, Doesn't Think Banning Straws in the Answer." *MIC*, July 27, 2018. www.mic.com/articles/190451/milo-cress-the-kid-who-started-the-straw-ban-movement-doesnt-think-banning-straws-is-the-answer.

Break Free From Plastic Movement. "Brand Audit Toolkit." 2022. www.breakfreefromplastic.org/brandaudittoolkit/.

———. "The #BrandAudit 2021 Report." 2022. www.breakfreefromplastic.org/brandaudit2021/.

———. "#BreakFreeFromPlastic Pollution Act." 2022. www.breakfreefromplastic.org/pollution-act/.

———. "The Coca-Cola Company and PepsiCo Named Top Plastic Polluters for the Fourth Year in a Row." October 25, 2021. www.breakfreefromplastic.org/2021/10/25/the-coca-cola-company-and-pepsico-named-top-plastic-polluters-for-the-fourth-year-in-a-row/.

———. *Missing the Mark: Unveiling Corporate False Solutions to the Plastic Pollution Crisis* (report). 2021. www.breakfreefromplastic.org/missing-the-mark-unveiling-corporate-false-solutions-to-the-plastic-crisis/.

———. "Single Use Plastic: Hidden Costs of Health & Environment in Bangladesh." November 4, 2019. www.breakfreefromplastic.org/2019/11/04/single-use-plastic-hidden-costs-of-health-environment-in-bangladesh/.

———. "Youth Leaders Reveal the #BrandAudit2021 Top Corporate Plastic Polluters." October 24, 2021. www.breakfreefromplastic.org/brandaudit2021/.

#breakfreefromplastic. "#breakfreefromplastic Toxic Tours." 2022. http://toxictours.org.

Briggs, Helen. "Plastic Pollution: Green Light for 'Historic' Global Treaty." *BBC*, March 2, 2022. www.bbc.com/news/science-environment-60590515.

Brock, Joe, and John Geddie. "Unilever's Plastic Playbook." Reuters, June 22, 2022. www.reuters.com/investigates/special-report/global-plastic-unilever/.

Brouwer, Daniel C. "Risibility Politics: Camp Humor in HIV/AIDS Zines." In *Public Modalities: Rhetoric, Culture, Media, and the Shape of Public Life*, edited by Daniel C.

Brouwer and Robert Asen, 219–39. Rhetoric, Culture, and Social Critique. Tuscaloosa: University of Alabama Press, 2010.

brown, adrienne maree. *We Will Not Cancel Us: Breaking the Cycle of Harm.* Emergent Strategy Series 4. Chico, CA: AK Press, 2020.

Brown, Brené. "Shame vs. Guilt" (blog), January 15, 2013. https://brenebrown.com/articles/2013/01/15/shame-v-guilt/.

Brown, Lydia X. Z. (@autistichoya). Verified. "Eco-Ableism Is Insisting." Twitter. March 3, 2022. https://twitter.com/autistichoya/status/1499568954490109955?s=11.

Brown, Sandy. "Ogilvy Gets BP Vote of Confidence." *AdWeek*, May 24, 2004. www.adweek.com/brand-marketing/ogilvy-gets-bp-vote-confidence-72697/.

Browne, Simone. *Dark Matters: On the Surveillance of Blackness.* Durham, NC: Duke University Press, 2015.

Brownstein, Michael, Daniel Kelly, and Alex Madva. "Individualism, Structuralism, and Climate Change." *Environmental Communication* 16, no. 2 (2021): 269–88. https://doi.org/10.1080/17524032.2021.1982745.

Brueck, Hillary. "California Just Became the First US State to Ban Plastic Straws in Restaurants —Unless Customers Ask." *Business Insider*, September 21, 2018. www.businessinsider.com/california-straw-ban-restaurants-what-you-need-to-know-2018-9.

———. "The Real Reason Why So Many Cities and Businesses Are Banning Plastic Straws Has Nothing to Do with Straws at All." *Business Insider*, October 22, 2018. www.businessinsider.com/plastic-straw-ban-why-are-there-so-many-2018-7.

Brulle, Robert J. "Networks of Opposition: A Structural Analysis of U.S. Climate Change Countermovement Coalitions 1989–2015." *Sociological Inquiry* 91, no. 3 (August 2021): 603–24. https://doi.org/10.1111/soin.12333.

Brulle, Robert J., Melissa Aronczyk, and Jason Carmichael, "Corporate Promotion and Climate Change: An Analysis of Key Variables Affecting Advertising Spending by Major Oil Corporations, 1986–2015." *Climatic Change* 159 (2020): 87–101.

Bsumek, Peter K., Steve Schwarze, Jennifer Peeples, and Jen Schneider. "Strategic Gestures in Bill McKibben's Climate Change Rhetoric." *Frontiers in Communication* 4 (August 19, 2019): 40. https://doi.org/10.3389/fcomm.2019.00040.

Buell, Frederick. *From Apocalypse to Way of Life: Environmental Crisis in the American Century.* New York: Routledge, 2003.

Bui, Nancy. "Slide: 12:28, Formosa Plastics Toxic Tour: Live Event; Author Attended." Stop Formosa Plastics. June 30, 2021. www.stopformosa.org/post/formosa-plastics-toxic-tour.

Bullard, Robert D. "Interview: The Politics of Race and Pollution." *Multinational Monitor*, June 1992. www.multinationalmonitor.org/hyper/issues/1992/06/mm0692_09.html.

———, ed. *Unequal Protection: Environmental Justice and Communities of Color.* San Francisco: Sierra Club Books, 1994.

Bullard, Robert D., Paul Mohai, Robin Saha, and Beverly Wright. *Toxic Wastes and Race at Twenty: 1987–2007: A Report Prepared for the United Church of Christ Justice and Witness Ministries. March 2007.* www.ucc.org/wp-content/uploads/2021/03/toxic-wastes-and-race-at-twenty-1987-2007_Part1.pdf.

Burford, Caitlyn, and Julie "Madrone" Kalil Schutten. "International Activists and the 'Blackfish Effect': Contemplating Captive Orcas' Protest Rhetoric through a Coherence

Frame." *Frontiers in Communication* 1 (January 12, 2017). https://doi.org/10.3389/fcomm
.2016.00016.

Burrows, Oliver. "The Plastic Bag: Kenya's National Flower?" *Capital News*, June 24, 2014.
www.capitalfm.co.ke/news/2014/06/the-plastic-bag-kenyas-national-flower/.

Büscher, Bram. *The Truth about Nature: Environmentalism in the Era of Post-Truth Politics
and Platform Capitalism.* Oakland: University of California Press, 2021.

Business & Human Rights Resource Center. "Formosa Plastics Lawsuit (Re Marine Pollution
in Vietnam, Filed in Taiwan)." June 11, 2019. www.business-humanrights.org/en/latest
-news/formosa-plastics-lawsuit-re-marine-pollution-in-vietnam-filed-in-taiwan/.

Business Insider. "Adrian Grenier Asks You to #StopSucking on Plastic Straws in New
Social Media Challenge." August 9, 2017. https://markets.businessinsider.com/news
/stocks/adrian-grenier-asks-you-to-stopsucking-on-plastic-straws-in-new-social-media
-challenge-1002243532.

———. "Responding to Trump's Plastic Straws, Bernie Sanders Supporters Launch 'Trump
Sucks' Metal Straws." August 5, 2019. https://markets.businessinsider.com/news/stocks
/responding-to-trump-s-plastic-straws-bernie-sanders-supporters-launch-trump
-sucks-metal-straws-1028416609.

Butler, Judith. *Bodies That Matter: On the Discursive Limits of Sex.* London: Routledge,
1993.

Butt, Nathalie, Frances Lambrick, Mary Menton, and Anna Renwick. "The Supply Chain of
Violence." *Nature Sustainability* 2, no. 8 (August 2019): 742–47. https://doi.org/10.1038
/s41893-019-0349-4.

Byaruhanga, Catherine, and Michael Onyiego. "Nashulai: The Community Trying to Conserve Kenya's Wildlife." *BBC News*, December 29, 2020. www.bbc.com/news/av/world
-africa-55477272.

Cadwalladr, Carole, and Duncan Campbell. "Revealed: Facebook's Global Lobbying against
Data Privacy Laws." *The Guardian*, March 2, 2019. www.theguardian.com/technology/2019
/mar/02/facebook-global-lobbying-campaign-against-data-privacy-laws-investment.

Calil, Juliano, Marce Gutiérrez-Graudiņš, Steffanie Munguía, and Christopher Chin.
Neglected: Environmental Justice Impacts of Marine Litter and Plastic Pollution (report).
Nairobi: United Nations Environment Programme (UNEP), 2021. https://wedocs.unep
.org/xmlui/bitstream/handle/20.500.11822/35417/EJIPP.pdf.

Calisto Friant, Martin, Walter J. V. Vermeulen, and Roberta Salomone. "A Typology of
Circular Economy Discourses: Navigating the Diverse Visions of a Contested Paradigm." *Resources, Conservation and Recycling* 161 (October 2020): 104917. https://doi
.org/10.1016/j.resconrec.2020.104917.

Calma, Justine. "Plastic Giant Coca-Cola Says People Want Its Plastic." *The Verge.* January 22, 2020. www.theverge.com/2020/1/22/21076868/plastic-bottle-coca-cola-davos
-world-economic-forum.

Care & Climate: Understanding the Policy Intersections. A Feminist Green New Deal Coalition Brief. April 2021. http://feministgreennewdeal.com/wp-content/uploads/2021/04
/FemGND-IssueBrief-Draft7-Apr15.pdf.

Carlson, Tucker. "Tucker Carlson: The Left Wants to Take Away Your Straws—but
Ignores Actual Environmental Problems." *Fox News*, January 3, 2019. www.foxnews

.com/opinion/tucker-carlson-the-left-wants-to-take-away-your-straws-but-ignores
-actual-environmental-problems.

Carpenter, Scott. "After Abandoned 'Beyond Petroleum' Re-Brand, BP's New Renewables Push Has Teeth." *Forbes*, August 4, 2020. www.forbes.com/sites/scottcarpenter/2020 /08/04/bps-new-renewables-push-redolent-of-abandoned-beyond-petroleum-rebrand /?sh=13d28f5d1ceb.

Carpenter, Zoë. "The Toxic Consequences of America's Plastics Boom." *The Nation*, March 14, 2019. www.thenation.com/article/archive/plastics-pollution-crisis-fracking -petrochemicals/.

Carrico, Amanda. "Climate Change, Behavior, and the Possibility of Spillover Effects: Recent Advances and Future Directions." *Current Opinion in Behavioral Science,* 42 (2021): 76–82. https://doi.org/10.1016/j.cobeha.2021.03.025.

Carrington, Damian. "Climate Denial Ads on Facebook Seen by Millions, Report Finds." *The Guardian*, October 8, 2020. www.theguardian.com/environment/2020/oct/08 /climate-denial-ads-on-facebook-seen-by-millions-report-finds.

———. "Microplastic Pollution Revealed 'Absolutely Everywhere' by New Research." *The Guardian*, March 6, 2019. www.theguardian.com/environment/2019/mar/07/micro plastic-pollution-revealed-absolutely-everywhere-by-new-research.

Carroll, Sam. "How This Man Is Helping to Solve Kenya's Waste Problem." SBS. May 15, 2017. www.sbs.com.au/topics/voices/culture/article/2017/05/12/how-man-helping-solve -kenyas-waste-problem.

Carrozza, Chiara, and Emanuele Fantini. "The Italian Water Movement and the Politics of the Commons. Water Alternatives." *Water Alternatives* 9, no. 1 (2016): 99–119.

Carson, Rachel. *The Sea Around Us*. 3rd ed. Oxford: Oxford University Press, 2018.

Carter, Lawrence. "Inside Exxon's Playbook: How America's Biggest Oil Company Continues to Oppose Action on Climate Change." *Unearthed*, June 30, 2021. https:// unearthed.greenpeace.org/2021/06/30/exxon-climate-change-undercover/.

Center for International Environmental Law (CIEL). "Formosa Plastics Group: A Serial Offender of Environmental and Human Rights (A Case Study)." October 2021. /www .ciel.org/reports/formosa-plastics-group-a-serial-offender-of-environmental-and -human-rights/.

———. "Fueling Plastics: Plastic Industry Awareness of the Ocean Plastics Problem." 2017. www.ciel.org/wp-content/uploads/2017/09/Fueling-Plastics-Plastic-Industry-Awareness -of-the-Ocean-Plastics-Problem.pdf.

———. "Plastic & Climate: The Hidden Costs of a Plastic Planet." 2019. www.ciel.org/wp -content/uploads/2019/05/Plastic-and-Climate-Executive-Summary-2019.pdf.

———. "UNEA 5.2: Historic Advances on Global Plastics Treaty, Chemicals." March 2, 2022. www.ciel.org/news/unea-5-2-historic-advances-on-global-plastics-treaty -chemicals/.

Chakrabarty, Dipesh. *The Climate of History in a Planetary Age*. Chicago: University of Chicago Press.

Chakraborty, Jayajit. "Proximity to Extremely Hazardous Substances for People with Disabilities: A Case Study in Houston, Texas." *Disability and Health Journal* 12, no. 1 (January 2019): 121–25. https://doi.org/10.1016/j.dhjo.2018.08.004.

———. "Unequal Proximity to Environmental Pollution: An Intersectional Analysis of People with Disabilities in Harris County, Texas." *Professional Geographer* 72, no. 4 (October 1, 2020): 521–34. https://doi.org/10.1080/00330124.2020.1787181.

Chalmin, Philippe. "The History of Plastics: From the Capitol to the Tarpeian Rock." Special issue, *Field Actions Science Reports*, no. 19 (March 1, 2019). http://journals.openedition.org/factsreports/5071.

Chan, Joan. "Albums, Sparrowclover." flickr. n.d. www.flickr.com/photos/193182308@N02/albums.

Charles, Dominic, Laurent Kimman, and Nakul Saran. "Plastic Waste Makers Index." Minderoo Foundation, 2021. https://cdn.minderoo.org/content/uploads/2021/05/27094234/20211105-Plastic-Waste-Makers-Index.pdf.

Charlier, Phillip. "Taiwan Company under Fire for Fish Kill in Vietnam." *Taiwan English News*, n.d. https://taiwanenglishnews.com/taiwan-company-under-fire-for-fish-kill-in-vietnam/.

Charlton, James I. *Nothing about Us without Us: Disability Oppression and Empowerment.* Berkeley: University of California Press, 2000.

Chatzidakis, Andreas, Jamie Hakim, Jo Littler, Catherine Rottenberg, and Lynne Segal. *The Care Manifesto: The Politics of Interdependence.* London: Verso, 2020.

Chávez, Karma R. *Palestine on the Air: A Supplement to the Journal of Civil and Human Rights.* Common Threads. Urbana: University of Illinois Press, 2019.

Chellel, Kit, and Ekow Dontoh. "West Africa Is Drowning in Plastic: Who Is Responsible?" *Bloomberg*, August 19, 2022. www.bloomberg.com/features/2022-coca-cola-nestle-west-africa-ghana-plastic-waste-recycling/#xj4y7vzkg.

Chen, Kuan-Hsing. *Asia as Method: Toward Deimperialization.* Durham, NC: Duke University Press, 2010.

Chen, Mel Y. "Toxic Animacies, Inanimate Affections." *GLQ: A Journal of Lesbian and Gay Studies* 17, nos. 2–3 (June 1, 2011): 265–86. https://doi.org/10.1215/10642684-1163400.

Chirindo, Kundai. "Bantu Sociolinguistics in Wangari Maathai's Peacebuilding Rhetoric." *Women's Studies in Communication* 39, no. 4 (October 2016): 442–59. https://doi.org/10.1080/07491409.2016.1228552.

Chokshi, Niraj. "How a 9-Year-Old Boy's Statistic Shaped a Debate on Straws." *New York Times*, July 19, 2018. www.nytimes.com/2018/07/19/business/plastic-straws-ban-fact-check-nyt.html.

Ciafone, Amanda. "If 'Thanda Matlab Coca-Cola' Then 'Cold Drink Means Toilet Cleaner': Environmentalism of the Dispossessed in Liberalizing India." *International Labor and Working-Class History* 81 (2012): 114–35. https://doi.org/10.1017/S0147547912000075.

Cirino, Erica. "Plastic Pollution: From Ship to Shore." *The Revelator*, May 17, 2017. https://therevelator.org/plastic-pollution-ship-shore/.

———. *Thicker Than Water: The Quest for Solutions to the Plastics Crisis.* Washington, DC: Island Press, 2021.

Clapp, Jennifer, and Linda Swanston. "Doing Away with Plastic Shopping Bags: International Patterns of Norm Emergence and Policy Implementation." *Environmental Politics* 18, no. 3 (May 2009): 315–32. https://doi.org/10.1080/09644010902823717.

Clark-Parsons, Rosemary, and Jessa Lingel. "Margins as Methods, Margins as Ethics: A Feminist Framework for Studying Online Alterity." *Social Media + Society* 6, no. 1 (2020): 1–11. https://doi.org/10.1177/2056305120913994.

Clean Up Kenya. "Statement from James Wakibia on Launch of 2021 PET Rubbish Report." September 1, 2021. https://cleanupkenya.org/statement-from-james-wakibia-on-launch -of-2021-pet-rubbish-report/.

Climate Justice Alliance: Communities United for a Just Transition. "About: Climate Justice Alliance." n.d. https://climatejusticealliance.org/about/.

Cloud, Dana. *Control and Consolation in American Culture and Politics: Rhetorics of Therapy*. London: Sage, 1997.

CNN. "Climate Crisis Town Hall with Sen. Elizabeth Warren (D-MA), Presidential Candidate. Aired 9:20–10p ET." September 4, 2019. https://transcripts.cnn.com/show/se/date /2019-09-04/segment/07.

The Coca-Cola Co. "The Coca-Cola Company and Bill Nye Demystify Recycling." April 5, 2022. YouTube video. www.youtube.com/watch?v=1HRadzzvQNY.

Collier, Kiah. "Retired Texas Shrimper Wins Record-Breaking $50 Million Settlement from Plastics Manufacturing Giant." *Texas Tribune*, December 3, 2019. www.texastribune.org /2019/12/03/texas-judge-approves-settlement-agreement-water-pollution-formosa/.

Collins Dictionary. "Why Single-Use? Etymology Corner—Collins Word of the Year 2018." November 7, 2018. https://blog.collinsdictionary.com/language-lovers/etymology-corner -collins-word-of-the-year-2018/.

Companies Market Cap. "Largest Taiwanese Companies by Market Capitalization." n.d. https://companiesmarketcap.com/taiwan/largest-companies-in-taiwan-by-market -cap/.

Conservation International. "Nature Is Speaking—Harrison Ford Is the Ocean." October 5, 2014. YouTube video. www.youtube.com/watch?v=rM6txLtoaoc.

Cook, John. *Cranky Uncle vs. Climate Change: How to Understand and Respond to Climate Science Deniers*. New York: Citadel Press, 2020.

———. "Seattle a Top Social Media City, but Lacking in Twitter Usage." *Puget Sound Business Journal* (blog), September 23, 2010. www.bizjournals.com/seattle/blog/techflash /2010/09/seattle_a_top_social_city_but_lacking_in_twitter_usage.html.

Corporate Responsibility Report 2005–2006. WPP. www.wpp.com/-/media/project/wpp /files/sustainability/reports/2005-06/wpp_csr_2006_aug06.pdf?la=en.

Cottle, Simon. "Stigmatizing Handsworth: Notes on Reporting Spoiled Space." *Critical Studies in Mass Communication* 11, no. 3 (September 1994): 231–56. https://doi.org/10 .1080/15295039409366900.

Cox, Robert. "Nature's 'Crisis Disciplines': Does Environmental Communication Have an Ethical Duty?" *Environmental Communication* 1, no. 1 (May 2007): 5–20. https://doi.org /10.1080/17524030701333948.

Craig, Jill. "Africa's Elephant Ivory: Sell or Destroy?" *VOA*, September 24, 2016. www .voanews.com/a/africa-seized-ivory-sell-destroy/3523240.html.

Cram, E. *Violent Inheritance: Sexuality, Land, and Energy in Making the North American West*. Environmental Communication, Power, and Culture 3. Oakland: University of California Press, 2022.

Cram, E., Martin P. Law, and Phaedra C. Pezzullo. "Cripping Environmental Communication: A Review of Eco-Ableism, Toxic Eco-Normativity, and Climate Justice Futures." *Environmental Communication* 16, no. 7 (2022): 851–63. https://doi.org/10.1080/17524032.2022.2126869.

Crenshaw, Kimberle. "Mapping the Margins: Intersectionality, Identity Politics, and Violence Against Women of Color." *Stanford Law Review* 43, no. 6 (July 1991): 1241–99.

Critical Past. "British Geologist George B. Reynolds and Workers Drilling for Oil and Striking Oi . . . HD Stock Footage." March 18, 2014. YouTube video. www.youtube.com/watch?v=Bcy9IlCtRA0.

The Daily Show. "If You Don't Know, Now You Know—Asian Nations Reject Western Trash." July 26, 2019. YouTube video. www.youtube.com/watch?v=-htnUTN4mHo.

Đặng Văn Tường. "Cá Voi Khổng Lồ Mắc Cạn ở Biển Nghệ An." May 24, 2016. YouTube video. www.youtube.com/watch?v=A58TIiyTt3g.

Đặng Văn Tường Channel. n.d. YouTube. www.youtube.com/channel/UCGlWIHbGy8rqPI_aH7Q3w9A.

Danovich, Tove, and Maria Godoy. "Why People with Disabilities Want Bans on Plastic Straws to Be More Flexible." *NPR*, July 11, 2018. www.npr.org/sections/thesalt/2018/07/11/627773979/why-people-with-disabilities-want-bans-on-plastic-straws-to-be-more-flexible.

Dasgupta, Debarshi, "India's Single-Use Plastic Ban to Start in July but Eradication Still Likely Long Way Off." *Straits Times*, May 30, 2022. www.straitstimes.com/asia/south-asia/indias-single-use-plastic-ban-to-start-in-july-but-eradication-still-likely-to-be-a-long-way-off.

Davis, Angela Y. *Are Prisons Obsolete?* n.p.: Seven Stories Press, 2003.

Davis, Brett. "Vietnam's Fishermen Stranded after Toxic Spill Destroys an Industry." *Forbes*, August 29, 2016. www.forbes.com/sites/davisbrett/2016/08/29/vietnams-fishermen-stranded-after-toxic-spill-destroys-an-industry/?fbclid=IwAR1Pcop2xlbonjGh5PXnrH_i46vM6OmdYowHkcF_mbIEwOarxWdNHX_zNWk&sh=52da8f848273.

de Onís, Catalina M. *Energy Islands: Metaphors of Power, Extractivism, and Justice in Puerto Rico.* Environmental Communication, Power, and Culture 1. Oakland: University of California Press, 2021.

de Onís, Catalina M., and Phaedra C. Pezzullo. "The Ethics of Embodied Engagement: Ethnographies of Environmental Justice." In *The Routledge Handbook of Environmental Justice*, edited by Ryan B. Holifield, Jayajit Chakraborty, and Gordon P. Walker, 231–40. London: Routledge, 2018.

de Wit, Wijnand, and Nathan Bigaud. "No Plastic in Nature: Assessing Plastic Ingestion from Nature to People." WWF. 2019. https://wwfint.awsassets.panda.org/downloads/plastic_ingestion_web_spreads.pdf.

"Declaration of Rights for Cetaceans: Whales and Dolphins." The Helsinki Group. May 22, 2010. www.cetaceanrights.org/.

DeFrane, Rae. "Single-Use Activism & the Disposable Human." *New Twenties*, July 23, 2021. https://thenewtwenties.ca/articles/singleuseactivism.

DeLuca, Kevin M. *Image Politics: The New Rhetoric of Environmental Activism.* London: Routledge, 2005.

———. "Trains in the Wilderness: The Corporate Roots of Environmentalism." *Rhetoric & Public Affairs* 4, no. 4 (2001): 633–52. https://doi.org/10.1353/rap.2001.0067.

Di Chiro, Giovanna. "Care Not Growth: Imagining a Subsistence Economy for All." *British Journal of Politics and International Relations*, 21, no. 2 (2019): 303–11. https://doi.org/10.1177/1369148119836349.

———. "Mobilizing 'Intersectionality' in Environmental Justice Research and Action in a Time of Crisis." In *Environmental Justice: Key Issues*, edited by Brendan Coolsaet, 316–33. London: Routledge, 2020.

———. "Polluted Politics? Confronting Toxic Discourse, Sex Panic, and Eco-Normativity." In *Queer Ecologies: Sex, Nature, Politics, Desire*, edited by Catriona Mortimer-Sandilands and Bruce Erickson, 199–230. Bloomington: Indiana University Press, 2010.

Di Ronco, Anna, James Allen-Robertson, and Nigel South. "Representing Environmental Harm and Resistance on Twitter: The Case of the TAP Pipeline in Italy." *Crime, Media, Culture: An International Journal* 15, no. 1 (March 2019): 143–68. https://doi.org/10.1177/1741659018760106.

Dickinson, Gregory M. "Big Tech's Tightening Grip on Internet Speech." *ProMarket*, August 9, 2021. www.promarket.org/2021/08/09/big-tech-internet-speech-common-carrier-clarence-thomas/.

Dinh, Van. "Minister of Natural Resources and Environment Tran Hong Ha: I Admit My Shortcoming." *Tuoi Tre*, April 28, 2016. https://tuoitre.vn/bo-truong-bo-tnmt-tran-hong-ha-toi-nhan-khuyet-diem-1092358.htm.

Dirty Money. Season 2, Episode 6, "Point Comfort." Aired March 11, 2020, on Netflix.

Disability Visibility Project. "About: Disability Visibility Project." n.d. https://disabilityvisibilityproject.com/about/.

———. "Hire Me!" n.d. https://disabilityvisibilityproject.com/hire-me/.

Disaster STS Network. "Formosa Plastics Global Archive." n.d. https://disaster-sts-network.org/content/formosa-plastics-archive/essay.

———. Home page. n.d. https://disaster-sts-network.org.

Discovering Alternative Straw Use for People with Disabilities: Survey Findings & Analysis. Research & Report by the Disability Organizing (DO) Network with support & collaboration from the Monterey Bay Aquarium & Central Coast Center for Independent Living. December 2018. http://disabilityorganizing.net/uploads/donet-straw-report-012319-ACCESSIBLE.pdf.

Dittoe PR. "Dittoe PR Takes Over Seattle with 'Strawless in Seattle' Campaign." October 26, 2017. https://dittoepr.com/dittoe-pr-takes-seattle-strawless-seattle-campaign/.

Domonoske, Camila. "The World Has Finally Stopped Using Leaded Gasoline: Algeria Used The Last Stockpile." *All Things Considered, NPR*, August 30, 2021. www.npr.org/2021/08/30/1031429212/the-world-has-finally-stopped-using-leaded-gasoline-algeria-used-the-last-stockp.

Dong, Vien. "Vietnamese Demand White House Probe of Mass Fish Deaths." *Voice of America (VOA) News*, May 3, 2016. www.voanews.com/a/vietnamese-demand-white-house-probe-mass-fish-deaths/3314548.html.

Doucette, Kitt. "The Plastic Bag Wars." *Rolling Stone*, July 25, 2011. www.rollingstone.com/politics/politics-news/the-plastic-bag-wars-243547/.

Douglas, Mary. *Purity and Danger: An Analysis of Concept of Pollution and Taboo*. Routledge Classics. London: Routledge, 2005.

The Dow Chemical Company. "A Compelling Investment." n.d. https://investors.dow.com
/en/investors/default.aspx.

———. "Golden Age of Inorganics." n.d. https://corporate.dow.com/en-us/about/company
/history/timeline/golden-age-of-inorganics.html.

Doyle, Julie. "Where Has All the Oil Gone? BP Branding and the Discursive Elimination of
Climate Change Risk." In *Culture, Environment and Eco-Politics*, edited by Nick Hef-
fernan and David Wragg, 200–225. Cambridge: Cambridge Scholars Press, 2011.

Doyle, Julie, Nathan Farrell, and Michael K. Goodman. *Celebrities and Climate Change*.
Vol. 1. Oxford: Oxford University Press, 2017. https://doi.org/10.1093/acrefore/978019
0228620.013.596.

"Dr. Vandana Shiva an Interview by Andy Opel: From Water Crisis to Water Culture."
Cultural Studies 22, nos. 3–4 (May 2008): 498–509. https://doi.org/10.1080/09502380
802012591.

Dsilva, Emmanuel. "Going Surgical on Plastics in Rwanda." *Down to Earth*, December 24,
2019. www.downtoearth.org.in/news/waste/going-surgical-on-plastics-in-rwanda-68446.

Duarte, Marisa Elena. "Connected Activism: Indigenous Uses of Social Media for Shaping
Political Change." *Australasian Journal of Information Systems* 21 (July 19, 2017): 1–12.
https://doi.org/10.3127/ajis.v21i0.1525.

Duignan, Brian. "Tucker Carlson." *Encyclopedia Britannica*. Accessed 14 July 2022.www
.britannica.com/biography/Tucker-Carlson.

Dunaway, Finis. *Seeing Green: The Use and Abuse of American Environmental Images*. Chicago: University of Chicago Press, 2015.

Duncan, Emily M., Alasdair Davies, Amy Brooks, Gawsia Wahidunnessa Chowdhury,
Brendan J. Godley, Jenna Jambeck, Taylor Maddalene, et al. "Message in a Bottle:
Open Source Technology to Track the Movement of Plastic Pollution." Edited by João
Miguel Dias. *PLOS ONE* 15, no. 12 (December 2, 2020): e0242459. https://doi.org/10.1371
/journal.pone.0242459.

Dunning, Barry. "Simply Refusing a Straw with Your Cocktail Isn't Going to Do Much
to Reduce the World's Plastic Consumption." *The Journal*, February 1, 2019. www.the
journal.ie/readme/opinion-tackling-the-global-plastic-problem-requires-action-to
-tackle-poverty-4462978-Feb2019/?utm_source=shortlink.

Dutta, Mohan J., and Mahuya Pal. "Theorizing from the Global South: Dismantling, Resisting, and Transforming Communication Theory." *Communication Theory* 30, no. 4
(December 22, 2020): 349–69. https://doi.org/10.1093/ct/qtaa010.

Duy, Ha. "Formosa Representative Nets Termination with 'Fish or Steel' Comment." *Vietnam Investment Review*, April 27, 2016. https://vir.com.vn/formosa-representative-nets
-termination-with-fish-or-steel-comment-41665.html.

Earth Justice and RISE St. James, "Decision Is a Major Victory for RISE St. James, Louisi-
ana Bucket Brigade, Healthy Gulf, No Waste Louisiana, Center for Biological Diversity,
Earthworks, the Sierra Club, and Others in a Years-Long Fight." Press release, Septem-
ber 15, 2022. https://myemail.constantcontact.com/Victory---Louisiana-Court-Vacates
-Air-Permits-for-Formosa-s-Massive-Petrochemical-Complex-in-Cancer-Alley----
-Sign-Up-For-Toxic-.html?soid=1123016499205&aid=zxFKEnm5Emo.

ECOLEX: The Gateway to Environmental Law. "Wildlife Conservation and Management Act 2013 (No. 47 of 2013)." n.d. www.ecolex.org/details/legislation/wildlife-conservation -and-management-act-2013-no-47-of-2013-lex-faoc134375/.

Econie, Alexis, and Michael L. Dougherty. "Contingent Work in the US Recycling Industry: Permatemps and Precarious Green Jobs." *Geoforum* 99 (February 2019): 132–41. https://doi.org/10.1016/j.geoforum.2018.11.016.

Editorial Board. "In Vietnam, Telling the Truth Is Criminal 'Propaganda.'" *Washington Post*, October 21, 2016. www.washingtonpost.com/opinions/in-vietnam-telling-the-truth -is-criminal-propaganda/2016/10/21/2a5745d2-923c-11e6-a6a3-d50061aa9fae_story.html ?fbclid=IwAR3lsL3TgQtInahlI_BRkskWC5_I86y7aYHonQ4dZ_2JFxKOzV-dtspr6og.

Edwards, Jason A. "Make America Great Again: Donald Trump and Redefining the U.S. Role in the World." *Communication Quarterly* 66, no. 2 (March 15, 2018): 176–95. https://doi.org/10.1080/01463373.2018.1438485.

Ehlers, Nadine, and Shiloh Krupar. *Deadly Biocultures: The Ethics of Life-Making*. Minneapolis: University of Minnesota Press, 2019.

Ehrenreich, Barbara. *Bright-Sided: How Positive Thinking Is Undermining America*. A Metropolitan Book. New York: Picador, 2010.

EJ Atlas. "Environmental Justice Atlas Database." n.d. https://ejatlas.org/.

Eldridge, Scott A., and Patrick Ferrucci, eds. *The Institutions Changing Journalism: Barbarians inside the Gate*. London: Routledge, 2022.

Ellcessor, Elizabeth. "'One Tweet to Make So Much Noise': Connected Celebrity Activism in the Case of Marlee Matlin." *New Media & Society* 20, no. 1 (January 2018): 255–71. https://doi.org/10.1177/1461444816661551.

Ellis, Katie. "#Socialconversations: Disability Representation and Audio Description on Marvel's Daredevil." In *Disability and Social Media: Global Perspectives*, edited by Katie Ellis and Mike Kent, 146–260. London: Routledge, Taylor & Francis Group, 2017.

Ellis, Katie, and Gerard Goggin. "Disability and Media Activism." In *The Routledge Companion to Media and Activism*, edited by Graham Meikle, 355–64. London: Routledge, 2018.

———. "Disability Media Participation: Opportunities, Obstacles and Politics." *Media International Australia* 154, no. 1 (February 2015): 78–88. https://doi.org/10.1177/1329878 X1515400111.

Enck, Judith. Foreword to *The New Coal: Plastics & Climate Change*. Beyond Plastics at Bennington College, 2021. https://static1.squarespace.com/static/5eda9126obbb7e7a4bf 528d8/t/616ef29221985319611a64e0/1634661022294/REPORT_The_New-Coal_Plastics _and_Climate-Change_10-21-2021.pdf.

Environmental Justice Foundation. "Climate Displacement in Bangladesh." 2021. https:// ejfoundation.org/reports/climate-displacement-in-bangladesh.

Eonnet, Estelle. "The Coca-Cola Company, PepsiCo and Nestlé Named Top Plastic Polluters for the Third Year in a Row." Break Free From Plastic Movement. December 2, 2020. www.breakfreefromplastic.org/2020/12/02/top-plastic-polluters-of-2020/.

Ephron, Nora, dir. *Sleepless in Seattle*. TriStar Pictures, 1993.

Eriksen, Marcus. "Opinion: I Thought I'd Seen It All Studying Plastics: Then My Team Found 2,000 Bags in a Camel." *Washington Post*, March 23, 2021. www.washingtonpost .com/opinions/2021/03/23/camels-plastic-bags-pollution-dubai/.

Eriksen, Marcus, Amy Lusher, Mia Nixon, and Ulrich Wernery. "The Plight of Camels Eating Plastic Waste." *Journal of Arid Environments* 185 (February 1, 2021): 104374. https://doi.org/10.1016/j.jaridenv.2020.104374.

Erni, John Nguyet. "Introduction: Affect and Critical Multiculturalism." In *Asia in Visuality, Emotions and Minority Culture: Feeling Ethnic*, edited by John Nguyet Erni, 1–10. Berlin: Springer, 2017.

———, ed. *Visuality, Emotions and Minority Culture*. Vol. 3, *The Humanities in Asia*. Berlin: Springer Berlin Heidelberg, 2017. https://doi.org/10.1007/978-3-662-53861-6.

Escobar, Arturo. *Designs for the Pluriverse: Radical Interdependence, Autonomy, and the Making of Worlds*. Durham, NC: Duke University Press, 2018.

———. "The Global Doesn't Exist." Great Transition Initiative Forum. August 2019. https://greattransition.org/gti-forum/global-local-escobar#endnote_5.

———. *Pluriversal Politics: The Real and the Possible*. Latin America in Translation. Durham, NC: Duke University Press, 2020.

———. *Territories of Difference: Place, Movements, Life, Redes*. New Ecologies for the Twenty-First Century. Durham, NC: Duke University Press, 2008.

The European Union. "Circular Economy: Commission Welcomes Council Final Adoption of New Rules on Single-Use Plastics to Reduce Marine Plastic Litter." May 21, 2019. https://ec.europa.eu/commission/presscorner/detail/en/IP_19_2631.

Everington, Keoni. "KFC Taiwan Stops Serving Plastic Straws." *Taiwan News*, June 12, 2019. www.taiwannews.com.tw/en/news/3722306.

Ewen, Stuart. *PR! A Social History of Spin*. 1st ed. New York: Basic Books, 1996.

Ewert, Benjamin. "Moving beyond the Obsession with Nudging Individual Behaviour: Towards a Broader Understanding of Behavioural Public Policy." *Public Policy and Administration* 35, no. 3 (2020), 337–60. https://doi.org/10.1177/0952076719889090

ExxonMobil. "Our History." February 9, 2023. https://corporate.exxonmobil.com/About-us/Who-we-are/Our-history.

———. "Press Release: ExxonMobil Earns $4.7 Billion in Second Quarter 2021," July 30, 2021. https://corporate.exxonmobil.com/News/Newsroom/News-releases/2021/0730_ExxonMobil-earns-4_7-billion-in-second-quarter-2021.

Fan, Mei-Fang, Chih-Ming Chiu, and Leslie Mabon. "Environmental Justice and the Politics of Pollution: The Case of the Formosa Ha Tinh Steel Pollution Incident in Vietnam." *Environment and Planning E: Nature and Space* 5, no. 1 (2020): 189–206. https://doi.org/10.1177/2514848620973164.

Feit, Steven, and Carroll Muffet. "Pandemic Crisis, Systemic Decline: Why Exploiting the COVID-19 Crisis Will Not Save the Oil, Gas, and Plastic Industries." Center for International Environmental Law (CIEL). 2020. www.ciel.org/wp-content/uploads/2020/04/Pandemic-Crisis-Systemic-Decline-April-2020.pdf.

Ferdman, Soraya. "4 Things We Learned about Plastic Straw Bans from People with Disabilities." *Mashable*, July 16, 2018. https://mashable.com/article/plastic-straw-ban-people-with-disabilities.

Fiesler, Casey, and Nicholas Proferes, "'Participant' Perceptions of Twitter Research Ethics." *Social Media + Society* 4, no. 1 (2018). https://doi.org/10.1177/2056305118763366.

Filloon, Whitney. "California Bans Restaurants From Automatically Giving Out Plastic Straws." Eater. September 21, 2018. www.eater.com/2018/9/21/17886256/california-straw -ban-plastic.

Fleetwood, Nicole R. "Failing Narratives, Initiating Technologies: Hurricane Katrina and the Production of a Weather Media Event." *American Quarterly* 58, no. 3 (2006), 767–89.

Flores, Lisa A. "Between Abundance and Marginalization: The Imperative of Racial Rhetorical Criticism." *Review of Communication* 16, no. 1 (2016): 4–24, https://doi.org/10 .1080/15358593.2016.1183871.

"Flying Toilets in Every Direction." United Nations Habitat. n.d. https://mirror.unhabitat .org/documents/media_centre/wwf2.pdf.

Food and Agriculture Organization of the United Nations. "Fishery and Aquaculture Country Profiles: The Socialist Republic of Viet Nam." 2021. www.fao.org/fishery/facp /vnm/en.

Footprint. "History of Straws: From Invention to Regulation" (blog), April 20, 2022. https://blog.footprintus.com/en/sustainability/history-of-straws-from-invention-to -regulation.

Forbes. "Global 500: #492 Formosa Plastics." n.d. www.forbes.com/companies/formosa -plastics/?list=global2000&sh=4b8061311b15.

———. "Global 2000: #749 Formosa Plastics." n.d. www.forbes.com/companies/formosa -plastics/?list=global2000&sh=609ed2631b15.

Ford, Dave. "COVID-19 Has Worsened the Ocean Plastic Pollution Problem." *Scientific American*, August 17, 2020. www.scientificamerican.com/article/covid-19-has -worsened-the-ocean-plastic-pollution-problem/.

Ford, Helen V., Nia H. Jones, Andrew J. Davies, Brendan J. Godley, Jenna R. Jambeck, Imogen E. Napper, Coleen C. Suckling, Gareth J. Williams, Lucy C. Woodall, and Heather J. Koldewey. "The Fundamental Links between Climate Change and Marine Plastic Pollution." *Science of the Total Environment* 806 (February 2022): 150392. https://doi.org/10 .1016/j.scitotenv.2021.150392.

Formosa Plastics Corporation and Subsidiaries, "Consolidated Financial Statements: For the Three Months Ended March 31, 2018 and 2017." 2018. www.fpc.com.tw/fpcwuploads /files/2018Q1-EN.pdf.

Formosa Plastics Group, "Introduction: Formosa Plastics Group." 2018. www.fpg.com.tw /uploads/images/media-center/ebook-top/FPG%20Introduction2018_en.pdf.

"Formosa Plastics Toxic Tour." Stop Formosa Plastics. n.d. www.stopformosa.org/post /formosa-plastics-toxic-tour.

"Formosa Toxic Waste Spill and Marine Life Disaster in Central Vietnam." Environmental Justice Atlas. n.d. https://ejatlas.org/conflict/formosa-toxic-waste-spill-and-marine-life -disaster-in-central-vietnam.

Fortun, Kim. "Essential2life." *Dialectical Anthropology* 34, no. 1 (March 2010): 77–86. https://doi.org/10.1007/s10624-009-9123-8.

Franta, Benjamin. "Early Oil Industry Disinformation on Global Warming." *Environmental Politics* 30, no. 4 (June 7, 2021): 663–68. https://doi.org/10.1080/09644016.2020 .1863703.

Fraser, Nancy. "Rethinking the Public Sphere: A Contribution to the Critique of Actually Existing Democracy." In *Habermas and the Public Sphere*, edited by Craig J. Calhoun, 109–42. Studies in Contemporary German Social Thought. Cambridge, MA: MIT Press, 1992.

———. "Transnationalizing the Public Sphere: On the Legitimacy and Efficacy of Public Opinion in a Post-Westphalian World." In *Transnationalizing the Public Sphere*, edited by Nancy Fraser and Kate Nash, 8–42. Cambridge, UK: Polity Press, 2014.

Freinkel, Susan. *Plastic: A Toxic Love Story*. New York: Houghton Mifflin Harcourt, 2011.

Fuller, Sascha, Tina Ngata, Stephanie B. Borrelle, and Trisia Farrelly. "Plastics Pollution as Waste Colonialism in Te Moananui." *Journal of Political Ecology* 29, no. 1 (2022): 534–60. https://doi.org/10.2458/jpe.2401.

Funding Universe. "Formosa Plastics Corporation History." n.d. www.fundinguniverse.com/company-histories/formosa-plastics-corporation-history/'.

Funes, Yessenia. "Criminalizing Protests Is a Dangerous Idea." *Gizmodo*, April 20, 2020. https://gizmodo.com/criminalizing-protests-is-a-dangerous-idea-1842624019.

GAIA and Break Free From Plastic. "Turning Back the Tide: GAIA and #BreakFreeFromPlastic Members Respond to Ocean Conservancy's Apology." July 15, 2022. www.no-burn.org/gaia-bffp-respond-to-oc-apology/.

GAIA Asia Pacific @ZeroWasteAsia. "#BreakFreeFromPlasticFriday." Twitter. September 8, 2022. https://twitter.com/zerowasteasia/status/1568104187975827456?s=11&t=3I5JFJZlGWj3irL-Wx__VQ.

Galoustian, Gisele. "Like Humans, Beluga Whales Form Social Networks beyond Family Ties." *Science Daily*, July 10, 2020. www.sciencedaily.com/releases/2020/07/200710212233.htm.

Gamba, Julien, Mohammed Rashed, Abbas Razaghpanah, Juan Tapiador, and Narseo Vallina-Rodriguez. "An Analysis of Pre-Installed Android Software." *41st IEEE Symposium on Security and Privacy*, May 7, 2019. https://doi.org/10.48550/ARXIV.1905.02713.

Garcia, Maria Fernanda. "La Nature—Zero Waste Store." Spotted by Locals. May 25, 2022. www.spottedbylocals.com/mexicocity/la-nature/.

Garcia-Rojas, Claudia. "(Un)Disciplined Futures: Women of Color Feminism as a Disruptive to White Affect Studies.' *Journal of Lesbian Studies* 21, no. 3 (2017): 254–71. https://doi.org/10.1080/10894160.2016.1159072.

Gardiner, Beth. "The Plastics Pipeline: A Surge of New Production Is on the Way." *Yale Environment 360*, December 19, 2019. https://e360.yale.edu/features/the-plastics-pipeline-a-surge-of-new-production-is-on-the-way.

Geneva Environment Network. "Geneva Beat Plastic Pollution Dialogues: Plastics and Human Rights." 2021. www.genevaenvironmentnetwork.org/events/geneva-beat-plastic-pollution-dialogues-plastics-and-human-rights/.

Geyer, Roland, Jenna R. Jambeck, and Kara Lavender Law. "Production, Use, and Fate of All Plastics Ever Made." *Science Advances* 3, no. 7 (July 7, 2017): e1700782. https://doi.org/10.1126/sciadv.1700782.

Ghani, Meera. "We Need a Culture of Care to Stop Climate Breakdown." Medium. October 9, 2018. https://meelaya.medium.com/we-need-a-culture-of-care-to-stop-climate-breakdown-91f739afbd39.

Ghosal, Anirudda. "Cups, Straws, Spoons: India Starts on Single-Use Plastic Ban." 104.5 WOKV. July 1, 2022. www.wokv.com/news/world/cups-straws-spoons/2UWKTPERO OI2NXAHKP5YX7KWIA/.

"Giant Steel Project in Ha Tinh to Start Operation Soon." Vietnam Net. April 23, 2016. http://english.vietnamnet.vn/fms/business/155339/giant-steel-project-in-ha-tinh-to -start-operation-soon.html.

Gilbert, Jeremy. "Against the Commodification of Everything: Anti-Consumerist Cultural Studies in the Age of Ecological Crises." *Cultural Studies* 22, no. 5 (September 2008): 551–66. https://doi.org/10.1080/09502380802245811.

Gillespie, Tarleton. *Custodians of the Internet: Platforms, Content Moderation, and the Hidden Decisions That Shape Social Media.* New Haven, CT: Yale University Press, 2018.

Gilmore, Ruth Wilson, Brenna Bhandar, and Alberto Toscano. *Abolition Geography: Essays towards Liberation.* New York: Verso, 2022.

Gingrich-Philbrook, Craig. "Autoethnography's Family Values: Easy Access to Compulsory Experiences." *Text and Performance Quarterly* 25, no. 4 (2005): 297–314. https:// doi.org/10.1080/10462930500362445.

Glad Network. "Disability-Inclusive Climate Action." n.d. https://gladnetwork.net/search /working-groups/disability-inclusive-climate-action.

Glennon, Robert. "The Unfolding Tragedy of Climate Change in Bangladesh." *Scientific American* (blog), April 21, 2017. https://blogs.scientificamerican.com/guest-blog/the -unfolding-tragedy-of-climate-change-in-bangladesh/.

Global Alliance for Incinerator Alternatives (GAIA). "Plastic Pollution and Waste Incineration." 2019. www.no-burn.org/wp-content/uploads/Plastic-x-Incineration-2019.pdf.

———. "The Zero Waste Master Plan: Companion Guide for Organizers." October 13, 2020. https://zerowasteworld.org/wp-content/uploads/GAIA_ZWMP_companion-guide _final_10.13.2020.pdf.

Global Alliance of Waste Pickers (@globalrec_org). Verified. "[BREAKING] It Is a Historical Moment for the #WastePickers Movement!" Twitter. March 2, 2022. https:// twitter.com/globalrec_org/status/1499043599677378568?s=11.

Global Environmental Justice Project, UC Santa Barbara. *Environmental Justice Struggles in Prisons and Jails around the World: The 2020 Annual Report of the Prison Environmental Justice Project.* 2020. https://gejp.es.ucsb.edu/sites/default/files/sitefiles/publication /Prison%20EJ%20Project%202020%20Report-compressed.pdf.

———. "What Is the GEJP?" n.d. https://gejp.es.ucsb.edu/.

Global High Seas Marine Preserve. "About Us: Global High Seas Marine Preserve." n.d. https://savingoceans.org/about-us.

Global Newswire. "Shale Gas Creating Renaissance in US Plastics Manufacturing." May 13, 2015. www.globenewswire.com/en/news-release/2015/05/13/1121802/0/en/Shale-Gas -Creating-Renaissance-in-U-S-Plastics-Manufacturing.html.

Global Witness. "About Us." n.d. www.globalwitness.org/en/about-us/.

Gómez, Eduardo J. "Coca-Cola's Political and Policy Influence in Mexico: Understanding the Role of Institutions, Interests and Divided Society." *Health Policy and Planning* 34, no. 7 (September 1, 2019): 520–28. https://doi.org/10.1093/heapol/czz063.

Gómez-Barris, Macarena. *The Extractive Zone: Social Ecologies and Decolonial Perspectives.* Dissident Acts. Durham, NC: Duke University Press, 2017.

Gordon, Constance. "Troubling 'Access': Rhetorical Cartographies of Food (In)Justice and Gentrification." PhD diss., University of Colorado, Boulder, 2018.

Gorsevski, Ellen W. "Wangari Maathai's Emplaced Rhetoric: Greening Global Peacebuilding." *Environmental Communication* 6, no. 3 (September 2012): 290–307. https://doi .org/10.1080/17524032.2012.689776.

Gössling, Stefan. "Celebrities, Air Travel, and Social Norms." *Annals of Tourism Research* 79 (2019). https://doi.org/10.1016/j.annals.2019.102775.

Graham, David A. "Trump Has No Shame." *The Atlantic*, September 24, 2019. www .theatlantic.com/ideas/archive/2019/09/trump-fears-only-consequences/598657/.

Gramsci, Antonio. *Prison Notebooks.* Vol. 1. Edited by Joseph A. Buttigieg. European Perspectives. New York: Columbia University Press, 2011.

———. *Prison Notebooks.* Vol. 2. Edited by Joseph A. Buttigieg. European Perspectives. New York: Columbia University Press, 2011.

Grandin, Greg. *The End of Myth: From the Frontier to the Border Wall in the Mind of America.* New York: Metropolitan Books, 2020.

Graziano, M., and K. Gillingham. "Spatial Patterns of Solar Photovoltaic System Adoption: The Influence of Neighbors and the Built Environment." *Journal of Economic Geography* 15, no. 4 (July 1, 2015): 815–39. https://doi.org/10.1093/jeg/lbu036.

Grech, Shaun, and Karen Soldatic. "Disability and Colonialism: (Dis)Encounters and Anxious Intersectionalities." *Social Identities* 21, no. 1 (January 2, 2015): 1–5. https://doi.org /10.1080/13504630.2014.995394.

The Green Belt Movement. "Who We Are." n.d. www.greenbeltmovement.org/.

Green Trees. *An Overview of the Marine Life Disaster in Vietnam* (report). 2019. https:// the88project.org/wp-content/uploads/2019/08/AnOverviewOfTheMarineLifeDisaster InVietnam-Final.pdf.

Greene, Ronald Walter, and Kevin Douglas Kuswa. "'From the Arab Spring to Athens, From Occupy Wall Street to Moscow': Regional Accents and the Rhetorical Cartography of Power." *Rhetoric Society Quarterly* 42, no. 3 (May 2012): 271–88. https://doi.org /10.1080/02773945.2012.682846.

Greenpeace. "The Climate Emergency Unpacked: How Consumer Goods Companies Are Fueling Big Oil's Plastic Expansion." 2021. www.greenpeace.org/usa/reports/the -climate-emergency-unpacked/.

———. "Throwing Away the Future: How Companies Still Have It Wrong on Plastic Pollution 'Solutions.'" 2019. https://drive.google.com/file/d/1vAcibVwP7yxxvJp3cZPshRF AvXmVxmYg/view.

Greenpeace International. "Story of a Plastic Bottle-Greenpeace." April 8, 2021. YouTube video. www.youtube.com/watch?v=CLeccbkBZzs.

GreenpeaceUSA. "Do you know what's REALLY behind the bottles of @CocaCola and @Pepsi?" Twitter. July 18, 2022. https://twitter.com/greenpeaceusa/status/15487292893 77341442.

Greenspan, Rachel E. "Meet the Man Who Popularized the Viral #Trashtag Challenge Getting People around the World Cleaning Up." *Time Magazine*, March 12, 2019. https:// time.com/5549019/trashtag-interview/.

Grenier, Adrian (@adriangrenier). Verified. "Kicking off #StrawlessInSeattle with the @Mariners and @LonelyWhale." Twitter. September 1, 2017. https://twitter.com/adrian grenier/status/903796886833913858.

Gribkoff, Elizabeth. "In Depth: First-of-Its Kind Testing Points to Dangers and Unknowns of PFAS in Clothing." Environmental Health Network. February 15, 2022. www.ehn.org /pfas-clothing-2656587709.html.

Grossberg, Lawrence. "On Postmodernism and Articulation: An Interview with Stuart Hall." In *Stuart Hall: Critical Dialogues in Cultural Studies*, edited by David Morley and Kuang-Hsing Chen, 131–150. London: Routledge, 1996.

———. *We Gotta Get out of This Place: Popular Conservatism and Postmodern Culture.* London: Routledge, 2013.

The Guardian. "Mother Mushroom: Vietnam Dissident and Blogger Arrives in US." October 17, 2018. www.theguardian.com/world/2018/oct/17/mother-mushroom-freed -vietnam-blogger-released-prison.

———. "Vietnam Declares US-Based Activist Group Is a Terrorist Organization." October 7, 2016. www.theguardian.com/world/2016/oct/07/vietnam-viet-tan-terrorists -dissent.

Guitiérrez Aguilar, Raquel. *Horizontes Comunitaros-Populares.* Madrid: Traficantes de Sueños, 2017.

Gumbs, Alexis Pauline. *Undrowned: Black Feminist Lessons from Marine Mammals.* Emergent Strategy Series. Chico, CA: AK Press, 2020.

Ha, Gwen. "Vietnamese Authorities Send 'Thugs' to Beat Activists." Translated by Viet Ha and Brooks Boliek. *Radio Free Asia*, February 14, 2017. www.rfa.org/english/news /vietnam/vietnamese-authorities-send-02142017143556.html.

Ha, Viet, and Brooks Boliek, trans. "Vietnam Cracks Down on Dissenters." *Radio Free Asia*, November 10, 2016. www.rfa.org/english/news/vietnam/vietnam-cracks-down -on-dissenters-11102016154306.html?fbclid=IwAR0CnprzQzvrkE18bTe6ErGMywwvF4 Xo2WQ19tOlq_NvuEtGTQLKEuz-qoA.

Haberman, Maggie, and Emily Cochrane. "Trump Mocks Al Franken for Quick Resignation Over Claims of Sex Misconduct." *New York Times*, October 4, 2018. www.nytimes .com/2018/10/04/us/politics/trump-al-franken-minnesota.html.

Hagmann, David, Emily H. Ho, and George Loewenstein. "Nudging Out Support for a Carbon Tax." *Nature Climate Change* 9 (2019): 484–89. https://doi-org.colorado.idm .oclc.org/10.1038/s41558-019-0474-0.

Hall, Stuart. "Notes on Deconstructing 'the Popular.'" In *People's History and Socialist Theory*, edited by Samuel Raphael, 185–92. London: Routledge & Kegan Paul, 1981.

Hall, Stuart, Chas Critcher, Tony Jefferson, John Clarke, and Brian Roberts. *Policing the Crisis: Mugging, the State, and Law and Order.* Critical Social Studies. London: Macmillan, 1978.

Hamraie, Aimi. *Building Access: Universal Design and the Politics of Disability.* Minneapolis: University of Minnesota Press, 2017.

Hanson, Carolyn. "Straw Wars: This Controversy Is Proof of Our Culture's Anti-Disability Bias." *Nylon*, August 7, 2018. www.nylon.com/articles/straws-unnecessary-people -disabilities.

Hao, Karen. "How to Poison the Data That Big Tech Uses to Surveil You." *MIT Technology Review*, March 5, 2021. www.technologyreview.com/2021/03/05/1020376/resist-big-tech-surveillance-data/.

Haraway, Donna. *Simians, Cyborgs, and Women: The Reinvention of Nature.* New York: Routledge, 1991.

———. *Staying with the Trouble: Making Kin in the Chthulucene.* Durham, NC: Duke University Press, 2016.

———. "Situated Knowledges: The Science Question in Feminism and the Privilege of Partial Perspective." *Feminist Studies* 14, no. 3 (1988): 575–99. https://doi.org/10.2307/3178066.

Harold, Christine. *Things Worth Keeping: The Value of Attachment in a Disposable World.* Minneapolis: University of Minnesota Press, 2020.

Hartnett, Stephen J. *A World of Turmoil: The United States, China, and Taiwan in the Long Cold War.* East Lansing: Michigan State University Press, 2021.

Hatfield, Joe Edward. "Toxic Identification: #Twinks4Trump and the Homonationalist Rearticulation of Queer Vernacular Rhetoric." *Communication Culture & Critique* 11 (2018) 147–61. https://doi.org/10.1093/ccc/tcx006.

Hauser, Gerard A. *Vernacular Voices: The Rhetoric of Publics and Public Spheres,* 2nd ed. Columbia: University of South Carolina Press, 2022.

Hautea, Samantha, Perry Parks, Bruno Takahashi, and Jing Zeng. "Showing They Care (Or Don't): Affective Publics and Ambivalent Climate Activism on TikTok." *Social Media + Society* 7, no. 2 (April 2021): 205630512110123. https://doi.org/10.1177/20563051211012344.

Heglar, Mary Annaïse. "Home Is Always Worth It." Medium, September 12, 2019. https://medium.com/@maryheglar/home-is-always-worth-it-d2821634dcd9.

———. "I Work in the Environmental Movement: I Don't Care If You Recycle." *Vox,* June 4, 2019. www.vox.com/the-highlight/2019/5/28/18629833/climate-change-2019-green-new-deal.

Hell on Wheels. "Podcast: What It Feels Like to Die." Episode 1, June 14, 2018. YouTube video. www.youtube.com/watch?v=iIAhoxnro_w.

Helvarg, David. *Blue Frontier: Dispatches from America's Ocean Wilderness.* 2nd ed. San Francisco: Sierra Club Books, 2006.

———. *The War against the Greens: The "Wise-Use" Movement, the New Right, and Anti-Environmental Violence.* San Francisco: Sierra Club Books, 1997.

Hendrickson, John. "Anthony Bourdain's Meal with Obama Was a Proud American Moment." *Rolling Stone,* June 8, 2018. www.rollingstone.com/culture/culture-news/anthony-bourdains-meal-with-obama-was-a-proud-american-moment-629690/.

Henn, Jamie. "Clean Creatives Pressure Wavemaker to Stop Greenwashing Chevron's Climate Record." Clean Creatives, February 11, 2021. https://cleancreatives.org/news/clean-creatives-pressure-wavemaker-to-stop-greenwashing-chevrons-climate-record.

Hern, Alex. "Facebook Says It Has Reached Net Zero Emissions." *The Guardian,* April 16, 2021. www.theguardian.com/technology/2021/apr/16/facebook-says-it-has-reached-net-zero-emissions.

Hernandez, Nicolas C., Cristi C. Horton, Danielle Endres, and Tarla Rai Peterson. "The Frontier Myth in U.S. Offshore Wind Energy Communication." *Frontiers in Communication* 4 (November 1, 2019): 57. https://doi.org/10.3389/fcomm.2019.00057.

Heumann, Judith E. *Being Heumann: An Unrepentant Memoir of a Disability Rights Activist*. Boston: Beacon Press, 2020.

Heumann, Judith E., and Kristen Joiner. *Rolling Warrior: The Incredible, Sometimes Awkward, True Story of a Rebel Girl on Wheels Who Helped Spark a Revolution*. Boston: Beacon Press, 2021.

Hickel, Jason (@jasonhickel). "Environmentalism without class struggle is using paper straws while the rich take 9 minute flights in their private jets." July 30, 2022. Twitter. https://twitter.com/jasonhickel/status/1553287090082926594?s=11&t=Wf8UzmgGGW Ic8bcYPGPL4g.

Higgins, Robert R. "Race, Pollution, and the Mastery of Nature:" *Environmental Ethics* 16, no. 3 (1994): 251–64. https://doi.org/10.5840/enviroethics199416315.

Hilary, Russ. "Coca-Cola, Criticized for Plastic Pollution, Pledges 25% Reusable Packaging." Reuters, February 15, 2022. www.reuters.com/business/sustainable-business/coca -cola-criticized-plastic-pollution-pledges-25-reusable-packaging-2022-02-10/.

Hinkel, Jason. "Quantifying National Responsibility for Climate Breakdown: An Equality-Based Attribution Approach for Carbon Dioxide Emissions in Excess of the Planetary Boundaries." *The Lancet* 4, no. 9 (September 1, 2020): E399–E404. https://doi.org/10 .1016/S2542-5196(20)30196-0.

"History: Taiwan." Taiwan government official website. n.d. www.taiwan.gov.tw/content _3.php.

Ho, Vivian. "'People Need Them': The Trouble with the Movement to Ban Plastic Straws." *The Guardian*, August 25, 2018. www.theguardian.com/us-news/2018/aug/25/plastic -straw-ban-california-people-with-disabilities.

Hodges, Heather E., and Galen Stocking. "A Pipeline of Tweets: Environmental Movements' Use of Twitter in Response to the Keystone XL Pipeline." *Environmental Politics* 25, no. 2 (March 3, 2016): 223–47. https://doi.org/10.1080/09644016.2015.1105177.

Hofman, Justin (@justinhofman). Verified. "It's a Photo That I Wish Didn't Exist." Instagram. September 12, 2017. www.instagram.com/p/BY8iyqxHx4r/?hl=en&taken-by= justinhofman.

Holden, Emily. "US Produces Far More Waste and Recycles Far Less of It Than Other Developed Countries." *The Guardian*, July 3, 2019. www.theguardian.com/us-news /2019/jul/02/us-plastic-waste-recycling.

Holmes, Paul. "Taking BP beyond Petroleum." *PRovoke Media*, April 18, 2001. www .provokemedia.com/latest/article/taking-bp-beyond-petroleum.

hooks, bell, and Amalia Mesa-Bains. *Homegrown: Engaged Cultural Criticism*. New York: Routledge, 2018.

Horton, Richard. "Offline: COVID-19 Is Not a Pandemic." *The Lancet* 396, no. 10255 (September 2020): 874. https://doi.org/10.1016/S0140-6736(20)32000-6.

Houreld, Katharine, and John Ndiso. "Kenya Imposes World's Toughest Law against Plastic Bags." Reuters, August 28, 2017. www.reuters.com/article/us-kenya-plastic/kenya -imposes-worlds-toughest-law-against-plastic-bags-idUSKCN1B8oNW.

Human Rights Watch. "Free Vietnam's Political Prisoners!" March 24, 3022. www.hrw.org /video-photos/interactive/2022/03/24/free-vietnams-political-prisoners.

———. *It's As If They're Poisoning Us: The Health Impact of Plastic Recycling in Turkey.* 2022. www.hrw.org/sites/default/files/media_2022/09/turkey0922web_0.pdf.

———. "Vietnam: Activist Facing Prison Term for Facebook Posts." November 14, 2019. www.hrw.org/news/2019/11/14/vietnam-activist-facing-prison-term-facebook-posts.

———. "Vietnam: New Wave of Arrests of Critics." January 27, 2017. www.hrw.org/news/2017/01/28/vietnam-new-wave-arrests-critics?fbclid=IwAR3ApAwVD8jAw6Ow_5u3fU2k5-7VruOh4ZChBYWv1lbo5XOSWGu1qUdNwQI.

Hung, Duc. "Dead Dolphin Washes up on Central Vietnam Coast." *VN Express International*, December 16, 2016. https://e.vnexpress.net/news/news/dead-dolphin-washes-up-on-central-vietnam-coast-3514688.html?fbclid=IwAR3pwoUQdpp59ZRSYUE3JjDBvGUWLdnY9QvkmIayBrzwsEBaJzsTHX-RrFc.

———. "Suspected Gas Leakage Kills Three Workers at Formosa Steel Plant." *VN Express International*, August 27, 2021. https://e.vnexpress.net/news/news/suspected-gas-leakage-kills-three-workers-at-formosa-steel-plant-4347204.html.

Hung, Duc, and Vuong Anh. "Taiwanese Steel Firm to Sack Executive over Remarks on Vietnam's Mass Fish Deaths." *VN Express International*, April 27, 2016. https://e.vnexpress.net/news/news/taiwanese-steel-firm-to-sack-executive-over-remarks-on-vietnam-s-mass-fish-deaths-3394574.html.

Hvistendahl, Mara. "Toxic Tiles: How Vinyl Flooring Made with Uyghur Forced Labor Ends Up at Big Box Stores." The Intercept. June 14, 2022. https://theintercept.com/2022/06/14/china-uyghur-forced-labor-pvc-home-depot/.

Imperial War Museums. "Why Whales Were Vital in the First World War." 2021. www.iwm.org.uk/history/why-whales-were-vital-in-the-first-world-war.

Ingraham, Chris. *Gestures of Concern*. Durham, NC: Duke University Press, 2020.

Institute Open Diplomacy. "Stepping up Climate Action: Conference with H. E Prof. Judi Wakhungu." July 9, 2021. YouTube video. www.youtube.com/watch?v=ffB5k-yF6Jg.

International Center for Not-for-Profit Law (ICNL). "US Protest Law Tracker." n.d. www.icnl.org/usprotestlawtracker/?location=&status=&issue=6&date=&type=.

International Institute for Sustainable Development. "Doubling Back and Doubling Down: G20 Scorecard on Fossil Fuel Funding." 2020. www.iisd.org/system/files/2020-11/g20-scorecard-report.pdf.

International Whaling Commission. "Welcome to the IWC Web Archive." n.d. https://archive.iwc.int/pages/home.php?login=true.

I-PEN: toxics-free (@ToxicsFree). "Festering Outrage." Twitter. October 4, 2016. https://twitter.com/ToxicsFree/status/783192491122122752.

IsupportbanplasticsKE. (@banplasticsKE.) Facebook page. n.d. Shared with permission. www.facebook.com/pg/banplasticsKE/photos/.

Ivanova, Irina. "States Take Aim at Ubiquitous 'Chasing Arrow' Symbol on Products That Aren't Recycled." *CBS News*, September 14, 2021. www.cbsnews.com/news/recycling-symbol-false-advertising-california-oregon-new-york/.

Ives, Dune. "The Gateway Plastic." Re:Wild. October 19, 2017. www.rewild.org/news/the-gateway-plastic.

———. "Hey, Could You #StopSucking?" Medium. n.d. https://medium.com/lonely-whale/hey-could-you-stopsucking-f4c9cd5b98e2.

Ives, Mike. "Outrage over Fish Kill in Vietnam Simmers 6 Months Later." *New York Times*, October 3, 2016. www.nytimes.com/2016/10/04/world/asia/formosa-vietnam-fish.html.

Jackson, Sarah J., Moya Bailey, and Brooke Foucault Welles. *#HashtagActivism: Networks of Race and Gender Justice.* Cambridge, MA: The MIT Press, 2020.

Jackson, Sarah J., and Brooke Foucault Welles. "Hijacking #myNYPD: Social Media Dissent and Networked Counterpublics." *Journal of Communication* 65, no. 6 (December 2015): 932–52. https://doi.org/10.1111/jcom.12185.

Jacoby, Karl. *Crimes against Nature: Squatters, Poachers, Thieves, and the Hidden History of American Conservation.* Berkeley: University of California Press, 2001.

Jacquet, Jennifer. *Is Shame Necessary? New Uses for an Old Tool.* 1st Vintage Books ed. New York: Vintage Books, a division of Penguin Random House LLC, 2016.

Jambeck, Jenna R., Roland Geyer, Chris Wilcox, Theodore R. Siegler, Miriam Perryman, Anthony Andrady, Ramani Narayan, and Kara Lavender Law. "Plastic Waste Inputs from Land into the Ocean." *Science* (2015): 347, 708–711.

Jambeck, Jenna R., and Kyle Johnsen. "Citizen-Based Litter and Marine Debris Data Collection and Mapping." *Computing in Science & Engineering* 17, no. 4 (July 2015): 20–26. https://doi.org/10.1109/MCSE.2015.67.

"Jamie Henn Profile." LinkedIn. n.d. www.linkedin.com/in/jamiehenn/.

Jarvis, Jeff. "Coca-Cola Live Positively." n.d. www.jefftjarvis.com/cocacola-live-positively.

"Jemez Principles for Democratic Organizing." EJnet.org: Web Resources for Environmental Justice Activists. 1991. www.ejnet.org/ej/jemez.pdf.

Jendukie. "BTS (방탄소년단)—Whalien 52 (Color Coded Lyrics Han|Rom|Eng)." n.d. YouTube video. www.youtube.com/watch?v=N6o-coKG67Y.

Jenner, Lauren C., Jeanette M. Rotchell, Robert T. Bennett, Michael Cowen, Vasileios Tentzeris, and Laura R. Sadofsky. "Detection of Microplastics in Human Lung Tissue Using μFTIR Spectroscopy." *Science of the Total Environment* 831 (July 20, 2022). https://doi.org/10.1016/j.scitotenv.2022.154907.

Jennings, Freddie J., Myria W. Allen, and Thuy Le Vu Phuong. "More Plastic Than Fish: Partisan Responses to an Advocacy Video Opposing Single-Use Plastics." *Environmental Communication* 15, no. 2 (February 17, 2021): 218–34. https://doi.org/10.1080/17524032.2020.1819363.

Johnson, Ayana Elizabeth. "Projects: Ayana Elizabeth Johnson." n.d. www.ayanaelizabeth.com/projects.

Johnson, Jenell. "Breaking Down: On Publicity as Capacity." *Rhetoric Society Quarterly* 50, no. 3 (May 26, 2020): 175–83. https://doi.org/10.1080/02773945.2020.1752128.

Johnson, Jenell, and Krista Kennedy. "Introduction: Disability, In/Visibility, and Risk." *Rhetoric Society Quarterly* 50, no. 3 (2020): 161–65. https://doi.org/10.1080/02773945.2020.1752126.

Johnson, Paul Elliott. *I the People: The Rhetoric of Conservative Populism in the United States.* Rhetoric, Culture, and Social Critique. Tuscaloosa: University of Alabama Press, 2022.

Jordan, Chris. "Midway: Message from the Gyre." *Lens Culture*, 2009. www.lensculture.com/articles/chris-jordan-midway-message-from-the-gyre#slideshow.

Jordan, Rob. "Do Plastic Straws Really Make a Difference?" *Stanford Earth Matters*, September 18, 2018. https://earth.stanford.edu/news/do-plastic-straws-really-make-difference#gs.1b8a01.

Just Transition Alliance. "Lifecycle of Plastics with an Environmental Justice Lens." Fact-sheet, 2020. http://jtalliance.org/wp-content/uploads/2020/02/Final-6.19-Lifecycle-of-Plastics-3.pdf.

Just Transition Alliance (@jtalliance). "#EarthDay." Twitter. April 22, 2022. https://twitter.com/jtalliance/status/1517506445876424710.

Justice, Tristan. "Joe Biden Says He's in Favor of Banning Plastic Straws." *The Federalist*, February 11, 2020. https://thefederalist.com/2020/02/11/joe-biden-says-hes-in-favor-of-banning-plastic-straws/.

The Justice Fleet. "The Justice Fleet." n.d. www.thejusticefleet.com/.

Justice For Formosa Victims. "About JFFV." n.d. https://jffv.org/our-mission/.

Kafer, Alison. *Feminist, Queer, Crip*. Bloomington: Indiana University Press, 2013.

Kahiu, Wanuri. *Pumzi*. Focus Features, 2009. https://vimeo.com/46891859.

Kaiser, Jaime. "What Comes after This Plastic-Filled Pandemic?" *Yes Magazine*, May 10, 2021. www.yesmagazine.org/issue/solving-plastic/2021/05/10/pandemic-plastic-what-comes-after?utm_term=Autofeed&utm_medium=Social&utm_source=Twitter#Echobox=1659335420.

Kämmerer, Annette. "The Scientific Underpinnings and Impacts of Shame." *Scientific American*, August 9, 2019. www.scientificamerican.com/article/the-scientific-underpinnings-and-impacts-of-shame/.

Kana, Lauriane Noelle Vofo. "Egyptian Environmental Group Builds 'World Biggest Plastic Pyramid.'" *Africanews*, September 19, 2022. www.africanews.com/2022/09/19/egyptian-environmental-group-builds-world-biggest-plastic-pyramid/.

Kandel, Denise. "Stages in Adolescent Involvement in Drug Use." *Science* 190, no. 4217 (November 28, 1975): 912–14. https://doi.org/10.1126/science.1188374.

Kandel, Denise, and R. Faust. "Sequence and Stages in Patterns of Adolescent Drug Use." *Archives of General Psychiatry* 32, no. 7 (July 1, 1975): 923. https://doi.org/10.1001/archpsyc.1975.01760250115013.

Kane, Caroline. "Kenya's Plastic Bag Ban: An Imperfect But Effective Policy. Fordham Environmental Law Review." *Fordham Environmental Law Review*, January 29, 2020. https://news.law.fordham.edu/elr/2020/01/29/kenyas-plastic-bag-ban-an-imperfect-but-effective-policy/.

Kane, Emy. "Lonely Whale: The Campaign." n.d. https://emy-kane-3kw8.squarespace.com/strawless-in-seattle.

Karasik, Rachel, Tibor Vegh, Zoie Diana, Janet Bering, Juan Caldas, Amy Pickle, Daniel Rittschoff, and John Virdin. "20 Years of Government Responses to the Global Plastic Pollution Program: The Plastics Policy Inventory." Duke Nicholas Institute for Environmental Policy Solutions, 2020. https://nicholasinstitute.duke.edu/sites/default/files/publications/20-Years-of-Government-Responses-to-the-Global-Plastic-Pollution-Problem-New_1.pdf.

Karmarkar, Uma R., and Bryan Bollinger. "BYOB: How Bringing Your Own Shopping Bags Leads to Treating Yourself and the Environment." *Journal of Marketing* 79, no. 4 (July 2015): 1–15. https://doi.org/10.1509/jm.13.0228.

Kaufman, Mark. "The Carbon Footprint Sham." *Mashable*, n.d. https://mashable.com/feature/carbon-footprint-pr-campaign-sham.

Kearney, Matt. "When Plastics Saved Turtles." *Researchers in Museums* (blog), May 25, 2019. https://blogs.ucl.ac.uk/researchers-in-museums/2019/05/25/when-plastics-saved-turtles/.

Keep America Beautiful. "Mission & History." n.d. https://kab.org/about/approach/mission-history/.

Kelly, Lynne. *Song for a Whale*. 1st ed. New York: Delacorte Press, 2019.

Kenner, Alison. *Breathtaking: Asthma Care in a Time of Climate Change*. Minneapolis; London: University of Minnesota Press, 2018.

Kenner, Robert. "Food, Inc." 2008. YouTube video. www.youtube.com/watch?v=zGrpgPQFU3A.

"Kenya Ivory Amnesty Ahead of Record-Breaking Tusk Burning." Phys.org. March 30, 2016. https://phys.org/news/2016-03-kenya-ivory-amnesty-record-breaking-tusk.html.

Kenya Plastics Pact. "Kenya Plastics Pact." n.d. https://kpp.or.ke.

Kidwell, Grant. "Amidst COVID-19, States Reversing Bans on Single-Use Plastics." ALEC. March 26, 2020. https://alec.org/article/amidst-covid-19-states-reversing-bans-on-single-use-plastics/.

Kiger, Patrick. "Jellyfish Invasion Shuts Down Nuclear Reactor." *National Geographic*, October 1, 2013. www.nationalgeographic.com/environment/article/jellyfish-invasion-shuts-down-nuclear-plant.

Kimmerer, Robin Wall. *Braiding Sweetgrass*. Minneapolis, MN: Milkweed Editions, 2013.

Kinefuchi, Etsuko. "Chapter 8: Wangari Maathai and Mottainai: Gifting 'Cultural Appropriation' with Cultural Empowerment." In *The Rhetorical Legacy of Wangari Maathai: Planting the Future*, edited by Eddah M. Mutua, Alberto González, and Anke Wolbert, 137–69. Transnational Communication and Critical/Cultural Studies. Lanham, MD: Lexington Books, 2018.

Kinkaid, Eden, Kelsey Emard, and Nari Senanayake. "The Podcast-as-Method? Critical Reflections on Using Podcasts to Produce Geographic Knowledge." *Geographical Review* 110, nos. 1–2 (January 2, 2020): 78–91. https://doi.org/10.1111/gere.12354.

Klein, Naomi. *This Changes Everything: Capitalism vs. the Climate*. London: Penguin Books, 2014.

Knorr, David. "7 Countries, States, and Cities That Have Banned Plastic Drinking Straws." Great Paper Straws. December 2, 2019. https://greatpaperstraws.com/7-countries-and-cities-that-have-banned-plastic-drinking-straws/.

Kothari, Ashish, Ariel Salleh, Arturo Escobar, Federico Demaria, and Alberto Acosta, eds. *Pluriverse: A Post-Development Dictionary*. New Delhi: Tulika Books and Authorsupfront, 2019.

Kuhn, Timothy. "Communicatively Constituting Organizational Unfolding through Counter-Narrative." In *Counter-Narratives and Organization*, edited by Sanne Frandsen, Timothy Kuhn, and Marianne Wolff Lundholt, 17–42. New York: Routledge, 2017.

———. "Negotiating the Micro-Macro Divide: Communicative Thought Leadership for Theorizing Organization." *Management Communication Quarterly* 26, no. 4 (2012): 543–84.

Kuo, Rachel. "Reflections on #Solidarity: Intersectional Movements in AAPI Communities." In *The Routledge Companion to Asian American Media*, edited by Lori Kido Lopez and Vincent N. Pham, 181–94. Routledge Companions. New York: Routledge, Taylor & Francis Group, 2017.

La Alianza México Sin Plástico (AMSP). "About Us?" n.d. https://alianzamexicosinplastico .org/.

Lã Trường Sơn. "Cá Voi Mắc Cạn Tại Biển Nam Định." October 29, 2018. YouTube video. www.youtube.com/watch?v=8cJZm81-QPI.

Lacey, Marc. "Flower of Africa: A Curse That's Blowing in the Wind." *New York Times*, April 7, 2005. www.nytimes.com/2005/04/07/world/africa/flower-of-africa-a-curse-thats -blowing-in-the-wind.html.

Laclau, Ernesto, and Chantal Mouffe, *Hegemony and Socialist Strategy: Towards a Radical Democratic Politics*. 2nd ed. London: Verso, 2001.

Lantern Books. "Wangari Maathai Talks about the Mottainai Campaign." 2009. YouTube video. www.youtube.com/watch?v=KMw-fP_GRP8.

Lau, Winnie W. Y., et al. [+ 28 authors]. "Evaluating Scenarios toward Zero Plastic Pollution." *Science*, July 23, 2020, 1455–61. https://doi.org/10.1126/science.aba9475.

Laughland, Oliver. "Multibillion-Dollar Louisiana Plastics Plant Put on Pause in a Win for Activists." *The Guardian*, August 28, 2021. www.theguardian.com/us-news/2021/aug/18 /louisiana-plastics-plant-toxic-emissions-cancer-alley.

Lauren. "A Zero Waste Guide to Mexico City." *Northern Lauren* (blog), n.d. https:// northernlauren.com/a-zero-waste-guide-to-mexico-city.

Lavers, Jennifer L., Alexander L. Bond, and Charles Rolsky. "Far from a Distraction: Plastic Pollution and the Planetary Emergency." *Biological Conservation* 272 (2022). https://doi .org/10.1016/j.biocon.2022.109655.

Law, Kara Lavender, Natalie Starr, Theodore R. Siegler, Jenna R. Jambeck, Nicholas J. Mallos, and George H. Leonard. "The United States' Contribution of Plastic Waste to Land and Ocean." *Science Advances* 6, no. 44 (October 30, 2020): eabd0288. https://doi.org /10.1126/sciadv.abd0288.

Learmouth, Imogen. "How the 'Carbon Footprint' Originated as a PR Campaign for Big Oil." *Thred*, September 23, 2020. https://thred.com/change/how-the-carbon-footprint -originated-as-a-pr-campaign-for-big-oil/.

The Leatherback Trust. "Plastic Straw Removed from Sea Turtle's Nostril (Short Version)." August 13, 2015. YouTube video.www.youtube.com/watch?v=d2J2qdOrW44.

Lebrecht, James, and Nicole Newnham. *Crip Camp*. Netflix, 2020.

LeBrón, Marisol. *Policing Life and Death: Race, Violence, and Resistance in Puerto Rico*. Oakland: University of California Press, 2019.

Lee, Ashley. "Invisible Networked Publics and Hidden Contention: Youth Activism and Social Media Tactics under Repression." *New Media & Society* 20, no. 11 (November 2018): 4095–4115. https://doi.org/10.1177/1461444818768063.

Lee, Jessica. "The Last Straw? Seattle Will Say Goodbye to Plastic Straws, Utensils with Upcoming Ban." *Seattle Times*, September 8, 2017. www.seattletimes.com/seattle-news /the-last-straw-seattle-will-say-goodbye-to-plastic-straws-utensils-with-upcoming -ban/.

LeMenager, Stephanie. *Living Oil: Petroleum Culture in the American Century*. Oxford: Oxford University Press, 2014.

Lepore, Jill. "The Right Way to Remember Rachel Carson." *New Yorker*, March 19, 2018. www.newyorker.com/magazine/2018/03/26/the-right-way-to-remember-rachel -carson.

Lerner, Sharon. "Africa's Exploding Plastic Nightmare." The Intercept. April 19, 2020. https://theintercept.com/2020/04/19/africa-plastic-waste-kenya-ethiopia/.

———. "Bottled Water Giant BlueTriton Admits Claims of Recycling and Sustainability Are 'Puffery.'" The Intercept. April 26, 2022. https://theintercept.com/2022/04/26 /plastic-recycling-bottled-water-poland-spring/.

———. "Leaked Audio Reveals How Coca-Cola Undermines Plastic Recycling Efforts." The Intercept. October 18, 2019. https://theintercept.com/2019/10/18/coca-cola-recycling -plastics-pollution/.

———. "Waste Only: How the Plastics Industry Is Fighting to Keep Polluting the World." The Intercept. July 20, 2019. https://theintercept.com/2019/07/20/plastics-industry -plastic-recycling/.

Lerner, Sharon (@fastlerner). Verified. "Destroy Life on Earth. Plant a Tree. Tweet about It." Twitter. September 2, 2021. https://twitter.com/fastlerner/status/1433497730282926087.

Leslie, Heather A., Martin J. M. van Velze, Sicco H. Brandsma, A. Dick Vethaak, Juan J. Garcia-Vallejo, and Marja H. Lamoree. "Discovery and Quantification of Plastic Particle Pollution in Human Blood." *Environment International* 163 (May 2022). https:// doi.org/10.1016/j.envint.2022.107199.

LEW. "Lonely Whale Foundation and Adrian Grenier Partner with Mariners, Sounders and Seahawks on 'Strawless in Seattle' September." *Green Sports Blog*, September 14, 2017. https://greensportsblog.com/lonely-whale-foundation-and-adrian-grenier-partner-with -mariners-sounders-and-seahawks-on-strawless-in-seattle-september/.

Liboiron, Max. "Plastics in the Gut: A Search for Sand on a Rocky Shoreline Upends Colonial Science." *Orion Magazine*, November 19, 2020. https://orionmagazine.org/article /plastics-in-the-gut/.

———. *Pollution Is Colonialism*. Durham, NC: Duke University Press, 2021.

———. "Waste Is Not 'Matter out of Place.'" *Discard Studies*, September 9, 2019. https:// discardstudies.com/2019/09/09/waste-is-not-matter-out-of-place/.

Liboiron, Max, and Josh Lepawsky. *Discard Studies: Wasting, Systems, and Power*. Durham, NC: Duke University Press, 2022.

Liebe, Ulf, Jennifer Gewinner, and Andreas Diekmann. "Large and Persistent Effects of Green Energy Defaults in the Household and Business Sectors." *Nature Human Behavior* 5 (2021): 576–85. https://doi.org/10.1038/s41562-021-01070-3.

Littler, Jo. "Making Fame Ordinary: Intimacy, Reflexivity, and 'Keeping It Real.'" *Mediactive*, no. 2 (January 2004): 8–25.

———. *Radical Consumption: Shopping for Change in Contemporary Culture*. Maidenhead, UK: McGraw-Hill/Open University Press, 2009.

Locker, Melissa. "Here's Why Elizabeth Warren Thinks Plastic Straw Bans Are Straw Dogs." Fast Company, September 5, 2019. www.fastcompany.com/90399831/heres-why -elizabeth-warren-thinks-plastic-straw-bans-are-straw-dogs.

Loeffelholz, Tracy Matsue. "The First Giant Step? Unpackage." *Yes! Magazine*, May 10, 2021. www.yesmagazine.org/issue/solving-plastic/2021/05/10/the-first-giant-step-unpackage.

Loepp, Don. "Hashtags, Real Problems, Symbols, and What They Mean for the Industry." *Plastics News*, October 11, 2017. www.plasticsnews.com/this-week-issue/archives?year =2017.

Lonely Whale. *Celebrating 5 Years*. 2020. https://static1.squarespace.com/static/5eb2 b8ab5c21c46cdd69f417/t/60380ff23de3d440cb7572dc/1614286834350/LonelyWhale _AnniversaryBrochure_digital.pdf.

———. "EP 1: Against the Current with Adrian Grenier." August 31, 2020. YouTube video. www.youtube.com/watch?v=y-Gqcm75Te4.

———. "#HydrateLike."June 4, 2019. YouTube video. www.youtube.com/watch?v=l1ahu BI4QFU&t=52s.

———. Mission. n.d. www.lonelywhale.org/.

———. "#StopSucking | Lonely Whale | For A Strawless Ocean." August 8, 2017. YouTube video. www.youtube.com/watch?v=Q91-23B8yCg.

———. "#StopSucking for a Strawless Ocean." n.d. https://lonely-whale-25l3.squarespace .com/global-team.

———. "Strawless In Seattle | "Straw Thief" Reveal | Lonely Whale | For A Strawless Ocean." September 26, 2017. Vimeo video. https://vimeo.com/235663203.

———. "Sucker Punch | Lonely Whale | For A Strawless Ocean." March 31, 2017. YouTube video. www.youtube.com/watch?v=rfFpz8KM-9E.

———. "Understanding Plastic Pollution." n.d. www.strawlessocean.org/faq.

———. "What We've Achieved So Far." 2018. https://web.archive.org/web/20180726194258 /https://www.strawlessocean.org/lonelywhale/.

Lonely Whale (@lonelywhale). Verified. "Last Night a Guy in a Turtle Costume." Instagram repost. October 31, 2018. www.instagram.com/p/BpmfB1kF5qr/.

Lonely Whale Foundation. "Show Me the Work: Strawless Ocean." n.d. https://show-methe.work/strawlessocean.

"Lonely Whale Foundation: #StopSucking by POSSIBLE." The Drum. August 2017. www .thedrum.com/creative-works/project/possible-lonely-whale-foundation-stopsucking.

"Lonely Whale (Philanthropy)." LinkedIn. n.d. www.linkedin.com/company/lonelywhale.

LOOP (Life Out of Plastic). "LOOP (Life Out of Plastic)." n.d. http://loop.pe/.

Lopez, Lori Kido. "Asian America Gone Viral: A Genealogy of Asian American YouTubers and Memes." In *The Routledge Companion to Asian American Media*, edited by Lori Kido Lopez and Vincent N. Pham, 157–69. New York: Routledge, 2017.

Los Angeles Times. "Thatcher Government to Sell 31.5%BP Stake in Biggest-Ever Stock Offer," August 21, 1987. www.latimes.com/archives/la-xpm-1987-08-21-fi-2366-story.html.

"Love Cans—Coca-Cola." Ogilvy. n.d. www.ogilvy.com/work/love-cans.

Lovett, Jamie. "Eddie Murphy Fans Are Furious at Bill Cosby: 'Have a Coke and a Smile and Shut the F--- Up.'" *Comicbook*, December 23, 2019. https://comicbook.com/tv -shows/news/eddie-murphy-snl-bill-cosby-statement-fans-react/.

Lugones, María. "Playfulness, 'World'-Travelling, and Loving Perception." *Hypatia* 2, no. 2 (1987): 3–19.

———. "Purity, Impurity, and Separation." *Signs* 19, no. 2 (1994): 458–79.

Lusher, Amy L., Valentina Tirelli, Ian O'Connor, and Rick Officer. "Microplastics in Arctic Polar Waters: The First Reported Values of Particles in Surface and Sub-Surface

Samples." *Scientific Reports* 5, no. 1 (December 2015): 14947. https://doi.org/10.1038 /srep14947.

Lustgarten, Abraham. "How Climate Change Is Contributing to Skyrocketing Rates of Infectious Disease." *ProPublica*, May 7, 2020. www.propublica.org/article/climate -infectious-diseases.

Luxon, Emily Matthews. "Economics-Oriented Discourse Strategies in Environmental Advocacy." *Environmental Communication* 13, no. 3 (April 3, 2019): 320–34. https://doi .org/10.1080/17524032.2019.1567569.

Ma, Wayne. "Facebook and Google Balance Booming Business with Censorship Pressure in Vietnam." *The Information*, December 10, 2019. www.theinformation.com /articles/facebook-and-google-balance-booming-business-with-censorship-pressure -in-vietnam.

Maathai, Wangari. *Unbowed: A Memoir*. 1st Anchor Books ed. New York: Anchor Books, 2007.

Madison, D. Soyini. *Acts of Activism: Human Rights as Radical Performance*. Cambridge: Cambridge University Press, 2010. https://doi.org/10.1017/CBO9780511675973.

———. *Critical Ethnography: Method, Ethics, and Performance*. 3rd ed. Los Angeles: Sage, 2020.

Mah, Alice. *Plastics Unlimited: How Corporations Are Fueling the Ecological Crisis and What We Can Do About It*. Cambridge, UK: Polity, 2022.

Mahdawi, Arwa. "Woke-Washing Brands Cash in on Social Justice: It's Lazy and Hypocritical." *The Guardian*, August 10, 2018. www.theguardian.com/commentisfree/2018 /aug/10/fellow-kids-woke-washing-cynical-alignment-worthy-causes.

Mahmud, Arshad. "Cash and Carry On." *The Guardian*, March 26, 2002. www.theguardian .com/society/2002/mar/27/guardiansocietysupplement7.

Makah. "The Makah Whaling Tradition." n.d. https://makah.com/makah-tribal-info /whaling/.

Maki, Alexander. "The Potential Cost of Nudges." *Nature Climate Change* 9, no. 439 (2019). https://doi-org.colorado.idm.oclc.org/10.1038/s41558-019-0491-z.

"Manipulating the Masses and Predicting the Future—Edward Bernays and W. Howard Chase." *Drilled Podcast*, written and produced by Amy Westervelt. Season 3, Episode 6. www.drilledpodcast.com/s3-the-mad-men-of-big-oil/.

Marez, Curtis. *Farmworker Futurism: Speculative Technologies of Resistance*. Berkeley: University of California Press.

"Marine Debris Tracker." n.d. https://debristracker.org/.

Massive Change Network. "Case Study: What Happens When a Global Company Is Designed for Perpetuity? Coca-Cola: Communicating Through Actions." 2021. www .massivechangenetwork.com/coca-cola-bruce-mau.

Mbaria, John, and Mordecai Ogada. *The Big Conservation Lie: The Untold Story of Wildlife Conservation in Kenya*. Auburn, WA: Lens & Pens, 2017.

Mbembe, Achilles. "Necropolitics." Translated by Libby Meintjes. *Public Culture* 15, no. 1 (2003): 11–40.

McCarthy, Joe. "Fed Up with Plastic, This Man Got Kenya to Ban It." *Global Citizen*, May 4, 2018. www.globalcitizen.org/en/content/kenya-plastic-ban-champion-james-wakibia/.

———. "Taiwan Announces Ban on All Plastic Bags, Straws, and Utensils." *Global Citizen*, February 22, 2018. www.globalcitizen.org/en/content/taiwan-ban-on-plastic-bags-straws-utensils-contain/.

McGuffey, James Coleman. "Copious Dwelling in a Sinking Landscape." In *Water, Rhetoric, and Social Justice: A Critical Confluence*, edited by Casey R. Schmitt, Christopher S. Thomas, and Theresa R. Castor, 193–214. Lanham, MD: Rowman and Littlefield, 2020.

McKibben, Bill. "Multiplication Saves the Day." *Orion Magazine*, November 11, 2008. https://orionmagazine.org/article/multiplication-saves-the-day/.

McQuay, Bill, and Christopher Joyce. "It Took a Musician's Ear to Decode the Complex Song in Whale Calls." *NPR*, August 6, 2015. www.npr.org/2015/08/06/427851306/it-took-a-musicians-ear-to-decode-the-complex-song-in-whale-calls.

McVeigh, Karen. "NGO Retracts 'Waste Colonialism' Report Blaming Asian Countries for Plastic Pollution." *The Guardian*, September 15, 2022. www.theguardian.com/environment/2022/sep/15/ocean-conservancy-ngo-retracts-2015-waste-colonialism-report-blaming-five-asian-countries-for-most-plastic-pollution.

———. "World Leaders Agree to Draw up 'Historic' Treaty on Plastic Waste." *The Guardian*, March 2, 2022. www.theguardian.com/environment/2022/mar/02/world-leaders-agree-draw-up-historic-treaty-plastic-waste.

Meekosha, Helen. "Decolonising Disability: Thinking and Acting Globally." *Disability & Society* 26, no. 6 (October 2011): 667–82. https://doi.org/10.1080/09687599.2011.602860.

"Meet James Wakibia." *Nairobi Ideas Podcast*. Mawazo Institute. September 13, 2019. https://mawazoinstitute.org/podcast-episodes/2019/9/18/episode-8-meet-james-wakibia?utm_source=google&utm_medium=website&utm_campaign=podcast_episodes.

Meiners, Joan. "Ten Years Later, BP Oil Spill Continues to Harm Wildlife—Especially Dolphins." *National Geographic*, April 17, 2020. www.nationalgeographic.com/animals/article/how-is-wildlife-doing-now--ten-years-after-the-deepwater-horizon.

Meiu, George Paul. "Panics over Plastics: A Matter of Belonging in Kenya." *American Anthropologist* 122, no. 2 (June 2020): 222–35. https://doi.org/10.1111/aman.13381.

Mendenhall, Emily. "The COVID-19 Syndemic Is Not Global: Context Matters." *The Lancet* 396, no. 10264 (November 2020): 1731. https://doi.org/10.1016/S0140-6736(20)32218-2.

Mendes, Sam, dir. *American Beauty*. Universal Pictures, 1999.

Mercieca, Jennifer. "A Field Guide to Trump's Dangerous Rhetoric." The Conversation. June 19, 2020. https://theconversation.com/a-field-guide-to-trumps-dangerous-rhetoric-139531.

Meredith, Sam. "BP Beats First-Quarter Estimates on Stronger Commodity Prices; Plans to Resume Share Buybacks." *CNBC*, April 27, 2021. www.cnbc.com/2021/04/27/bp-earnings-q1-2021.html.

Messe Frankfurt. "Coca-Cola Introduces a World-First: A Coke Bottle Made with Plastic from the Sea." Press release, October 11, 2019. https://rosmould.ru.messefrankfurt.com/moscow/en/rosmould/press/exhibition-news/press-folder/recycledbottle.html.

Messina, Chris. "The Hashtag Turns 13." Medium. September 2, 2020. https://medium.com/chris-messina/the-hashtag-turns-13-edd4c93b5685.

"Mia Ives-Rublee, Disability Policy Director and Public Speaker." LinkedIn. n.d. www.linkedin.com/in/mia-ives-rublee.

Miller, Barbara M., and Julie Lellis. "Audience Response to Values-Based Marketplace Advocacy by the Fossil Fuel Industries." *Environmental Communication* 10, no. 2 (March 3, 2016): 249–68. https://doi.org/10.1080/17524032.2014.993414.

Miller, Toby. *A COVID Charter, a Better World.* New Brunswick, NJ: Rutgers University Press, 2021.

Millions Missing. "Millions Missing." 2022. https://millionsmissing.meaction.net/millions missing-2022/.

Milstein, Tema. "The Performer Metaphor: 'Mother Nature Never Gives Us the Same Show Twice.'" *Environmental Communication* 10, no. 2 (March 3, 2016): 227–48. https://doi .org/10.1080/17524032.2015.1018295.

Mina, An Xiao. "Batman, Pandaman and the Blind Man: A Case Study in Social Change Memes and Internet Censorship in China." *Journal of Visual Culture* 13, no. 3 (2014): 359–75. https://doi.org/10.1177/1470412914546576.

Mingus, Mia. "Access Intimacy, Interdependence and Disability Justice." *Leaving Evidence* (blog), April 12, 2017. https://leavingevidence.wordpress.com/2017/04/12/access -intimacy-interdependence-and-disability-justice/.

———. "The Four Parts of Accountability." *Leaving Evidence* (blog), December 18, 2019. https://leavingevidence.wordpress.com/2019/12/18/how-to-give-a-good-apology-part-1 -the-four-parts-of-accountability/.

Minh, Gia. "Formosa Spill Still Roils Public Opinion in Vietnam." Translated by Viet Ha and Brooks Boliek. *Radio Free Asia*, December 7, 2016. www.rfa.org/english/news /vietnam/formosa-spill-still-roils-12072016125227.html?fbclid=IwAR0UtQy_RzyBSw WiFoAiO5lBjxNtH-pF-ttL569IZhEnS--vobKBZ9pqYqg%20%5C.

Ministerio Del Medio Ambiente. "Campaña Ciudadana Chao Bombillas." September 12, 2018. YouTube video. www.youtube.com/watch?v=9gnmoPQBOpM&list=PLoooEno xiEY6rowXRBDY71nRLrABhZKLM&index=4.

Ministry of Economy, Trade, and Industry. "The Containers and Packaging Recycling Law." n.d. www.meti.go.jp/policy/recycle/main/english/pamphlets/pdf/the_containers _e.pdf.

Mintor, Adam. "Plastics Straws Aren't the Problem." *Bloomberg*, June 7, 2018. www .bloomberg.com/opinion/articles/2018-06-07/plastic-straws-aren-t-the-problem #xj4y7vzkg.

"Missing the Mark: Unveiling Corporate False Solutions to the Plastic Pollution Crisis." #BreakFreeFromPlastic. n.d. www.breakfreefromplastic.org/missing-the-mark -unveiling-corporate-false-solutions-to-the-plastic-crisis/.

Mitchell, Stuart. "Lonely Whale Foundation Launch New #StopSucking Campaign." *Ethical Marketing News*, September 17, 2017. https://ethicalmarketingnews.com/lonely -whale-foundation-launch-new-stopsucking-campaign.

MKTO. "Classic." By Evan "Kidd" Bogart, Andrew Goldstein, Emanuel Kiriakou, and Lindy Robbins. Track 2 on *MKTO*. Columbia, 2014.

Momoland and CHROMANCE. "Wrap Me in Plastic." By DJ Stanfill, Johannes-Chane Becker, and Marcus Layton. Single. Columbia, 2021.

Monbiot, George. "Capitalism Is Killing the Planet—It's Time to Stop Buying into Our Own Destruction." *The Guardian*, October 30, 2021. www.theguardian.com/environment

/2021/oct/30/capitalism-is-killing-the-planet-its-time-to-stop-buying-into-our-own
-destruction.

Montalvo, Andrés Flores, and Fairuz O. Loutfi Olivares. "Policy Mechanisms to Reduce
Single-Use Plastic Waste: Review of Available Options and Their Applicability in Mex-
ico." WRI México. n.d. https://wrimexico.org/sites/default/files/Factsheets_Single-Use
_Plastic_Waste_WRI_Mexico_2020%20digital.pdf.

Moore, Mark P. "Rhetorical Criticism of Political Myth: From Goldwater Legend to Rea-
gan Mystique." *Communication Studies* 42, no. 3 (September 1991): 295–308. https://doi
.org/10.1080/10510979109368343.

Morais, Jandira, Glen Corder, Artem Golev, Lynda Lawson, and Saleem Ali. "Global
Review of Human Waste-Picking and Its Contribution to Poverty Alleviation and a
Circular Economy." *Environmental Research Letters* 17 (2022). https://iopscience.iop
.org/article/10.1088/1748-9326/ac6b49/pdf.

Morgan, Scott. "First Stage of Taiwan's Plastic Straw Ban to Begin on July 1, 2019." *Taiwan
News*, June 8, 2018. www.taiwannews.com.tw/en/news/3452133.

"Most Innovative Companies: Lonely Whale." Fast Company. n.d. www.fastcompany.com
/company/lonely-whale.

Movement Generation. "Movement Generation, Just Transition Framework Resources."
n.d. https://movementgeneration.org/movement-generation-just-transition-framework
-resources/.

Mukherjee, Roopali, and Sarah Banet-Weiser, eds. *Commodity Activism: Cultural Resistance in Neoliberal Times*. Critical Cultural Communication. New York: New York
University Press, 2012.

Muttitt, Greg, and Sivan Kartha. "Equity, Climate Justice and Fossil Fuel Extraction: Principles for a Managed Phase Out." *Climate Policy* 20, no. 8 (September 13, 2020): 1024–
42. https://doi.org/10.1080/14693062.2020.1763900.

Myhal, Natasha, and Clint Carroll. "Indigenous Optimism in the Colonialcene." In
Anthropological Optimism: The Power of What Could Go Right, edited by Anna J. Wil-
low, 88–103. London: Routledge, 2023.

Na'puti, Tiara R. "Archipelagic Rhetoric: Remapping the Marianas and Challenging Mili-
tarization from 'A Stirring Place.'" *Communication and Critical/Cultural Studies* 16,
no. 1 (January 2, 2019): 4–25.

———. "Oceanic Possibilities for Communication Studies." *Communication and Critical/
Cultural Studies* 17, no. 1 (January 2, 2020): 95–103. https://doi.org/10.1080/14791420
.2020.1723802.

Na'puti, Tiara R., and Jöelle M. Cruz. "Mapping Interventions: Toward a Decolonial and
Indigenous Praxis across Communication Subfields." *Communication, Culture and
Critique* 15 (2022): 1–20.

National Environment Management Authority (NEMA). "Court Upholds Plastic Bag
Ban." June 28, 2018. www.nema.go.ke/index.php?option=com_content&view=article&
id=225:judges-upholds-plastic-bags-ban&catid=10&Itemid=375.

———. "2 Years on: Say No to Plastic Bags." n.d. www.nema.go.ke/index.php?option=com
_content&view=article&id=296:2-years-on-say-no-to-plastic-bags&catid=2&Itemid
=451.

NBC News. "As Mexico City's Plastic Bag Ban Takes Effect, Some Rethink Old Ways of Carrying Things." January 2, 2020. www.nbcnews.com/news/latino/mexico-city-s -plastic-bag-ban-takes-effect-some-rethink-n1109436.

NCSL (National Conference of State Legislatures). "State Plastic Bag Legislation." 2021. www.ncsl.org/environment-and-natural-resources/state-plastic-bag-legislation.

NEMA Kenya (@NemaKenya). "The Kenya Gazette: Legal Notice on Carrier Bags." n.d. www.nema.go.ke/images/Docs/Awarness%20Materials/Gazette_legal_Notice_on _carrier_bags.pdf.

———. "Today Morning @NemaKenya Together with DCI Officers." Twitter. May 29, 2019. https://twitter.com/NemaKenya/status/1133720468257759234?ref_src=twsrc%5Etfw% 7Ctwcamp%5Etweetembed%7Ctwterm%5E1133720468257759234%7Ctwgr%5E%7Ctw con%5Es1_&ref_url=https%3A%2F%2Fwww.newsmoto.co.ke%2Fnema-arrest-a-major -distributor-of-banned-plastic-bags%2F.

———. "Two People Were Arrested in Syokimau, Machakos." Twitter. June 22, 2022. https://twitter.com/nemakenya/status/1274964078964936704?lang=en.

———. Verified. "Nema Arrested Three People Using the Banned Plastic Bags." Twitter. October 15, 2020. https://twitter.com/nemakenya/status/1316620935055773697.

The New Twenties. "Writers' Co-Op." n.d. https://thenewtwenties.ca/who/#raedefrane.

New York Times. "Y. C. Wang, Billionaire Who Led Formosa Plastics, Is Dead at 91." October 16, 2008. www.nytimes.com/2008/10/17/business/17wang.html.

News Moto. "NEMA Arrest a Major Distributor of Banned Plastic Bags." May 29, 2019. www.newsmoto.co.ke/nema-arrest-a-major-distributor-of-banned-plastic-bags/.

Ngata, Tina. "An Unconquerable Tide." *Kia Mau: Resisting Colonial Fictions* (blog), February 18, 2018. https://tinangata.com/2018/02/18/2494/.

Ngoc, Duc. "Scaffolding Collapse in Formosa Kills 14, Injures 18." *Nguoi Lao Dong*, March 25, 2015. http://nld.com.vn/thoi-su-trong-nuoc/sap-gian-giao-o-formosa-14-nguoi-chet-18 -nguoi-bi-thuong-20150325234933354.htm.

Nguyên, Linh Chi. "Tons of Fish Are Dying Mysteriously in Vietnam." *Global Voices*, June 7, 2016. https://globalvoices.org/2016/06/07/tons-of-fish-are-dying-mysteriously -in-vietnam-whales-too/.

Nguyen, Nam, and Gia Ming. "Formosa Steel Goes on the Offensive over Fish Kill in Vietnam." Translated by Brooks Boliek. *Radio Free Asia*, April 26, 2016. www.rfa.org /english/news/vietnam/formosa-steel-goes-04262016161331.html.

Nguyen, Quang Dung. "From Cyberspace to the Streets. Emerging Environmental Paradigm of Justice and Citizenship in Vietnam." *Newsletter: International Institute for Asian Studies* 86 (Summer 2020). www.iias.asia/the-newsletter/article/cyberspace -streets-emerging-environmental-paradigm-justice-and-citizenship.

Nguyen, Thieu-Dang, and Simone Datzberger. *The Environmental Movement in Vietnam: A New Frontier of Civil Society Activism?* The Transnational Institute. May 2018. www .tni.org/files/publication-downloads/tni-authoritarianism-vietnams-environmental -movement.pdf.

Nguyen, Truc. "Victims of 2016 Vietnam Marine Life Disaster Submit Lawsuit against Formosa Plastics." *New Bloom*, June 19, 2019. https://newbloommag.net/2019/06/19 /formosa-vietnam-lawsuit/.

Nichols, Mike, dir. *The Graduate*. Embassy Pictures, 1967.

Nielson, Kristian S., Kimberly A. Nicholas, Felix Creutzig, Thomas Dietz, and Paul C. Stern. "The Role of High-Socioeconomic-Status People in Locking in or Rapidly Reducing Energy-Driven Greenhouse Gas Emissions." *Nature Energy* 6 (2021): 1011–16. https://doi.org/10.1038/s41560-021-00900-y.

Nishime, LeiLani, and Kim D. Hester Williams. "Afterword: Collective Struggle, Collective Ecologies." In *Racial Ecologies*, edited by LeiLani Nishime and Kim D. Hester Williams. Seattle: University of Washington Press, 2018.

Njuguna, John Kariuki. "The Efficacy of the Ban on Use of Plastic Bags in Kenya." *Journal of CMSD* 2, no. 1 (2018): 91–101. https://journalofcmsd.net/wp-content/uploads/2018/05/Dealing-with-the-plastic-menace-John-Kariuki-17th-May-2018.pdf.

Noor, Dharna. "Exxon's Secret Assist from the World's Top PR Firm." *Gizmodo*, September 22, 2021. https://gizmodo.com/exxon-s-secret-assist-from-the-world-s-top-pr-firm-1847721598.

NowThis Earth. "The Connection between Fracking and Plastics: One Small Step." July 15, 2021. YouTube video. www.youtube.com/watch?v=vpdkJz4UBGk.

Nuñez, Erica (@Erica__Nunez). "The #PlasticsTreaty Process Cannot Be Driven." Twitter. March 15, 2022. Shared with permission. https://twitter.com/Erica__Nunez/status/1503722866445012992.

Nuojua, Sohvi, Sabine Pahl, and Richard Thompson. "Ocean Connectedness and Consumer Responses to Single-Use Packaging." *Journal of Environmental Psychology* 81 (2022): 101814. https://doi.org/10.1016/j.jenvp.2022.101814.

Nyathi, Brian, and Chamunorwa Aloius Togo. "Overview of Legal and Policy Framework Approaches for Plastic Bag Waste Management in African Countries." Edited by John Yabe. *Journal of Environmental and Public Health* 2020 (October 31, 2020): 1–8. https://doi.org/10.1155/2020/8892773.

Obama, President Barrack. "Remarks by President Obama after Meeting Vietnamese Civil Society Leaders: Hanoi, Vietnam. JW Marriot Hotel Hanoi." Obama White House Archives. May 24, 2016. https://obamawhitehouse.archives.gov/the-press-office/2016/05/24/remarks-president-obama-after-meeting-vietnamese-civil-society-leaders.

O'Byrne, Megan. "Rhetorical Critic(ism)'s Body: Affect and Fieldwork on a Plane of Immanence." *Southern Communication Journal* 79, no. 4 (2014): 293–310. http://dx.doi.org/10.1080/1041794X.2014.906643.

Ocean Conservancy. "Stemming the Tide Statement of Accountability." July 10, 2022. https://oceanconservancy.org/trash-free-seas/take-deep-dive/stemming-the-tide/.

———. "Trash Free Seas." 2022. https://oceanconservancy.org/trash-free-seas/plastics-in-the-ocean/global-ghost-gear-initiative/.

Ochieng, Omedi. *Groundwork for the Practice of the Good Life: Politics and Ethics at the Intersection of North Atlantic and African Philosophy*. London: Routledge, Taylor & Francis Group, 2017.

O'Connor, John. "The Promise of Environmental Democracy." In *Toxic Struggles: The Theory and Practice of Environmental Justice*, edited by Richard Hofrichter, 47–57. Philadelphia: New Society Publishers, 1993.

O'Dell, Cary. "Songs of the Humpback Whale." Library of Congress. 2010. www.loc.gov /static/programs/national-recording-preservation-board/documents/humpback %20whales.pdf.

Odonkor, Elsie, and Katherine Gilchrist. "Why Gender Is at the Heart of Transforming the Plastics Value Chain." World Economic Forum. May 26, 2021. www.weforum.org /agenda/2021/05/gender-women-plastics-ghana/.

OED Online. "Single-use." n.d. www.oed.com/view/Entry/180129.

———. "Throwaway." n.d. www.oed.com/view/Entry/201417.

Office of the Deputy Prime Minister (UK). "EEA Glossary." http://glossary.eea.europa.eu /EEAGlossary.

Office of the Historian. "56: Joint Resolution by the Congress." January 29, 1955. https:// history.state.gov/historicaldocuments/frus1955-57v02/d56.

"Ogilvy: Coca-Cola Love Story." WPP. 2021. www.wpp.com/featured/work/2019/05/ogilvy ---coca-cola-love-story.

Ogilvy, David. *Confessions of an Advertising Man.* London: Southbank Publishing, 1963.

Okune, Angela. "Open Ethnographic Archiving as Feminist, Decolonizing Practice." *Catalyst: Feminism, Theory, Technoscience* 6, no. 2 (2020): 1–24. https://doi.org/10.28968/cftt .v6i2.33041.

Ondieki, Elvis. "Inside Kenya's Criminal Justice System: Only the Tough Survive." *Nation: Kenya Edition,* January 28, 2017. https://nation.africa/kenya/life-and-style/lifestyle/ inside-kenya-s-criminal-justice-system-only-the-tough-survive--354188.

One Planet Network. "Reducing Plastic Pollution: Campaigns That Work." March 15, 2021. www.oneplanetnetwork.org/knowledge-centre/resources/reducing-plastic-pollution -campaigns-that-work.

Ono, Kent A., and John M. Sloop. "The Critique of Vernacular Discourse." *Communication Monographs* 62 (1995): 19–46. https://doi.org/10.1080/03637759509376346.

"Open Letter to Ocean Conservancy Regarding the Report: Stemming the Tide." October 2, 2015. www.no-burn.org/wp-content/uploads/Open_Letter_Stemming_the_Tide _Report_2_Oct_15.pdf.

Oprah. "Oprah's Favorite Things 2018." 2018. www.oprah.com/gift/oprahs-favorite-things -2018-full-list-izola-cocktail-straws_1?editors_pick_id=75233.

Orellana, Marcos. *Report of the Special Rapporteur on the Implication for Human Rights of the Environmentally Sound Management and Disposal of Hazardous Substances and Wastes: The Stages of the Plastics Cycle and Their Impacts on Human Rights.* July 22, 2021. https://undocs.org/A/76/207.

Osman, Osman Mohamed. "Will Kenya's War on Plastic Be Successful This Time?" *CNN,* August 31, 2017. www.cnn.com/2017/08/31/africa/kenya-plastic-ban/index.html.

Osnes, Beth, Maxwell Boykoff, and Patrick Chandler. "Good-Natured Comedy to Enrich Climate Communication." *Comedy Studies* 10, no. 2 (2019): 224–36, https://doi.org/10 .1080/2040610X.2019.1623513.

Ott, Brian L., and Greg Dickinson. *The Twitter Presidency: Donald J. Trump and the Politics of White Rage.* NCA Focus on Communication Studies. New York: Routledge, 2019.

Oxfam. *Confronting Carbon Inequality* (Report). September 21, 2020. https://oxfamilibrary .openrepository.com/bitstream/handle/10546/621052/mb-confronting-carbon-inequality -210920-en.pdf.

———. "Ten Richest Men Double Their Fortunes in Pandemic While Incomes of 99 Percent of Humanity Fall." Press release, January 17, 2022. www.oxfam.org/en/press -releases/ten-richest-men-double-their-fortunes-pandemic-while-incomes-99-percent -humanity.

Package Recovery Organization Europe. "The Green Dot Trademark." n.d. www.pro-e.org /the-green-dot-trademark.

Paddock, Richard C. "Toxic Fish in Vietnam Idle a Local Industry and Challenge the State." *New York Times*, June 8, 2016. www.nytimes.com/2016/06/09/world/asia/vietnam-fish -kill.html.

Pang, Natalie, and Pei Wen Law. "Retweeting #WorldEnvironmentDay: A Study of Content Features and Visual Rhetoric in an Environmental Movement." *Computers in Human Behavior* 69 (April 2017): 54–61. https://doi.org/10.1016/j.chb.2016.12.003.

"Panoramic View of Mass Fish Death in Central Vietnam." Vietnam Net. April 20, 2016. http://english.vietnamnet.vn/fms/special-reports/155823/panoramic-view-of-mass-fish -death-in-central-vietnam.html.

Papacharissi, Zizi. *Affective Publics: Sentiment, Technology, and Politics*. Oxford: Oxford University Press, 2014. https://doi.org/10.1093/acprof:oso/9780199999736.001.0001.

———. "Affective Publics and Structures of Storytelling: Sentiment, Events and Mediality." *Information, Communication & Society* 19, no. 3 (March 3, 2016): 307–24. https://doi .org/10.1080/1369118X.2015.1109697.

Papacharissi, Zizi, and Maria de Fatima Oliveira. "Affective News and Networked Publics: The Rhythms of News Storytelling on #Egypt." *Journal of Communication* 62, no. 2 (April 2012): 266–82. https://doi.org/10.1111/j.1460-2466.2012.01630.x.

Pariser, Eli. "Eli Pariser." n.d. www.elipariser.org/.

Park, Lisa Sun-Hee, and David N. Pellow. *The Slums of Aspen: Immigrants vs. the Environment in America's Eden*. Nation of Newcomers : Immigrant History as American History. New York: New York University Press, 2011.

Parscale, Brad (@parscale). Verified. "I'm so over Paper Straws. #LiberalProgress." Twitter. July 18, 2019. https://twitter.com/parscale/status/1151899687349301248.

Pattee, Emma. "Forget about Your Carbon Footprint. Let's Talk about Your Climate Shadow." *MIC*, October 12, 2021. www.mic.com/impact/forget-your-carbon-footprint -lets-talk-about-your-climate-shadow.

———. "Leading Climate Scientist Katharine Hayhoe: 'You Have the Ability to Use Your Voice.'" *The Guardian*, November 10, 2021. www.theguardian.com/environment/2021 /nov/10/katharine-hayoe-climate-change-interview.

Paystrup, P. "Plastics as a 'Natural Resource': Perspective by Incongruity for an Industry in Crisis." In *The Symbolic Earth: Discourse and Our Creation of the Environment*, edited by James G. Cantrill and Christine L. Oravec, 176–97. Lexington: University Press of Kentucky, 1996.

PBS Frontline. Episode 14, "Plastic Wars." Aired on March 31, 2020, on PBS. www.pbs.org /wgbh/frontline/documentary/plastic-wars/.

Pellow, David N., and Robert J. Brulle, eds. *Power, Justice, and the Environment: A Critical Appraisal of the Environmental Justice Movement*. Urban and Industrial Environments. Cambridge, MA: MIT Press, 2005.

Pellow, David Naguib. "Political Prisoners and Environmental Justice." *Capitalism Nature Socialism* 29, no. 4 (October 2, 2018): 1–20. https://doi.org/10.1080/10455752.2018.1530835.

———. *Resisting Global Toxics: Transnational Movements for Environmental Justice*. Urban and Industrial Environments. Cambridge, MA: MIT Press, 2007.

———. *Total Liberation: The Power and Promise of Animal Rights and the Radical Earth Movement*. Minneapolis: University of Minnesota Press, 2014.

———. "Toward a Critical Environmental Justice Studies: Black Lives Matter as an Environmental Justice Challenge." *Du Bois Review: Social Science Research on Race* 13, no. 2 (2016): 221–36. https://doi.org/10.1017/S1742058X1600014X.

———. *What Is Critical Environmental Justice?* Cambridge, UK: Polity Press, 2017.

The Peninsula. "Vietnamese Blogger Arrested for Criticizing Government." November 3, 2016. www.thepeninsulaqatar.com/article/03/11/2016/Vietnamese-blogger-arrested-for -criticizing-government?fbclid=IwAR26kyzdYDqE2MtxOm9JsoOmFKOGG5y6Uni _i3jvDidZY-MJ1vNQM2rjJsc.

Perez, Sarah. "Facebook Blocked in Vietnam over the Weekend Due to Citizen Protests." *TechCrunch*, May 17, 2016. https://techcrunch.com/2016/05/17/facebook-blocked-in -vietnam-over-the-weekend-due-to-citizen-protests/?fbclid=IwAR0nZ2NZlRVbGad WVmfufXOghqW6-0ZfuffHjPPrUphgsoHrBj1dNfuzniA.

Perry, Katy. "Firework." By Katy Perry and Ester Dean. Track 3 on *Teenage Dream*. Capitol, 2010.

Petsko, Emily. "Q&A: Susan Freinkel, Author of 'Plastic: A Toxic Love Story,' on Recycling Myths and the Problem with Single-Use Plastics." *Oceana* (blog), January 27, 2020. https://oceana.org/blog/qa-susan-freinkel-author-%E2%80%9Cplastic-toxic-love-story %E2%80%9D-recycling-myths-and-problem-single-use.

Pezzullo, Phaedra C. "Afterword: Decentralizing and Regenerating the Field." In *Text + Field: Innovations in Rhetorical Method*, edited by Sara L. McKinnon, Robert Asen, Karma R. Chávez, and Robert Glenn Howard, 177–88. University Park: The Pennsylvania State University Press, 2016.

———. "Articulating 'Sexy' Anti-Toxic Activism on Screen: The Cultural Politics of A Civil Action and Erin Brockovich." *Environmental Communication Yearbook* 3, no. 1 (January 2006): 21–48. https://doi.org/10.1207/s15567362ecy0301_2.

———. "Between Crisis and Care: Projection Mapping as Creative Climate Advocacy." *Journal of Environmental Media* 1, no. 1 (January 1, 2020): 59–77. https://doi.org/10.1386 /jem_00006_1.

———. "Contaminated Children: Debating the Banality, Precarity, and Futurity of Chemical Safety." *Resilience: A Journal of the Environmental Humanities* 1, no. 2 (2014). https:// doi.org/10.5250/resilience.1.2.004.

———. "Contextualizing Boycotts and Buycotts: The Impure Politics of Consumer-Based Advocacy in an Age of Global Ecological Crises." *Communication and Critical/Cultural Studies* 8, no. 2 (June 2011): 124–45. https://doi.org/10.1080/14791420.2011.566276.

———. "CU Boulder Alumna Emy Kane, Lonely Whale, on Plastic Bans & Ocean Advocacy." *Communicating Care* (podcast), April 13, 2022. https://communicatingcare.buzz sprout.com/1924467/10086170-cu-boulder-alumna-emy-kane-lonely-whale-on-plastic -bans-ocean-advocacy.

———. "Environment." In *Oxford Research Encyclopedia of Communication*, edited by Dana L. Cloud. Oxford: Oxford University Press, 2017. https://doi.org/10.1093/acrefore /9780190228613.013.575.

———. "Environmental Justice and Climate Justice." In *The Routledge Handbook of Environmental Movements*, 1st ed., edited by Maria Grasso and Marco Giugni, 229–44. London: Routledge, 2021. https://doi.org/10.4324/9780367855680.

———. "Hello from the Other Side: Popular Culture, Crisis, and Climate Activism." *Environmental Communication* 10, no. 6 (November 2016): 803–6. https://doi.org/10.1080 /17524032.2016.1209325.

———. "Her Excellency Prof. Judi Wakhungu on Kenya's Ivory & Rhino Horn Ban, as Well as the Nation's Plastic Bag Ban." *Communicating Care* (podcast), February 16, 2022. https://communicatingcare.buzzsprout.com/1924467/9927685-her-excellency-prof -judi-wakhungu-on-kenya-s-ivory-rhino-horn-ban-as-well-as-the-nation-s-plastic -bag-ban.

———. "Introduction: Environmental Communication in China and Beyond." In *Green Communication and China: On Crisis, Care, and Global Futures*, 1st ed., edited by Jingfang Liu and Phaedra C. Pezzullo, xiii–xiiv. US-China Relations in an Age of Globalization. East Lansing: Michigan State University Press, 2020.

———. "James Wakibia on Kenya's Plastic Ban & Hashtag Activism." *Communicating Care* (podcast), February 2, 2022. https://communicatingcare.buzzsprout.com/1924467 /9927627-james-wakibia-on-kenya-s-plastic-bag-ban-hashtag-activism.

———. "Joe Andenmatten, Director of CU Boulder's Disability Services, on Accessibility." *Communicating Care* (podcast), May 11, 2022. https://communicatingcare.buzzsprout .com/1924467/10515396-joe-andenmatten-director-of-cu-boulder-s-disability-services -on-accessibility.

———. "José Toscano Bravo, Exec. Director of the Just Transition Alliance, on Just Transitions and Plastics." *Communicating Care* (podcast), March 16, 2022. https:// communicatingcare.buzzsprout.com/1924467/10086143-jose-toscano-bravo-exec -director-of-the-just-transition-alliance-on-just-transitions-and-plastics.

———. "Mia Ives-Rublee: Disability Justice, Coalition Work and Environmental Futures." *Communicating Care* (podcast), April 27, 2022. https://communicatingcare.buzz sprout.com/1924467/10515342-mia-ives-rublee-disability-justice-coalition-work-and -environmental-futures.

———. "Michelle Gabrieloff-Parish on Art and Environmental Justice." *Communicating Care* (podcast), May 25, 2022. https://communicatingcare.buzzsprout.com/1924467 /10515430-michelle-gabrieloff-parish-on-art-and-environmental-justice.

———. "Nancy Bui on the 2016 Vietnam Marine Life Kill, & #StopFormosaPlastic." *Communicating Care* (podcast), March 2, 2022. https://communicatingcare.buzz sprout.com/1924467/10086120-nancy-bui-on-the-2016-vietnam-marine-life-kill-stop formosaplastic.

———. "On Bats, Breathing, and Bella Vita Verde: Reflections on Environmental Communication during a Global Pandemic." In *Communication in the 2020s: How Communication Studies Makes Sense of Our Times*, edited by Christina S. Beck, 155–64. New York: Routledge, 2022.

———. "Performing Critical Interruptions: Stories, Rhetorical Invention, and the Environmental Justice Movement." *Western Journal of Communication* 65, no. 1 (March 2001): 1–25. https://doi.org/10.1080/10570310109374689.

———. "Resisting 'National Breast Cancer Awareness Month': The Rhetoric of Counterpublics and Their Cultural Performances." *Quarterly Journal of Speech* 89, no. 4 (2003): 345–65.

———. "Sharir Hossain on the Bangladesh Plastic Bag Ban, The First in the World." *Communicating Care* (podcast), March 30, 2022. https://communicatingcare.buzzsprout.com/1924467/10086155-sharir-hossain-on-the-bangladesh-plastic-bag-ban-the-first-in-the-world.

———. *Toxic Tourism: Rhetorics of Pollution, Travel, and Environmental Justice*. Tuscaloosa: University of Alabama Press, 2007.

Pezzullo, Phaedra C., and J. Robert Cox. *Environmental Communication and the Public Sphere*. 6th ed. Los Angeles: Sage, 2021.

Pezzullo, Phaedra C., and Catalina M. de Onís. "Rethinking Rhetorical Field Methods on a Precarious Planet." *Communication Monographs* 85, no. 1 (January 2, 2018): 103–22. https://doi.org/10.1080/03637751.2017.1336780.

Pezzullo, Phaedra C., and Ronald Sandler. "Conclusion: Working Together and Working Apart." In *Environmental Justice and Environmentalism: The Social Justice Challenge to the Environmental Movement*, edited by Ronald Sandler and Phaedra C. Pezzullo, 309–20. Cambridge, MA: MIT Press, 2007.

Pharr, Suzanne. *In a Time of the Right: Reflections on Liberation*. Little Rock, AR: Women's Project, 1996.

Phengsitthy, Nyah. "Plastic Reduction, Responsibility Bill Passed in California." *Bloomberg Law*, June 30, 2022. https://news.bloomberglaw.com/environment-and-energy/plastic-reduction-responsibility-bill-passed-in-california.

Piatek, Simon J. "#sandiegofire—the First Successful Hashtag." March 31, 2021. www.sjpiatek.com/research/sandiegofire-the-first-successful-hashtag/.

Picchi, Aimee. "Ikea: No More Plastic Straws or Single-Use Plastics by 2020." *CBS News*, June 7, 2018. www.cbsnews.com/news/ikea-no-more-plastic-straws-or-single-use-plastics-by-2020/.

Piepzna-Samarasinha, Leah Lakshmi. *Care Work: Dreaming Disability Justice*. Vancouver, BC: Arsenal Pulp Press, 2018.

Plasticbaglaws.org. Home page. n.d. www.plasticbaglaws.org/.

PlasticPollutionCoalition, "Yelp Adds 'Plastic-Free Packaging' and More Eco-Friendly Business Attributes." April 13, 2022. www.plasticpollutioncoalition.org/blog/2022/4/13/yelpaddsplasticfreepackaging.

Pompeo, Ellen (@ellenpompeo). Verified. "@theellenshow Did You Know Humans Use 500 Million Plastic Straws a Day?" Instagram. August 17, 2017. www.instagram.com/p/BX54GSPlLnI/?utm_source=ig_embed&ig_rid=0a8a1bc9-f9f7-4ce7-be8e-2ed586075d31.

Poole, Eva, and Elizabeth Giraud. "Right-Wing Populism and Mediated Activism: Creative Responses and Counter-narratives Special Collection." *Open Library of Humanities* 5, no. 1 (2019): 31. https://doi.org/10.16995/olh.438.

Postmes, Tom, and Suzanne Brunsting. "Collective Action in the Age of the Internet: Mass Communication and Online Mobilization." *Social Science Computer Review* 20, no. 3 (August 2002): 290–301. https://doi.org/10.1177/089443930202000306.

PotatoMcFry. "Doreen the Whale—Kate Micucci." April 18, 2012. YouTube video, www.youtube.com/watch?v=xNv-I6vgYqY.

Prashad, Vijay. *The Darker Nations: A People's History of the Third World.* New York: The New Press, 2007.

——. *The Poorer Nations: A Possible History of the Global South.* New York: Verso, 2012.

"Principles of Environmental Justice." EJnet.org: Web Resources for Environmental Justice Activists. 1991. www.ejnet.org/ej/principles.html.

"The Principles of Working Together." EJnet.org: Web Resources for Environmental Justice Activists. 1991. www.ejnet.org/ej/workingtogether.pdf.

Pritchett, Liam. "Leonardo DiCaprio's New Documentary Is about the World's Loneliest Whale." LIVEKINDLY. July 9, 2021. www.livekindly.co/leonardo-dicaprios-documentary-worlds-loneliest-whale/.

PRWeek. "PR Week Awards 2001." February 19, 2001. www.prweek.com/article/1238330/pr-week-awards-2001-prweek-award-winners-2000-2001.

Quito, Anne. "The Bendy Plastic Straw Was Originally Used in Hospitals and Vital for People with Disabilities." *Quartz*, July 9, 2018. https://qz.com/1324094/how-the-starbucks-plastic-straw-ban-affects-people-with-disabilities/.

——. "Vital Tubes: The Bendy Plastic Straw Was Originally Used in Hospitals and Vital for People with Disabilities." *Quartz*, July 9, 2018. https://qz.com/1324094/how-the-starbucks-plastic-straw-ban-affects-people-with-disabilities/.

Quizar, Jessi. "A Logic of Care and Black Grassroots Claims to Home in Detroit." *Antipode*, May 1, 2022. https://doi.org/10.1111/anti.12842.

Radical Philosophy. "Interview: Forgetting Vietnam Trinh T Minh-Ha and Lucie Kim-Chi Mercier." December 2018. www.radicalphilosophy.com/article/forgetting-vietnam.

Rage, Raju. "Access Intimacy and Institutional Ableism: Raju Rage on the Problem with 'Inclusion.'" Disability Arts Online. April 17, 2020. https://disabilityarts.online/magazine/opinion/access-intimacy-and-institutional-ableism-raju-rage-on-the-problem-with-inclusion/.

Ragusa, Antonio, Valentina Notarstefano, Alessandro Sveltao, Alessia Belloni, Girogia Gioacchini, Christine Blondeel, Emma Zucchelli, Caterina De Luca, Sara D'Avino, Alessandra Gulotta, Oliana Carnevali, and Elisabetta Giorgini. "Raman Microspectroscopy Detection and Characterisation of Microplastics in Human Breastmilk." *Polymers* 14, no. 13 (2022): 2700. https://doi.org/10.3390/polym14132700.

Ragusa, Antonio, Alessandro Svelato, Criselda Santacroce, Piera Catalano, Valentina Notarstefano, Oliana Carnevali, Fabrizio Papa, Mauro Ciro Antonio Rongioletti, Federico Baiocco, Simonetta Draghi, Elisabetta D'Amore, Denise Rinaldo, Maria Matta, and Elisabetta Giorgini. "Plasticenta: First Evidence of Microplastics in Human Placenta," *Environment International* 146 (2021). https://doi.org/10.1016/j.envint.2020.106274.

Ramirez, Chris. "Formosa Plastics Will Pay $2.8 Million for Air Pollution Fines at Its Point Comfort Plant." *Corpus Christi Caller Times*, September 15, 2021. www.caller.com/story

/news/2021/09/15/formosa-pays-2-8-m-air-pollution-fines-its-point-comfort-plant /8338634002/?utm_source=ActiveCampaign&utm_medium=email&utm_content= Plastic+Pollution%3A&utm_campaign=Plastic+Pollution+Email.

Ran, Yu. "CSR Special: Coca-Cola Launches 'Live Positively' Campaign." *China Daily*, April 20, 2010. www.chinadaily.com.cn/cndy/2010-04/20/content_9750200.htm.

Rapoza, Kenneth. "China Quits Recycling U.S. Trash as Sustainable Start-Up Makes Strides." *Forbes*, January 10, 2021. www.forbes.com/sites/kenrapoza/2021/01/10/china -quits-recycling-us-trash-as-sustainable-start-up-makes-strides/?sh=4213ab3d5a56.

Ray, Sarah Jaquette. *A Field Guide to Climate Anxiety: How to Keep Your Cool on a Warming Planet*. Oakland, California: University of California Press, 2020.

———. "Risking Bodies in the Wild: The 'Corporeal Unconscious' of American Adventure Culture." *Journal of Sport and Social Issues* 33, no. 3 (August 2009): 257–84. https://doi .org/10.1177/0193723509338863.

Ray, Sarah Jaquette, and Jay Sibara, eds. *Disability Studies and the Environmental Humanities: Toward an Eco-Crip Theory*. Lincoln: University of Nebraska Press, 2017.

Reich, Michael R. *Toxic Politics: Responding to Chemical Disasters*. Ithaca, NY: Cornell University Press, 1991.

Renegar, Valerie R., and Stacey K. Sowards. "Contradiction as Agency: Self-Determination, Transcendence, and Counter-Imagination in Third Wave Feminism." *Hypatia* 24, no. 2 (2009): 1–20. https://doi.org/10.1111/j.1527-2001.2009.01029.x.

Rentschler, Carrie A. "Bystander Intervention, Feminist Hashtag Activism, and the Anti-Carceral Politics of Care." *Feminist Media Studies* 17, no. 4 (July 4, 2017): 565–84. https:// doi.org/10.1080/14680777.2017.1326556.

Report: U.S. Plastic Waste Exports May Violate Basel Convention. The Maritime Executive. March 14, 2021. www.maritime-executive.com/article/report-u-s-plastic-waste-exports -may-violate-basel-convention.

Revkin, Andrew C. "A Song of Solitude." *New York Times*, December 26, 2004. www .nytimes.com/2004/12/26/weekinreview/a-song-of-solitude.html.

Ribeiro-Broomhead, John, and Neil Tangri. "Zero Waste and Economic Recovery: The Job Creation Potential of Zero Waste Solutions." Global Alliance for Incinerator Alternatives. February 11, 2021. https://doi.org/10.46556/GFWE6885.

Ripoll, Daniela. "'Live Positively': The Lessons Taught by Advertisements of Coca-Cola Company." *Comunicacao, Midia E Consumo* 11, no. 31 (2014). http://revistacmc.espm .br/index.php/revistacmc/article/viewFile/756/pdf_22.

Robinson, Britany. "A Whale of a Controversy: Animal Rights Groups and the Makah Nation Tangle over Traditional Whaling." *Sierra*, May 11, 2021. www.sierraclub.org /sierra/whale-controversy.

Rodriguez, Ashley. "Alice Wong Says #suckitableism." *Boss Barista*, August 23, 2018. http:// bossbarista.com/bossbarista/051.

Romero, Dennis. "Want a Plastic Straw in California? You'll Have to Specifically Ask for One." *NBC News*, September 20, 2018. www.nbcnews.com/science/environment/want -plastic-straw-california-you-ll-have-specifically-ask-one-n911651.

Römpke, Anne-Kristin, Immo Fritsche, and Gerhard Reese. "Get Together, Feel Together, Act Together: International Personal Contact Increases Identification with Humanity and Global Collective Action: XXXX." *Journal of Theoretical Social Psychology* 3, no. 1 (January 2019): 35–48. https://doi.org/10.1002/jts5.34.

Rosenberg, Eli. "Someday, a Turtle May End up with a Trump-Branded Straw in Its Nose. Here's Why." *Washington Post*, July 19, 2019. www.washingtonpost.com/politics /2019/07/20/some-day-turtle-may-end-up-with-trump-branded-straw-its-nose-heres -why/.

Roston, Eric, Leslie Kaufman, and Hayley Warren. "How the World's Richest People Are Driving Global Warming." *Bloomberg*, March 23, 2022. www.bloomberg.com/graphics /2022-wealth-carbon-emissions-inequality-powers-world-climate/.

Rothenberg, David. "Nature's Greatest Hit: The Old and New Songs of the Humpback Whale." *The Wire*, September 2014. www.thewire.co.uk/in-writing/essays/nature_s -greatest-hit_the-old-and-new-songs-of-the-humpback-whale.

Rourke, Alison. "Greta Thunberg Responds to Asperger's Critics: 'It's a Superpower.'" *The Guardian*, September 2, 2019. www.theguardian.com/environment/2019/sep/02/greta -thunberg-responds-to-aspergers-critics-its-a-superpower.

Rushing, Janice Hocker. "Mythic Evolution of 'The New Frontier' in Mass Mediated Rhetoric." *Critical Studies in Mass Communication* 3, no. 3 (September 1986): 265–96. https:// doi.org/10.1080/15295038609366655.

Said, Edward W. *Orientalism*. 1st ed. New York: Pantheon Books, 1978.

———. "Orientalism Reconsidered." *Race & Class* 27, no. 2 (October 1985): 1–15. https://doi .org/10.1177/030639688502700201.

Sanchez, Rudy. "The History of Plastic: The Theft of the Recycling Symbol." *Dieline* (blog), April 22, 2020. https://thedieline.com/blog/2020/4/22/the-history-of-plastic-the-theft -of-the-recycling-symbol?

Sandler, Ronald, and Phaedra C. Pezzullo, eds. *Environmental Justice and Environmentalism: The Social Justice Challenge to the Environmental Movement*. Urban and Industrial Environments. Cambridge, MA: MIT Press, 2007.

"Saving Sea Turtles in Viet Nam by Touching the Hearts of Local Communities." IUCN. August 4, 2017. www.iucn.org/news/viet-nam/201708/saving-sea-turtles-viet-nam -touching-hearts-local-communities.

Schaltegger, Megan. "Donald Trump Sold $200,000 Worth Of Plastic Straws Because 'Liberal Paper Straws Don't Work.'" *Delish*, July 24, 2019. www.delish.com/food-news /a28496591/donald-trump-selling-plastic-straws/.

Scheible, Jeff. *Digital Shift: The Cultural Logic of Punctuation*. Minneapolis: Minnesota University Press, 2015.

Scheidel, Arnim, Daniela Del Bene, Juan Liu, Grettel Navas, Sara Mingorría, Federico Demaria, Sofía Avila, et al. "Environmental Conflicts and Defenders: A Global Overview." *Global Environmental Change* 63 (July 2020): 102104. https://doi.org/10.1016/j .gloenvcha.2020.102104.

Schiffer, Zoe. "'Filter Bubble' Author Eli Pariser on Why We Need Publicly Owned Social Networks." *The Verge*, November 12, 2019. www.theverge.com/interface/2019/11/12 /20959479/eli-pariser-civic-signals-filter-bubble-q-a.

Schlosberg, Deia, dir. *The Story of Plastic*. The Story of Stuff Project, 2019. www.storyofstuff .org/movies/the-story-of-plastic-documentary-film/.

Schmidt, Laura, Melissa Mialon, Cristin Kearns, and Eric Crosbie. "Transnational Corporations, Obesity and Planetary Health." *The Lancet: Planetary Health* 4, no. 7 (July 2020): e266–67. https://doi.org/10.1016/S2542-5196(20)30146-7.

Schneider, Jennifer J., Steve Schwarze, Peter K. Bsumek, and Jennifer Ann Peeples. *Under Pressure: Coal Industry Rhetoric and Neoliberalism.* Palgrave Studies in Media and Environmental Communication. London: Palgrave Macmillan, 2016.

Schutten, Julie "Madrone" Kalil, and Caitlyn Burford. "'Killer' Metaphors and the Wisdom of Captive Orcas." *Rhetoric Society Quarterly* 47, no. 3 (May 27, 2017): 257–63. https:// doi.org/10.1080/02773945.2017.1309911.

Schwandt, Hilary M., Joanna Skinner, Adel Takruri, and Douglas Storey. "The Integrated Gateway Model: A Catalytic Approach to Behavior Change." *International Journal of Gynecology & Obstetrics* 130, no. S3 (August 2015). https://doi.org/10.1016/j.ijgo.2015 .05.003.

Schwarz, Steve, Jennifer Peeples, Jen Schneider, and Pete Bsumek. "The Hypocrite's Trap: Inside the Coal Industry's Rhetorical Playbook." *Salon*, January 12, 2017. www.salon .com/2017/01/12/the-hypocrites-trap-inside-the-coal-industrys-rhetorical-playbook _partner/.

Science History Institute. "Plastics: American as Apple Pie." n.d. https://digital.science history.org/works/9i35s67.

———. "Science Matters: The Case of Plastics." n.d. www.sciencehistory.org/the-history -and-future-of-plastics.

Seattle Public Utilities. "Straws & Utensils." n.d. www.seattle.gov/documents/Departments /SPU/Services/Recycling/EnglishSPUFlyer-LetterStrawsandUtensilsAM.pdf.

Sebeelo, Tebogo B. "Hashtag Activism, Politics and Resistance in Africa: Examining #This-Flag and #RhodesMustFall Online Movements." *Insight on Africa* 13, no. 1 (January 2021): 95–109. https://doi.org/10.1177/0975087820971514.

See Mia Paddle (@SeeMiaRoll). "As @SFdirewolf says, #SuckItAbleism." Twitter. July 30, 2022. https://twitter.com/seemiaroll/status/1553380992601268224?s=11&t=CQBQTfuv YzPHamvIxMpFuw

Seigworth, Gregory J. "Capaciousness." *Capacious: Journal for Emerging Affect Inquiry* 1, no. 1 (2017): i–v.

Seigworth, Gregory J., and Melissa Gregg. "An Inventory of Shimmers." In *The Affect Theory Reader*, edited by Melissa Gregg and Gregory J. Seigworth. Durham, NC: Duke University Press, 2010. https://doi.org/10.1215/9780822393047.

Seigworth, Gregory J., and Matthew Tiessen. "Mobile Affects, Open Secrets, Global Illiquidity: Pockets, Pools, and Plasma." *Theory Culture & Society* 29, no. 6 (2012): 47–44. https://doi.org/10.1177/0263276412444473.

Shome, Raka. "Thinking Culture and Cultural Studies—from/of the Global South." *Communication and Critical/Cultural Studies* 16, no. 3 (July 3, 2019): 196–218. https://doi.org /10.1080/14791420.2019.1648841.

Shorty Awards. "#STOPSUCKING Silver Distinction in Social Good Campaign." 2018. https://shortyawards.com/10th/strawless-ocean-campaign.

Silver, Laura, and Courtney Johnson. "Majorities in Sub-Saharan Africa Own Mobile Phones, but Smartphone Adoption Is Modest." Pew Research Center. October 9,

2018. /www.pewresearch.org/global/2018/10/09/majorities-in-sub-saharan-africa-own-mobile-phones-but-smartphone-adoption-is-modest/.

Simpson, Leanne. *As We Have Always Done: Indigenous Freedom through Radical Resistance*. Indigenous Americas. Minneapolis: University of Minnesota Press, 2017.

SinoPec Holdings. "Monthly Net Profits." August 2021. www.sinopac.com/en/financeInfo/20170217124543537000000000000395.html.

———. "Our Story: SinoPec Holdings." n.d. www.sinopac.com/en/about/20170308183544 18900000000000053.html.

Sins Invalid. "10 Principles of Disability Justice." September 17, 2015. www.sinsinvalid.org/blog/10-principles-of-disability-justice.

Skager, Kelsey. "When Were Plastic Grocery Bags Invented? A History." *Quality Logo Products Blog*, July 2, 2021. www.qualitylogoproducts.com/blog/the-history-of-plastic-bags/.

Skill, Karin, Sergio Passero, and Marie Francisco. "Assembling Amazon Fires through English Hashtags. Materializing Environmental Activism within Twitter Networks." *Computer Supported Cooperative Work (CSCW)* 30, nos. 5–6 (December 2021): 715–32. https://doi.org/10.1007/s10606-021-09403-6.

Skoric, Marko M., Qinfeng Zhu, Debbie Goh, and Natalie Pang. "Social Media and Citizen Engagement: A Meta-Analytic Review." *New Media & Society*, 18, no. 9 (2016): 1817–39. https://doi.org/10.1177/1461444815616221.

Smith, Dominic. "Kenya Burns Largest Ever Ivory Stockpile to Highlight Elephants' Fate." *The Guardian*, April 30, 2016. www.theguardian.com/environment/2016/apr/30/kenya-to-burn-largest-ever-ivory-stockpile-to-highlight-elephants-fate.

Smith, Toby Maureen. *The Myth of Green Marketing: Tending Our Goats at the Edge of Apocalypse*. Toronto: University of Toronto Press, 1998. https://go.exlibris.link/y5shrlCd.

Sofar Sounds. "52 Hertz Whale—The Whale of Youth | Sofar Bratislava." March 5, 2017. YouTube video. www.youtube.com/watch?v=oxun_ey_nAg.

Sontag, Susan. *Under the Sign of Saturn*. 1st Picador USA ed. New York: Picador USA/Farrar, Straus and Giroux, 2002.

Soto, Rev. Theresa I. "What Plastic Activists Need to Know about Disability Justice." *Greenpeace USA* (blog), July 26, 2018. www.greenpeace.org/usa/single-use-plastic-multi-issue-lives/.

Southwest Pennsylvania Environmental Health Project. "The Health behind Plastics Cracker Plants." *Environmental Health Project* (blog), December 10, 2019. www.environmentalhealthproject.org/post/the-health-behind-plastics-cracker-plants#:~:text=The%20pollutants%20emitted%20from%20the,emissions%20contribute%20to%20climate%20change.

Sowards, Stacey K. *¡Sí, Ella Puede! The Rhetorical Legacy of Dolores Huerta and the United Farm Workers*. 1st ed. Inter-America Series. Austin: University of Texas Press, 2019.

Squires, Catherine R. "Rethinking the Black Public Sphere: An Alternative Vocabulary for Multiple Public Spheres." *Communication Theory* 12, no. 4 (November 2002): 446–68. https://doi.org/10.1111/j.1468-2885.2002.tb00278.x.

Squirmy and Grubs. "Should We Ban Plastic Straws? / Squirmy and Grubs." , February 19, 2019. YouTube video. www.youtube.com/watch?v=zydt1olmQgg.

Stafford, Richard, and Peter J. S. Jones. "Viewpoint—Ocean Plastic Pollution: A Convenient but Distracting Truth?" *Marine Policy* 103 (May 2019): 187–91. https://doi.org/10.1016/j.marpol.2019.02.003.

Starbird, Kate, and Leysia Palen. "'Voluntweeters': Self-Organizing by Digital Volunteers in Times of Crisis." *CHI '11: Proceedings of the SIGCHI Conference on Human Factors in Computing Systems* (May 7, 2011): 1071–80. https://doi.org/10.1145/1978942.1979102.

Starbucks. "Follow Up to Starbucks Sustainability News." July 13, 2018. https://stories.starbucks.com/press/2018/follow-up-to-starbucks-sustainability-news/.

———. "Starbucks Commitment to Access and Disability Inclusion." July 28, 2015. https://stories.starbucks.com/press/2015/starbucks-commitment-to-access-and-disability-inclusion/.

Statistica.com. "Extreme Poverty Rate in Kenya from 2017–2021." Reuters, April 7, 2022. www.reuters.com/business/exxon-chevron-paid-their-ceos-over-22-mln-each-last-year-filings-2022-04-07/.

Stauber, John, and Sheldon Rampton. *Toxic Sludge Is Good for You! Lies, Damn Lies and the Public Relations Industry.* Monroe, ME: Common Courage Press, 1995.

Steele, Jason. "The History of Credit Cards."Creditcards.com. July 11, 2022. www.creditcards.com/statistics/history-of-credit-cards/.

The Story of Stuff Project. *Glass, Metal, Plastic: The Story of New York's Canners.* Film. www.storyofstuff.org/movies/plastic/glass-metal-plastic/.

———. "The Story of Bottled Water." March 17, 2010. YouTube video. www.youtube.com/watch?v=Se12y9hSOMo.

Stossel, John. "John Stossel: The Absurd Hysteria around Plastic Straws." *Fox News,* July 19, 2018. www.foxnews.com/opinion/john-stossel-the-absurd-hysteria-around-plastic-straws.

Stouffer, Lloyd. "Plastics Packaging: Today and Tomorrow." In *1963 National Plastics Conference,* section 6-A, 1–3. The Society of the Plastics Industry, Inc., 1963. https://discardstudies.files.wordpress.com/2014/07/stoffer-plastics-packacing-today-and-tomorrow-1963.pdf.

Strasser, Susan. *Waste and Want: A Social History of Trash.* New York: Macmillan, 1999.

Striphas, Ted. *Algorithmic Culture before the Internet.* New York: Columbia University Press, 2023.

Strong, Matthew. "Taiwanese Tycoon Needs 90 Minutes to Find NT$400 Million for Bail." *Taiwan News,* August 18, 2017. www.taiwannews.com.tw/en/news/3234818.

Sullivan, Laura. "How Big Oil Misled the Public into Believing Plastic Would Be Recycled." *NPR,* September 11, 2020. www.npr.org/2020/09/11/897692090/how-big-oil-misled-the-public-into-believing-plastic-would-be-recycled.

Sultana, Farhana. "Critical Climate Justice." *Geographical Journal* 188, no. 1 (2022): 118–24.

Sultana, Marium. "Polybag Ban Fails to Discourage Use." *Dhaka Tribune,* May 21, 2021. www.dhakatribune.com/bangladesh/environment/2021/05/21/polybag-ban-fails-to-discourage-use.

Sun, Kang, and Mohan J. Dutta. "Meanings of Care: A Culture-Centered Approach to Left-Behind Family Members in the Countryside of China." *Journal of Health Communication* 21, no. 11 (November 2016): 1141–47. https://doi.org/10.1080/10810730.2016.1225869.

Sun, Yu-Huay. "Fish Death Crisis Prompts Vietnam Waste Water Probe." *Bloomberg*, May 8, 2016. www.bloomberg.com/news/articles/2016-05-04/fish-death-crisis-prompts -vietnam-to-probe-waste-water-pipes.

Supran, Geoffrey, and Naomi Oreskes. "Rhetoric and Frame Analysis of ExxonMobil's Climate Change Communications." *One Earth* 4, no. 5 (May 2021): 696–719. https://doi .org/10.1016/j.oneear.2021.04.014.

Sutton, Jeannette, Leysia Palen, and Irina Shklovski. "Backchannels on the Front Lines: Emergent Uses of Social Media in the 2007 Southern California Wildfires." *Proceedings of the 5th International ISCRAM Conference* (May 2008).

Sze, Julie. *Environmental Justice in a Moment of Danger.* Oakland: University of California Press, 2020.

———. *Fantasy Islands: Chinese Dreams and Ecological Fears in an Age of Climate Crisis.* Oakland: University of California Press, 2015.

Tabuchi, Hiroko. "The World Is Awash in Plastic: Nations Plan a Treaty to Fix That." *New York Times*, March 2, 2022. www.nytimes.com/2022/03/02/climate/global-plastics -recycling-treaty.html.

Tabuchi, Hiroko, Michael Corkery, and Carlos Mureithi. "Big Oil Is in Trouble; Its Plan: Flood Africa with Plastic." *New York Times*, August 30, 2020. www.nytimes.com/2020 /08/30/climate/oil-kenya-africa-plastics-trade.html.

Takahashi, Bruno, Edson C. Tandoc Jr., and Christine Carmichael. "Communicating on Twitter during a Disaster: An Analysis of Tweets During Typhoon Haiyan in the Philippines." *Computers in Human Behavior* 50 (2015) 392–98. https://doi.org/10.1016/j.chb .2015.04.020.

Tannenbaum, Emily. "Every Social Media Platform Donald Trump Is Banned from Using (So Far)." *Glamour*, June 4, 2021. www.glamour.com/story/donald-trump-social-media -bans-twitter-facebook.

Tanzer, Andrew. "Y. C. Wang Gets up Very Early in the Morning." *Forbes*. July 15, 1985. Downloaded from LexisNexis.

Taylor, Diana. *Disappearing Acts: Spectacles of Gender and Nationalism in Argentina's "Dirty War".* Durham, NC: Duke University Press, 1997.

———. *¡Presente! The Politics of Presence.* Durham, NC: Duke University Press, 2020.

Teki Paper Bags. "Why? Teki Paper Bags." n.d. www.tekipaperbags.com/#Timefora change.

Thaker, Jagadish. "Environmentalism of the Poor." In *The Handbook of International Trends in Environmental Communication*, edited by Bruno Takahashi, Julia Metag, and Jagadish Thaker, 193–205. New York: Routledge, 2022.

Thiel, Cassandra L., Matthew Eckelman, Richard Guido, Matthew Huddleston, Amy E. Landis, Jodi Sherman, Scott O. Shrake, Noe Copley-Woods, and Melissa M. Bilec. "Environmental Impacts of Surgical Procedures: Life Cycle Assessment of Hysterectomy in the United States." *Environmental Science & Technology* 49, no. 3 (February 3, 2015): 1779–86. https://doi.org/10.1021/es504719g.

Thompson, Derek. "The Amazing History and the Strange Invention of the Bendy Straw." *The Atlantic*, November 22, 2011. www.theatlantic.com/business/archive/2011/11/the -amazing-history-and-the-strange-invention-of-the-bendy-straw/248923/.

Thompson, Richard C., Shanna H. Swan, Charles J. Moore, and Frederick S. vom Saal. "Our Plastic Age." *Philosophical Transactions of the Royal Society B: Biological Sciences* 364, no. 1526 (July 27, 2009): 1973–76. https://doi.org/10.1098/rstb.2009.0054.

Thuy, Minh. "Over 3,000 Chinese Laborers Working without Permit in Vung Ang." *Tien Phong*, October 9, 2014. www.tienphong.vn/xa-hoi/hon-3000-lao-dong-trung-quoc-lam-viec-khong-phep-tai-vung-ang-768996.tpo#epi_web.

Tinh, Ha. "Taiwanese Firm Exec Makes Shocking Remarks over Vietnam's Environmental Disaster." *Thanh Nien News*, April 26, 2016. www.thanhniennews.com/society/taiwanese -firm-exec-makes-shocking-remarks-over-vietnams-environmental-disaster-61560.html.

Tombleson, Bridget, and Katharina Wolf. "Rethinking the Circuit of Culture: How Participatory Culture Has Transformed Cross-Cultural Communication." *Public Relations Review* 43, no. 1 (March 2017): 14–25. https://doi.org/10.1016/j.pubrev.2016.10.017.

Toner, Kathleen. "He's Doing the 'Dirty Work' to Keep Plastic out of the Ocean." *CNN*, October 17, 2019. www.cnn.com/2019/10/17/world/cnnheroes-afroz-shah-afroz-shah -foundation/index.html.

Tong, Linh. "Vietnam Fish Deaths Cast Suspicion on Formosa Steel Plant." *The Diplomat*, April 30, 2016. https://thediplomat.com/2016/04/vietnam-fish-deaths-cast-suspicion -on-formosa-steel-plant/.

Towns, Armond. *On Black Media Philosophy*. Oakland: University of California Press, 2022.

Townsend, Robert, dir. *Eddie Murphy: Raw*. Paramount Pictures, 1987.

Tran, Tony. "Vietnamese Diasporic Films and the Construction of Dysfunctional Transnational Families: The Rebel and Owl and the Sparrow." In *The Routledge Companion to Asian American Media*, edited by Lori Kido Lopez and Vincent N Pham, 211–22. New York: Routledge, 2019.

Trang, Doan. "Timeline: The Formosa Environmental Disaster." *The Vietnamese*, November 8, 2017. www.thevietnamese.org/2017/11/timeline-the-formosa-environmental-disaster/.

Transnational Resource and Action Center (TRAC). "Greenhouse Gangsters vs. Climate Justice." 1999. www.corpwatch.org/sites/default/files/Greenhouse%20Gangsters.pdf.

Truelove, Heather Barnes, Amanda R. Carrico, Elke U. Weber, Kaitlin Toner Raimi, and Michael P. Vandenbergh. "Positive and Negative Spillover of Pro-Environmental Behavior: An Integrative Review and Theoretical Framework." *Global Environmental Change* 29 (November 2014): 127–38. https://doi.org/10.1016/j.gloenvcha.2014.09.004.

Truelove, Heather Barnes, Kam Leung Yeung, Amanda R. Carrico, Ashley J. Gillis, and Kaitlin Toner Raimi. "From Plastic Bottle Recycling to Policy Support: An Experimental Test of Pro-environmental Spillover." *Journal of Environmental Psychology* 46 (2016): 55–66. http://dx.doi.org/10.1016/j.jenvp.2016.03.004.

"The Truth behind Trash: The Scale and Impact of the International Trade in Plastic Waste." Environmental Investigation Agency and Rethink Plastic. September 2021. https://rethinkplasticalliance.eu/wp-content/uploads/2021/09/EIA_UK_Plastic_Waste _Trade_Report.pdf.

Tsiaoussidis, Alex. "MrBeast Smashes Team Seas Goal as Donation Drive Breaks $30 Million." Dexerto. January 2022. www.dexerto.com/entertainment/mrbeast-smashes-team -seas-goal-donation-drive-breaks-30-million-1731360/.

Tsing, Anna L., Jennifer Deger, Alder Keleman Saxena, and Feifei Zhou. "Plastics Saturate US, Inside and Out." In *Feral Atlas: The More-Than-Human Anthropocene*. Stanford University Press, 2020. https://doi.org/10.21627/2020fa.

Tsing, Anna Lowenhaupt. *Friction: An Ethnography of Global Connection*. Princeton, NJ: Princeton University Press, 2005.

Tuck, Eve. "Suspending Damage: A Letter to Communities," *Harvard Educational Review* 79, no. 3 (2009): 409–27. https://doi.org/10.17763/haer.79.3.n0016675661t3n15.

Tufekci, Zeynep. *Twitter and Tear Gas: The Power and Fragility of Networked Protest*. New Haven, CT: Yale University Press, 2017.

Twitter. "Permanent Suspension of @realDonaldTrump." Twitter Blog, January 8, 2021. https://blog.twitter.com/en_us/topics/company/2020/suspension.

Unghetto, Mathieu. *PLASTIC*. Epic Records, 2020.

Union of Concerned Scientists. "The Hidden Costs of Fossil Fuels." August 30, 2016. www.ucsusa.org/resources/hidden-costs-fossil-fuels.

UN News. "Climate Change Recognized as 'Threat Multiplier': UN Security Council Debates Its Impact on Peace." January 25, 2019. news.un.org/en/story/2019/01/01031322.

United Nations (UN). "Disability-Inclusive Disaster Risk Reduction and Emergency Services," n.d. www.un.org/development/desa/disabilities/issues/disability-inclusive-disaster-risk-reduction-and-emergency-situations.html.

———. "USA: Environmental Racism in 'Cancer Alley' Must End—Experts." Press release, March 2, 2021. www.ohchr.org/en/press-releases/2021/03/usa-environmental-racism-cancer-alley-must-end-experts?LangID=E&NewsID=26824.

United Nations Environment Programme (UNEP). "About Montreal Protocol." n.d. www.unep.org/ozonaction/who-we-are/about-montreal-protocol.

———. "Blue Awakening as Latin America and Caribbean States Say No to Plastic." October 22, 2018. www.unep.org/news-and-stories/story/blue-awakening-latin-america-and-caribbean-states-say-no-plastic.

———. "Double Trouble: Plastics Found to Emit Potent Greenhouse Gases." August 24, 2018. www.unep.org/news-and-stories/story/double-trouble-plastics-found-emit-potent-greenhouse-gases?_ga=2.238947395.2146704485.1629400532-264916678.1623774169.

———. "Draft Resolution on an Internationally Legally Binding Instrument on Plastic Pollution." 2022. https://wedocs.unep.org/bitstream/handle/20.500.11822/37395/UNEA5.2%20Global_Agreement_Explanatory%20note%20and%20Resolution%2027%20October.pdf?sequence=1&isAllowed=y.

———. *Drowning in Plastics: Marine Litter and Plastic Waste Vital Graphics* (Report). 2021. https://wedocs.unep.org/xmlui/bitstream/handle/20.500.11822/36964/VITGRAPH.pdf.

———. "From Birth to Ban: A History of the Plastic Shopping Bag." December 20, 2021. www.unep.org/news-and-stories/story/birth-ban-history-plastic-shopping-bag.

———. "Historic Day in the Campaign to Beat Plastic Pollution: Nations Commit to Develop a Legally Binding Agreement." Press release, March 2, 2022. www.unep.org/news-and-stories/press-release/historic-day-campaign-beat-plastic-pollution-nations-commit-develop.

———. "Kenya Bans Single-Use Plastics in Protected Areas." June 5, 2020. www.unep.org/news-and-stories/story/kenya-bans-single-use-plastics-protected-areas.

———. "Latin America Wakes up to the Problem of Plastic Straws." August 29, 2018. www .unep.org/news-and-stories/story/latin-america-wakes-problem-plastic-straws.

———. "Meet James Wakibia, the Campaigner behind Kenya's Plastic Bag Ban." May 4, 2018. www.unep.org/news-and-stories/story/meet-james-wakibia-campaigner-behind -kenyas-plastic-bag-ban.

———. "Megacities' War on Plastic Bags." February 12, 2020. YouTube video. www.youtube .com/watch?v=JNvkGK7Kyhg.

———. "Reframing Tourism to Address Plastic Pollution." November 25, 2021. www.unep .org/news-and-stories/story/reframing-tourism-address-plastic-pollution.

———. "Selection, Design and Implementation of Economic Instruments in the Solid Waste Management Sector in Kenya: The Case of Plastic Bags." 2005. https://wedocs.unep.org /bitstream/handle/20.500.11822/8655/Selection-Design-Implementation-of-Economic -Instruments-Solid-Waste-Management-Kenya.pdf?sequence=3&isAllowed=y.

———. "Single-Use Plastics: A Roadmap for Sustainability." 2018. https://wedocs.unep.org /bitstream/handle/20.500.11822/25496/singleUsePlastic_sustainability.pdf?isAllowed= y&sequence=1.

———. "Visual Feature: Beat Plastic Pollution." n.d. www.unep.org/interactives/beat -plastic-pollution/.

United States Environmental Protection Agency (EPA). "Plastics: Material-Specific Data." 2021. www.epa.gov/facts-and-figures-about-materials-waste-and-recycling/plastics -material-specific-data.

US Energy Information Administration. "How Much Oil Is Used to Make Plastic?" Frequently Asked Questions. www.eia.gov/tools/faqs/faq.php?id=34&t=6.

Valeur Magazine. "Strawless In Seattle or Lonely Whale or For A Strawless Ocean." December 30, 2018. YouTube video. www.youtube.com/watch?v=wpFYQRyTDCs.

Valle, Sabrina. "Exxon, Chevron Paid Their CEOs over \$22 Mln Each Last Year-Filings." Reuters, April 7, 2022. www.reuters.com/business/exxon-chevron-paid-their-ceos-over -22-mln-each-last-year-filings-2022-04-07/.

van Doorn, Jenny, and Tim Kurz. "The Warm Glow of Recycling Can Make Us More Wasteful." *Journal of Environmental Psychology* 77 (2021): 101672. https://doi.org/10 .1016/j.jenvp.2021.101672.

Van Gelder, Sarah, and Vandana Shiva. "Earth Democracy: An Interview with Vandana Shiva." *Yes Magazine*, January 1, 2003. www.yesmagazine.org/issue/democracy/2003/01 /01/earth-democracy-an-interview-with-vandana-shiva.

Van Truong, Tuyen, Melissa Marschke, Tuan Viet Nguyen, Georgina Alonso, Mark Andrachuk, and Phuong Le Thi Hong. "Household Recovery from Disaster: Insights from Vietnam's Fish Kill." *Environmental Hazards* 21, no. 1 (January 1, 2022): 1–16. https://doi .org/10.1080/17477891.2021.1873098.

Vasquez, Krystal. "Environmental Injustice and Disability: Where Is the Research?" *Environmental Health News*, August 23, 2021. www.ehn.org/disability-justice-2654683649 .html.

Vats, Anjali. "Cooking Up Hashtag Activism: #PaulasBestDishes and Counternarratives of Southern Food." *Communication and Critical/Cultural Studies* 12, no. 2 (April 3, 2015): 209–13. https://doi.org/10.1080/14791420.2015.1014184.

———. "(Dis)owning Bikram: Decolonizing Vernacular and Dewesternizing Restructur-
ing in the Yoga Wars." *Communication and Critical/Cultural Studies* 13, no. 4 (2016):
325–45.

Venhoeven, Leonie A., Jan Willem Bolderdijk, and Linda Steg. "Why Going Green Feels
Good." *Journal of Environmental Psychology* 71 (October 2020): 101492. https://doi.org
/10.1016/j.jenvp.2020.101492.

Viet Diaspora Stories. "Justice for Formosa's Victims Presents RED SEA: Vietnam's Modern
Disaster." July 31, 2021. YouTube video. www.youtube.com/watch?v=Io7RsnomGXk.

Việt Nam News. "Formosa Steel to Boost Investment." June 22, 2012. https://vietnamnews
.vn/economy/226473/formosa-steel-to-boost-investment.html.

———. "Ministry Opposes Steel Region Plans." June 28, 2014. https://vietnamnews.vn
/economy/256780/ministry-opposes-steel-region-plans.html.

———. "PM Hails Efforts at Major Steel Plant Construction Launch." December 3, 2012.
vietnamnews.vn/industries/233539/pm-hails-efforts-at-major-steel-plant-construction
-launch.html.

Vineberg, Dan. "We Choose Fish." Medium, November 28, 2016. https://medium.com/
@danvineberg/we-choose-fish-40c882503472#.109ysynmf.

Vuong, Brenda. "In the Wake of Formosa Plastics: Rebuilding Coastal Communities Texas,
Taiwan, Vietnam." Disaster STS Network at UC Irvine. March 4, 2022. disaster-sts
-network.org/content/wake-formosa-plastics-rebuilding-coastal-communities-texas
-taiwan-vietnam-030422/essay.

Wachira, James. "Wangari Maathai's Environmental Afrofuturist Imaginary in Wanuri
Kahiu's *Pumzi*." *Critical Studies in Media Communication* 37, no. 4 (August 7, 2020):
324–36. https://doi.org/10.1080/15295036.2020.1820543.

Wakibia, James. "Ban Plastic Bags in Kenya Campaign." June 21, 2016. https://jwakibia
.medium.com/ban-plastic-bags-in-kenya-campaign-2d1c621e0bfd#.n7f3jbdr5.

———. "A Deaf Women Community in Ethiopia's Capital, Addis Ababa, Is Using Sign Language to Fight Plastic Pollution as They Create Jobs from Their Paper Bags Enterprise."
Medium, June 20, 2021. https://jwakibia.medium.com/a-deaf-women-community-in
-ethiopias-capital-addis-abab-is-using-sign-language-to-fight-plastic-ae6daa26f6ee.

———. "James Wakibia." Medium. n.d. https://jwakibia.medium.com/.

———. "Officially Closing My #BanPlasticsKE Campaign." Medium, August 27, 2017.
https://jwakibia.medium.com/officially-closing-my-banplasticske-campaign-4bfe
8ae04882.

Wakibia, James (@JamesWakibia). "Am Always Pissed Off When Somebody Tells Me."
Twitter. May 20, 2015. Shared with permission. https://twitter.com/JamesWakibia
/status/601060189228441600.

———. "Why #BanPlasticKE Picture Says It All." Twitter. May 20, 2015. Shared with permission. https://twitter.com/JamesWakibia/status/601230424287875072.

Walker, Heather R. "#CripTheVote: How Disabled Activists Used Twitter for Political
Engagement during the 2016 Presidential Election." *Participations: Journal of Audience
and Reception Studies* 17, no. 1 (May 2020).

"Walking the Talk in Vanuatu, the First Country in the World to Ban Plastic Straws."
SPREP. May 15, 2018. www.sprep.org/news/walking-talk-vanuatu-first-country-world
-ban-plastic-straws.

Wang, Jiu-Liang, dir. *Plastic China*. Journeyman Pictures, 2016. www.cnex.tw/plasticchina.

Wang, Rong, John A. Dearing, Peter G. Langdon, Enlou Zhang, Xiangdong Yang, Vasilis Dakos, and Marten Scheffer. "Flickering Gives Early Warning Signals of a Critical Transition to a Eutrophic Lake State." *Nature* 492, no. 7429 (December 20, 2012): 419–22. https://doi.org/10.1038/nature11655.

Waters, Mark, and Melanie Mayron, dirs. *Mean Girls*. Paramount Pictures, 2004.

Wathuti, Elizabeth (@lizwathuti). Verified. "The poor have a voice too and the rich can also be arrested! No one is above the law. We hold everyone accountable!" Twitter. February 18, 2020. https://twitter.com/lizwathuti/status/1229676265231323136.

Watts, Eric King. "Coda: Food, Future, Zombies." In *Voice and Environmental Communication*, edited by Stephen P Depoe and Jennifer Peeples, 257–63. London: Palgrave Macmillan, 2014.

———. "'Voice' and 'Voicelessness' in Rhetorical Studies." *Quarterly Journal of Speech* 87, no. 2 (May 2001): 179–96. https://doi.org/10.1080/00335630109384328.

Watts, Jonathan. "Climatologist Michael E. Mann: 'Good People Fall Victim to Doomism; I Do Too Sometimes.'" *The Guardian*, February 27, 2021. www.theguardian.com /environment/2021/feb/27/climatologist-michael-e-mann-doomism-climate-crisis -interview.

———. "Eight Months on, Is the World's Most Drastic Plastic Bag Ban Working?" *The Guardian*, April 25, 2018. www.theguardian.com/world/2018/apr/25/nairobi-clean-up -highs-lows-kenyas-plastic-bag-ban.

The Webby Awards. "BP Carbon Footprint Calculator." 2007. https://winners.webbyawards .com/2007/websites/general-websites/lifestyle/143421/bp-carbon-footprint-calculator.

Weinstein, Dana, Adam Amel Rogers, and Erica L. Rosenthal. *Flip the Script: Can Hollywood Help Us Imagine a Future without Plastic?* Report supported by the Plastic Pollution Coalition. https://static1.squarespace.com/static/55b29de4e4b088f33db802c6/t /6182f0838998131cadb657fb/1635971216312/FlipTheScriptReport_November2021.pdf.

Werft, Meghan. "Eliminating Plastic Bags in Rwanda Saved Lives and the Economy." *Global Citizen*, September 22, 2015. www.globalcitizen.org/fr/content/how-eliminating -plastic-bags-in-rwanda-saves-liv-2/.

West, Isaac. "PISSAR's Critically Queer and Disabled Politics." *Communication and Critical/ Cultural Studies* 7, no. 2 (June 2010): 156–75. https://doi.org/10.1080/14791421003759174.

———. *Transforming Citizenships: Transgender Articulations of the Law*. New York: New York University Press, 2013.

Westervelt, Amy. "Big Oil's 'Wokewashing' Is the New Climate Science Denialism." *The Guardian*, September 9, 2021. www.theguardian.com/environment/2021/sep/09/big-oil -delay-tactics-new-climate-science-denial.

Wetherbee, Gregory, Austin Baldwin, and James Ranville. "It Is Raining Plastic." USGS. n.d. https://pubs.usgs.gov/of/2019/1048/ofr20191048.pdf.

White, Edwards. "Vietnamese Netizens Defiant Despite Looming Prison Threats." *News Lens*, October 17, 2016. https://international.thenewslens.com/article/51631?fbclid= IwAR2bf9NXQ9aha3bbDQ4dLeEyzla4lzgegtKGHJc_aCcPk-7ji8aJxpcGNQ8.

Whiting, Kate. "This Is How Humans Have Affected Whale Populations over the Years." World Economic Forum. October 28, 2019. www.weforum.org/agenda/2019/10/whales -endangered-species-conservation-whaling/.

Whyte, Kyle Powys, and Chris Cuomo. *Ethics of Caring in Environmental Ethics*. Edited by Stephen M. Gardiner and Allen Thompson. Vol. 1. Oxford: Oxford University Press, 2016. https://doi.org/10.1093/oxfordhb/9780199941339.013.22.

Wiens, Brianna I. "Virtual Dwelling: Feminist Orientations to Digital Communities." In *Networked Feminisms: Activist Assemblies and Digital Practices*, edited by Shana Mac-Donald, Michelle MacArthur, Milena Radzikowska, and Brianna I. Wiens, 85–108. Lanham, MD: Lexington Book, 2021.

Wilkins, Matt. "More Recycling Won't Solve Plastic Pollution." *Observations* (blog), *Scientific American*, July 6, 2018. https://blogs.scientificamerican.com/observations/more-recycling-wont-solve-plastic-pollution/.

Williams, Raymond. *Keywords: A Vocabulary of Culture and Society*. New York: Oxford University Press, 1976.

———. *Marxism and Literature*. Oxford: Oxford University Press, 1977.

Wilson, Jason. "Eco-Fascism Is Undergoing a Revival in the Fetid Culture of the Extreme Right." *The Guardian*, March 19, 2019. www.theguardian.com/world/commentisfree/2019/mar/20/eco-fascism-is-undergoing-a-revival-in-the-fetid-culture-of-the-extreme-right.

Wilson, Mark. "How a Student Designed the Recycling Logo, and Got a Measly $2,500." Fast Company, July 6, 2012. www.fastcompany.com/1670164/how-a-student-designed-the-recycling-logo-and-got-a-measly-2500.

Wilz, Kelly. *Resisting Rape Culture through Pop Culture: Sex after #MeToo*, Lanham, MD: Lexington Books, 2021.

Winters, Joseph. "California Passes Nation's Toughest Plastic Reduction Bill." Grist. July 1, 2022. https://grist.org/regulation/california-new-legislation-fights-plastic-pollution/.

Wired News Report. "A Rebel Movement's Life on the Web." *Wired*, March 6, 1998. www.wired.com/1998/03/a-rebel-movements-life-on-the-web/.

Wong, Alice. "Disability Visibility Project: Ingrid Tischer, Part 2." *Disability Visibility Blog*, November 12, 2014. https://disabilityvisibilityproject.com/2014/11/12/disability-visibility-project-ingrid-tischer-part-2/.

———. "The Last Straw." Eater. July 19, 2018. www.eater.com/2018/7/19/17586742/plastic-straw-ban-disabilities.

———. "The Rise and Fall of the Plastic Straw: Sucking in Crip Defiance." *Catalyst: Feminism, Theory, Technoscience* 5, no. 1 (April 1, 2019): 1–12. https://doi.org/10.28968/cftt.v5i1.30435.

World Economic Forum. "The New Plastics Economy: Rethinking the Future of Plastics." 2016. www3.weforum.org/docs/WEF_The_New_Plastics_Economy.pdf.

World Wildlife Fund. "The Lifecycle of Plastics." n.d. www.wwf.org.au/news/blogs/the-lifecycle-of-plastics#gs.auozth.

Wu, Sarah. "Q&A: Seattle's Plastic Straw Ban Now in Effect; Here's What You Need to Know." *Seattle Times*, July 1, 2018. www.seattletimes.com/seattle-news/q-heres-what-you-need-to-know/.

Yang, Guobin, and Craig Calhoun. "Media, Civil Society, and the Rise of a Green Public Sphere in China." In *China's Embedded Activism: Opportunities and Constraints of a Social Movement*, edited by Peter Ho and Richard Edmonds, 69–88. London: Routledge, 2008.

Yoder, Kate. "Footprint Fantasy: Is It Time to Forget about Your Carbon Footprint?" Grist. August 26, 2020. https://grist.org/energy/footprint-fantasy/?utm_content=bufferc3467 &utm_medium=social&utm_source=twitter.com&utm_campaign=buffer.

Zeman, Joshua, dir. *The Loneliest Whale: The Search for 52*. Bleecker Street, 2021.

Zhang, Emma, Minji Kim, Lezlie Rueda, Chelsea Rochman, Elizabeth VanWormer, James Moore, and Karen Shapiro. "Association of Zoonotic Protozoan Parasites with Microplastics in Seawater and Implications for Human and Wildlife Health." *Scientific Reports*, 12 no. 6532 (2022). https://doi.org/10.1038/s41598-022-10485-5.

Zhang, Junjie, Lei Wang, Leonardo Trasande, and Kurunthachalam Kannan. "Occurrence of Polyethylene Terephthalate and Polycarbonate Microplastics in Infant and Adult Feces." *Environmental Science & Technology Letters* 8, no. 11 (November 9, 2021): 989–94. https://doi.org/10.1021/acs.estlett.1c00559.

Zhu, Xia. "The Plastic Cycle—An Unknown Branch of the Carbon Cycle." *Frontiers in Marine Science* 7 (January 14, 2021): 609243. https://doi.org/10.3389/fmars.2020.609243.

INDEX

Founded in 1893,
UNIVERSITY OF CALIFORNIA PRESS
publishes bold, progressive books and journals
on topics in the arts, humanities, social sciences,
and natural sciences—with a focus on social
justice issues—that inspire thought and action
among readers worldwide.

The UC PRESS FOUNDATION
raises funds to uphold the press's vital role
as an independent, nonprofit publisher, and
receives philanthropic support from a wide
range of individuals and institutions—and from
committed readers like you. To learn more, visit
ucpress.edu/supportus.

www.ingramcontent.com/pod-product-compliance
Lightning Source LLC
Chambersburg PA
CBHW020843270326
41928CB00006B/524